The Neuroscience of Tinnitus

The Neuroscience of Tinnitus

Jos J. Eggermont

OXFORD
UNIVERSITY PRESS

OXFORD
UNIVERSITY PRESS

Great Clarendon Street, Oxford OX2 6DP
United Kingdom

Oxford University Press is a department of the University of Oxford.
It furthers the University's objective of excellence in research, scholarship,
and education by publishing worldwide. Oxford is a registered trade mark of
Oxford University Press in the UK and in certain other countries

First edition published 2012

Impression: 1

British Library Cataloguing in Publication Data

Data available

Library of Congress Cataloging-in-Publication Data

Library of Congress Control Number: 2012933691

ISBN 978–0–19–960560–6

Printed in Great Britain
on acid-free paper by
CPI Group (UK) Ltd, Croydon, CR0 4YY

Contents

Preface

Tinnitus (ringing in the ears) is a prevalent and often debilitating disorder with approximately 15% of people (incorporating ages from childhood to the elderly) perceiving it continuously, and in about 3% of the population it seriously affects the quality of life. There is currently no cure for it; however, there are promising ways to alleviate it. The most common cause of tinnitus is hearing loss, and its prevalence has surged in recent years as a result from the various large-scale military actions in the Middle East. The veterans of these wars now join occupational- and recreational-noise victims of tinnitus, such as duck hunters, factory workers, classical musicians, and rock stars as visible examples of this often debilitating symptom. Because of the rising costs for compensation and rehabilitation especially facing the Veterans Administration, and the need for adequate treatments, tinnitus has recently become a hot topic of neuroscience research.

Strong advances have been made in the area of behavioral animal models for tinnitus, in the understanding of human brain imaging aspects of tinnitus, and in addressing the long-range resting-state connectivity changes in the human brain that accompany tinnitus. Tinnitus is not just an auditory system disorder; major influences of the limbic and striatal systems on the psychological aspects of tinnitus have emerged as well. Especially the interaction between the auditory system and those "emotional" systems is currently actively explored.

Continued investigation of the three major animal models of tinnitus: salicylate-induced, noise trauma induced, and that induced by interactions of the somatosensory with the auditory system, has further delineated the relative roles of cochlear activity versus central auditory system changes. Evidence for the role of neural synchrony changes in tinnitus originates both from human electroencephalographic and magnetoencephalographic studies as well as from neuron pair-correlation studies in animals.

This monograph examines tinnitus from the viewpoint of a neuroscientist with a long background in translational research. Therefore the book focuses on the understanding of the mechanisms that underlie tinnitus and is based on data-driven approaches to characterize its properties in humans and in animal models. It is hoped that a better and coherent understanding of the findings from the various neuroscience methods, ranging from brain imaging, electrophysiology, and quantifying the subjective aspects of tinnitus, to molecular biology and genetic aspects, will lead to more and better science-driven approaches to alleviate tinnitus and ultimately produce a cure.

Since tinnitus most often results from an auditory dysfunction, the book is aimed first of all at the auditory neuroscientist, the audiologist, and the research-minded otolaryngologist. However, there are more and more indications that tinnitus is also a neurological disorder, and therefore this book is equally geared towards neurologists and neurosurgeons who currently contribute considerably to this field. The clinical psychologist and psychiatrist who bear the brunt of managing the tinnitus patient will also find much needed information. Although the material will generally require auditory and neuroscience knowledge at the entry postgraduate level, most parts will be accessible to undergraduate neuroscience students and the interested academic in general.

A single-author book about tinnitus is an important vehicle to provide a critical overview of what has resulted from human and animal research, what data fit together, and where the many discrepancies are. Multiauthored books, and I have contributed to quite a few, offer very specialized knowledge about the individual authors' expertise areas, but the individual chapters are typically not well integrated and the respective authors do not tend to criticize each other. My background in both human and animal research on numerous aspects of hearing, deafness, and tinnitus gives me a good viewpoint to do this. I hope that where my expertise about certain aspects of tinnitus research is somewhat limited it will be compensated by the integrated presentation.

For the backbone for the book I used three well-researched animal models of tinnitus, the salicylate (Chapter 6), noise trauma (Chapter 7), and somatic models (Chapter 8), and in addition, the hypersynchrony model (Chapter 9) that is aimed at the integration of these three models. The reasons to use these models are found in human research, which is extensively described in Chapters 3 and 4. Since tinnitus complaints increase with age, I also incorporate a chapter (Chapter 10) on the mechanisms of aging, the interaction between age-related and noise-induced hearing loss, and how this affects the increasing prevalence of tinnitus in old age. An extensive discussion of the pros and cons of behavioral animal models that are employed to decide whether an animal has tinnitus is presented in Chapter 5.

Chapter 11 revisits most aspects discussed in previous chapters and advocates the role that modified brain networks may play in generating and maintaining tinnitus. This is the most speculative chapter of the book. Epidemiology and etiology (Chapter 2), and evidence-based treatment of tinnitus (Chapter 12) have their own place at the beginning and end of this book. However, because of the neuroscience emphasis of the book, the treatment or management aspects of tinnitus are limited to the science underlying these treatments, and the, so far, few positive outcomes of clinical trials. Some 1000 references have provided the information presented here. I believe that the book is up to date as of the time of submitting it, but with the current rapid pace of new findings it is quite possible that exciting findings will be published before this book is on the shelf. I hope that it nevertheless will be a useful reference work and introduction for a fruitful start or continuation in tinnitus research.

I am grateful to various colleagues who read all or part of earlier versions of the book; Manny Don, Arnaud Noreña and Martin Pienkowski read the entire book and were very critical when needed and offered many important suggestions for improvement. Larry Roberts and Jim Kaltenbach read selected chapters (of their choice) and set me on the right track when I was wandering aimlessly. My wife, Mary, read all the chapters and pointed out the many stylistic and grammatical weeds that needed to be pulled. I followed most of the suggestions made by the readers, but my own point of view has prevailed in many places. As one of the readers said: it is a view of tinnitus as seen from the cortex. I do not disagree, but I have extensively treated the bottom-up aspects in the generation of tinnitus. I firmly believe that tinnitus is a network disorder where large parts of the brain, cortical as well as subcortical, are involved. If any of the components in that network is at fault the entire feedback system fails, and spontaneous neural activity becomes audible. The big issue still is how the myriad of animal research findings relates to what is experienced as tinnitus in humans, in other words will it help to understand how we hear?

Calgary, September 2011

List of abbreviations

2-DG	2-deoxyglucose
5-HT	serotonin
AAF	anterior auditory field (cat)
ABR	auditory brainstem response
AC	auditory cortex
ACC	anterior cingulate cortex
ACh(R)	acetylcholine (receptor)
ACTH	adrenocorticotropin hormone
AEF	auditory evoked magnetic field
AEP	auditory evoked potential
AI	primary auditory cortex (cat)
AII	secondary auditory cortex (cat)
AMPA	α-amino-3-hydroxyl-5-methyl-4-isoxazole-propionate
ANF	auditory nerve fiber
Aqp	aquaporin
ARHL	age-related hearing loss
ASR	acoustic startle reflex
ASSR	auditory steady-state response
AT	acoustic trauma
ATP	adenosine-triphosphate
AVCN	anteroventral cochlear nucleus
BA	Brodmann area
BBN	broad band noise
BDNF	brain-derived neurotrophic factor
BM	basilar membrane
BOLD	blood oxygenation level-dependent
CaM	Ca^{2+}-calmodulin
CAP	compound action potential
CaRE	calcium-response element
CB	calbindin
CBT	cognitive behavioral therapy
CBF	cerebral blood flow
CF	characteristic frequency
ChAT	choline acetyltransferase
CI	cochlear implant
CM	cochlear microphonic
CN	cochlear nucleus
CNS	central nervous system
CR	calretinin
CREB	cAMP-response element binding protein
CRF	corticotropin-releasing factor
CS	conditioned stimuli
CSD	current source density
DA	dopamine
dB	decibel
dB(A)	decibel A-weighted
DCN	dorsal cochlear nucleus
DFN	X-chromosome linked hereditary hearing loss
DFNA	autosomal dominant hereditary hearing loss
DFNB	autosomal recessive hereditary hearing loss
DG	deoxyglucose
DLPFC	dorsal lateral prefrontal cortex
DPOAE	distortion product otoacoustic emissions
DRG	dorsal root ganglia
DTI	diffusion tensor imaging
EAE	enhanced acoustic environment
EEG	electroencephalography
EP	endocochlear potential
EPSC	excitatory postsynaptic current
EPSP	excitatory postsynaptic potential
FA	fractional anisotropy
FDG	fluorodeoxyglucose
fMRI	functional magnetic resonance imaging
FTC	frequency tuning curve
GABA	gamma-aminobutyric acid
GAD	glutamate decarboxylase
GDNF	glial cell derived neurotophic factor
GLU	glutamate
Gly	glycine
GlyR	glycine receptor
GST	glutathione-S-transferase
HG	Heschl's gyrus
HL	hearing level
HPA	hypothalamic–pituitary–adrenal axis

HPLC	high-performance liquid chromatography
HRP	horse radish peroxidase
IC	inferior colliculus
ICC	central nucleus of the inferior colliculus
ICD	dorsal cortex of the inferior colliculus
ICX	external nucleus of the inferior colliculus
IEG	immediate early gene
IHC	inner hair cell
IPSC	inhibitory postsynaptic current
IRN	immunoreactive neuron
ISI	inter spike interval
LC	locus coeruleus
LFP	local field potential
LORETA	low-resolution electromagnetic tomography
LSO	lateral superior olive
LTD	long-term depression
LTP	long-term potentiation
MAP	mitogen activated protein
MEG	magnetoencephalography
MEMRI	manganese-enhanced magnetic resonance imaging
mEPSC	miniature excitatory postsynaptic current
MGB	medial geniculate body
mGluR	metabotropic glutamate receptor
mHG	medial Heschl's gyrus
MLR	middle latency response
MMN	mismatch negativity
MNTB	medial nucleus of the trapezoid body
MRI	magnetic resonance imaging
mRNA	messenger ribonucleic acid
MSN	medullary somatosensory nuclei
MSO	medial superior olive
MSU	multiple single unit
NA	noradrenaline
NAc	nucleus accumbens
NAT	N-acetyl transferase
NHP	Nottingham Health Profile
NIHL	noise-induced hearing loss
NLL	nucleus of the lateral lemniscus
NMDA	N-methyl-D-aspartic acid
NSH	National Study of Hearing
NSHL	non-syndromic hearing loss
OAE	otoacoustic emission
OCA	octopus cell area
OHC	outer hair cell
OR	odds ratio
OSHA	Occupational Safety and Health Administration
PAC	primary auditory cortex
PCR	polymerase chain reaction
PET	positron emission tomography
PMCA	plasma-membrane Ca^{2+}-ATPase
PnC	nucleus reticularis pontis caudalis
PSP	postsynaptic potential
PV	parvalbumin
PVCN	posteroventral cochlear nucleus
rCBF	regional cerebral blood flow
RI	residual inhibition
RLF	rate-level function
ROI	region of interest
ROS	reactive oxygen species
SA	silence–activity
SAC	secondary auditory cortex
SD	standard deviation
SF-36	Short Form 36 Health Survey Questionnaire
SFR	spontaneous firing rate
SGC	spiral ganglion cell
SGN	spiral ganglion neuron
SIP	schedule-induced polydipsia
sIPSC	spontaneous inhibitory postsynaptic current
SL	sensation level
SNHL	sensorineural hearing loss
SOAE	spontaneous otoacoustic emission
SOC	superior olivary complex
SP	summating potential, substance P
Sp5	spinal trigeminal nucleus
SPECT	single-photon emission computed tomography
SPL	sound pressure level
SSR	steady state response
STA	spike-triggered average

TBI	traumatic brain injury
TCD	thalamocortical dysrhythmia
TCQ	Tinnitus Cognitions Questionnaire
TCSQ	Tinnitus Coping Style Questionnaire
TG	trigeminal ganglion
TEM	transmission electron microscopy
TEOAE	transient-evoked otoacoustic emission
TEQ	Tinnitus Effect Questionnaire
THI	Tinnitus Handicap Inventory
THQ	Tinnitus Handicap Questionnaire
TH/SS	Tinnitus Handicap/Support
TMJ	temporomandibular joint
trkB	tyrosine kinase B
TRN	thalamic reticular nucleus
TRPV	transient receptor potential vanilloid
TRQ	Tinnitus reaction questionnaire
TRT	tinnitus retraining therapy
TQ	Tinnitus Questionnaire
TSI	Tinnitus Severity Index
TTF	trauma tone frequency
TTS	temporary threshold shift
US	unconditioned stimuli
VAS	visual analog scale
VCN	ventral cochlear nucleus
VGLUT	vesicular glutamate transporter
vmPFC	ventral medial prefrontal cortex
VNLL	ventral nucleus of the lateral lemniscus
VNS	vagus nerve stimulation
VS	vestibular schwannoma

Chapter 1

What is tinnitus?

Tinnitus can be defined as the conscious perception of a sound that is not generated by any source outside of the body. If there are physical sound sources inside the body, e.g. pulsating blood vessels, then the tinnitus is called objective because it may be audible to another person equipped with a sensitive microphone and amplifier. In the absence of any physical sound source, in- or outside the body, the tinnitus is called subjective.

Tinnitus, commonly known as "ear noise" or "ringing in the ears," is a sensation of a sound located in the head, in both ears or in one ear. Pliny the Elder (23–79 CE) seems to have coined the term tinnitus (Latin for "ringing") although he did not have it himself (Morgenstern, 2005). The noise can be a whistling, a hissing, or a roaring sound. In fact most ongoing spontaneous sound percepts except voices or music qualify as tinnitus. Hearing voices or music is usually considered to be a hallucination, but not less real than tinnitus. However, its cause is at least partly different albeit that the brain areas involved may be largely the same (Cope and Baguley, 2009; Chapter 11).

Martin Luther (1483–1546), one of the initiators of the Reformation, suffered from tinnitus. At the age of 44, about 10 years after he had cut himself loose from the Roman Catholic Church, he was stricken with "great pains in the head and a violent buzzing in the ear. He felt convinced he was going to die." In that same year he wrote: "When I try to work, my head becomes filled with all sorts of whizzing, buzzing, thundering noises, and if I did not leave off in the instant, I should faint away." But he knew how to alleviate his tinnitus to a bearable percept: "We should live high and drink wine when we are not well." One wonders though if his rebellion against the Roman Catholic church or his problems with his faith in general were at the origin of all his suffering: "I have not got beyond more than three or four words of the Psalms when buzz, buzz! The noise begins again and often I am near falling off my chair with the pain." The latter may be related to the vertigo (dizziness) from which he also suffered. He may have had Ménière's disease that is defined by the symptoms of vertigo, roaring tinnitus, fullness in the ear, and hearing loss. It is known that stress can trigger a vertigo attack in Ménière patients (Horner, 2003; citations are from Morgenstern, 2005).

Beethoven's (1770–1827) deafness is well known, especially since it did not affect his ability to write rousing and sensitive music throughout his entire life. At the age of 29, Beethoven began to lose his hearing. From the very beginning of his suffering from hearing loss, he complained of noises in his ears and head: "my ears hum and buzz continuously day and night. I can tell you that I lead a miserable existence" (cited in Morgenstern, 2005).

These highly personal narratives illustrate that tinnitus can often be debilitating.

1.1 **Objective tinnitus**

Objective tinnitus is commonly the result of blood flow that becomes audible, e.g. when a vessel is pulsating against the auditory nerve at the site where the nerve enters the brain (De Ridder, 2010). It can also be a vessel pulsating in the middle ear and thus changing the middle ear pressure periodically and so producing an audible sound. Pulsating tinnitus often is the result of the sound made by turbulent blood flow that is transmitted to the inner ear. Hearing one's own heart murmur causes complaints of pulsating tinnitus.

There are some very early reports (1910–1930) in the *Proceedings of the Royal Society of Medicine* that illuminate the various forms of objective tinnitus. I quote them fully:

> Miss G., aged 13, sought advice for a "purring" noise in her right ear, which she had noticed for at least two years, but which had become louder during the past twelve months after an attack of "influenza." Hearing in both ears is normal, and no pathological conditions are to be seen in the meatus or in the nasopharynx. The general health is excellent. She has menstruated once – about four months ago – but this seems to have exerted no influence for good or evil upon the aural symptom. Listening through the otoscope, it may be noted that (1) The "purring" sound – very like a haemic murmur – is best heard when the head is erect, and it is synchronous with the pulse. (2) It is diminished by either very light pressure over the right carotid artery, or by much firmer pressure on the left. Tilting the head towards the left shoulder has also a similar effect. The pulsating murmur can be heard in a quiet room if the observer's ear be placed close to the patient's right ear, and is just perceptible if a stethoscope be applied to the temporal bone above the pinna (Tilley, 1910).
>
> A. D., Male, aged 45, complains of a "scraping noise" in his right ear, which began about twelve years ago and has been almost constant ever since. The patient attributes it to a chill which followed a long cycle run, and he states that the character of the noise has varied very little since he first noticed it. Hearing in both ears is practically normal, and there is no marked pathological condition in the meatus, nose, or nasopharynx. Sleep is much disturbed and his general health is impaired. The murmur can be heard if the observer places his ear close to the right ear of the patient. With the otoscope the murmur is very distinctly heard and is synchronous with the pulse. Pressure applied over the carotid artery controls it. Listening with the stethoscope it may be noted: (1) That the murmur is heard very distinctly over the right mastoid process and over the right temporal region; (2) that it is also heard distinctly over the same areas on the left side, although the patient does not hear it in the left ear; (3) that it is heard over any part of the cranium, diminishing as one approaches the middle line. Sneezing, coughing, or inflation of the middle ear causes a momentary cessation of the sound (Sharp, 1910).
>
> A boy, aged 9, complained a few days ago of a "ticking" in his ears; his mother noted that she could hear the noise herself. There is an easily audible ticking in the left ear, very clearly heard through an otoscope; it is not synchronous with the pulse, and is much less audible when the boy's mouth is open; a similar noise is to be heard in the right ear, but much less loud. The boy had retracted membranes, and large tonsils, and adenoids. The noise is presumably due to irregular contractions of the tensor tympani (Mollison, 1916).
>
> Mrs. K., aged 48, has complained of various kinds of head noises, worse always in left ear, since September, 1926. She now definitely states she hears a sound like the "cawing of a rook" in the left ear. The noises are very distressing to her and cause much sleeplessness and mental anxiety. Her systolic blood-pressure is 150. By induction one can actually hear this noise, which is synchronous with the pulse and can be definitely controlled by pressure over the left common carotid. Mr. Graham Brown said that he had actually heard the so-called "cawing of a rook" in hospital, by placing his ear close to the patient's face, at the same time occluding the other ear. The noise was said

> to be preying on the patient's mind to such an extent that even the possibility of suicide had arisen, so that the question of treatment had become urgent. At his (the speaker's) suggestion the patient had worn during the night a broad band tied round the neck, to this was attached a small pad which pressed gently on the carotid. This had given her some relief, but he would be glad of suggestions as to further treatment (Brown, 1927).

Thus, children as well as adults can suffer from pulsating (often called pulsatile) tinnitus that has numerous causes, including benign intracranial hypertension, glomus tumors and atherosclerotic carotid arteries. Objective tinnitus can often be treated by surgical intervention.

1.2 **Subjective tinnitus**

Subjective tinnitus can only be perceived by the sufferer. Someone who does not have tinnitus has no idea what it sounds like, what it feels like, and what it means to have it. That makes it very difficult to communicate what one is experiencing, but the famous people cited in the beginning of this chapter had a way to do this effectively.

Some further distinctions in subjective tinnitus are sometimes made. Transient or reversible tinnitus is often distinguished from chronic tinnitus because it typically does not need treatment and may have a source that is different from that of chronic tinnitus. Very short transient tonal tinnitus, accompanied by fullness in the ear and transient mild hearing loss, is experienced by nearly everyone. It is not clear what the underlying mechanism is, but it combines three of the symptoms that define Ménière's disease (see section 1.1).

Acute application of salicylate (the active ingredient in aspirin) in large doses typically produces a reversible form of tinnitus. Only after prolonged salicylate use (such as prescribed in the past as an anti-inflammatory agent and painkiller in case of arthritis) can the tinnitus become chronic. This chronic tinnitus is likely caused by the development of a permanent hearing loss or diffuse spiral ganglion cell loss as a result of this continued aspirin use (Chapter 6). Tinnitus accompanied by hearing loss, and resulting from damage to the hair cells in the inner ear, is nearly always chronic and generally irreversible. The distinction of peripheral versus central tinnitus is, in my opinion, similar to that for transient versus chronic tinnitus (Eggermont, 2003). I hold that all chronic tinnitus is central in origin, but of course is often triggered by hearing loss or, in substantial numbers, also by somatic injuries such as whiplash resulting from car accidents (Pang, 1971; Levine, 1999).

1.3 **How does tinnitus sound?**

The best-documented artist of the 16th century is Michelangelo (1475–1564). His output in every field during his long life was extensive and much has been recovered, including a large volume of correspondence, sketches, and reminiscences. He described passionately in one of his poems that he heard cricket sounds: "A spider's web is hidden in one ear, in the other a cricket sings throughout the night; I do not sleep, and snore with catarrhal breath" (cited in Morgenstern, 2005).

The famous Czech composer Smetana (1824–1884) explained in a letter to a violinist why the violin part had a high-sustained harmonic sound at the end of the piece: "I felt I must describe the onset of my deafness, and tried to represent this with the sustained E of the first violin in the finale of my quartet" (Clapham, 1972).

These artistic descriptions of tinnitus, accurate as they may be for the individual case, have been extended early on by McNaughton Jones (1890) who studied 260 cases of tinnitus aurium:

> The following were the noises I have recorded as complained of by patients. The sound resembling buzzing; sea roaring; trees agitated; singing of kettle; bellows; bee humming; noise of shell; horse out of breath, puffing; thumping noise; continual beating; crackling sounds in the head; train; vibration of a metal; whistle of an engine; steam engine puffing; furnace blowing; constant hammering; rushing water; sea waves; drumming; rain falling; booming; railway whistling; distant thunder; chirping of birds; kettle boiling; waterfall; mill wheel; music; bells.

Around the same time, Sexton (1880) a physician who suffered from tinnitus himself describes it very aptly:

> The writer has been the subject of moderate singing in the ears for a long time, and has had, therefore, frequent opportunities for its study. When he is at work, or in the midst of a noise, the singing is unobserved; but when everything is quiet, and his attention is not drawn to any particular thing, the phenomenon is seldom out of mind. Sometimes, when first lying down in bed at night, the tinnitus increases, as it also does when he awakes suddenly. If he have been riding in a noisy conveyance, or standing near an engine letting off steam, or by the surf at the sea-side, there is afterwards for a time a continuance of the same sounds, the tinnitus seeming to assume the phase of the latest impression made on the organ of hearing. Patients from the different walks of life, describing the symptoms of tinnitus, are very likely to compare it to some sound with which they have long been familiar, as the surf, a waterfall, the wind in the trees, escaping steam, motions of machinery, etc.

Statistical detail is provided by two more recent representative and well-documented studies that give extensive illustrations of tinnitus quality. Eighteen hundred tinnitus cases were studied by Meikle and Taylor-Walsh (1984) and more than 500 people with tinnitus were investigated by Stouffer and Tyler (1990a). The various characterizations of their tinnitus, as reported by the patients, are shown in Table 1.1. Whereas the distributions for unilateral and bilateral tinnitus are very similar in the two studies, it is clear that the sound of ocean waves is a characterization used by persons living close to the Pacific (the Oregon study by Meikle and Taylor-Walsh), but not by land-locked people in Iowa (Stouffer and Tyler). It is also obvious that cricket sounds appear more familiar in Iowa compared to Oregon. This suggests that characterization of tinnitus is typically done by referral to known sounds that resemble it. The Iowa study people were obviously more adept with this referral or encouraged more strongly to assign a label to the tinnitus sound (~75% assigned a label to it compared to only ~40% in the Oregon study). Although both studies find a ringing sound the most common description, reminiscent of the English description of tinnitus as ringing in the ears, otherwise there is not much in common for these two large surveys in the characterization of how tinnitus sounds.

After reading these studies, one may wonder why there are so many descriptions of tinnitus sounds. The problem may be partly that it is difficult to describe unfamiliar sounds.

Table 1.1 The sound of tinnitus

	Meikle and Taylor-Walsh (1984) N = 1800	Stouffer and Tyler (1990a) N = 500
Bilateral, ear-localized	52%	44%
Unilateral	37%	34%
In the head	11%	22%
Ringing	25%	37.5%
Clear tone (humming)	4.4%	5.3%
Chirping (cricket like)	1.2%	8.5%
Whistling	1%	6.6%
Hissing	4.3%	7.8%
Ocean waves	1.7%	–
Buzzing	1.4%	11.2%

To investigate this, low level but audible pure tones were presented to tinnitus patients and they were asked to describe what they heard. It appeared that the description of these simple tonal sounds was not an easy task, and sometimes tinnitus patients even labeled pure tones as noise (Wahlström and Axelsson, 1995). This may be related to their hearing loss of sensory origin, which distorts sound perception. In this study, pure tones of different frequencies were labeled differently by the patients. For instance, a 4-kHz tone was only listed in 34% of cases as a tone, in 26% as a hissing sound, in 18% as a roaring sound, and in the remaining 22% as whistling, squeaking, etc. An 8-kHz tone was not perceived as having any tonal quality: 48% of the cases described it as rushing, 16% as beeping, 12% as ringing, the remainder as whistling, a cricket sound, a dentist drill, etc. Buzzing sounds were identified for a low-pitched 250-Hz tone (10%), and also for a high-pitched 4-kHz narrow-band noise (12%). Roaring sounds were heard to 250-Hz tones (18%) and 250-Hz narrow-band noise (8%).

No correlation was found between the person's labeling of their own tinnitus sound and the earlier given descriptions of the external sound. One gets the impression that the characterization of external sounds, and likely also internal percepts such as tinnitus, is not an easy task for people with tinnitus who generally also have a hearing loss that can produce distortion of sound. So the proliferation of labels (Table 1.1) to describe what tinnitus sounds like becomes understandable when it is realized how difficult it is to assign labels to simple unfamiliar sounds such as pure tones and filtered noise. Huss and Moore (2005) provide a potential explanation for the noisiness of tones; they found that people with hearing loss rate tones with frequencies that fall in dead regions of the cochlea generally as noise-like.

Normal-hearing people without tinnitus also have difficulties in characterizing low-level, tinnitus-like sounds (Tucker et al., 2005; Del Bo et al., 2008). After staying in a silent and soundproof room for 4 min, they reported hearing mostly buzzing and humming sounds, followed by squeaking and whirling sounds, the sound of ocean waves, clicking, heartbeat sound, crickets, flowing water, etc. These are sounds of the same quality as those characterizing tinnitus as listed in Table 1.1.

1.4 Is tinnitus a form of pain?

Early studies had already pointed to the similarity of severe tinnitus and central neuropathic pain that occurs without stimulation of pain receptors (Tonndorf, 1987; Møller, 1997). For instance, perception of auditory stimuli is often abnormal in tinnitus patients, and perception of nociceptive stimuli is often abnormal in people with central pain. Many individuals with severe tinnitus often have hyperacusis (Chapter 3) and individuals with central pain often have hyperalgesia. The similarity between these two forms of enhanced sensitivity and excessive reaction to normal sound (hyperacusis) and normal touch (hyperalgesia) is striking. Hyperalgesia is dependent on N-methyl-D-aspartic acid (NMDA) receptor-mediated activity and the loss of inhibitory control (Dickenson, 1996). It is likely, but so far not demonstrated, that hyperacusis has the same neural correlates.

The auditory extralemniscal or non-tonotopic pathways involve the dorsal and medial thalamus whereas the lemniscal or tonotopic pathways involve the ventral part of the auditory thalamus. The medial and dorsal medial geniculate body (MGB) project directly to the lateral nucleus of the amygdala in the limbic system (LeDoux, 1991) and this may explain the emotional components that often accompany severe chronic tinnitus. Chronic pain is, in part, an emotion (Chapman, 1996) and tinnitus is also, in part, definitely an emotion. The findings that limbic structures are more active in response to sound stimulation in some patients with tinnitus (Lockwood et al., 1998) support the notion that the extralemniscal auditory system is involved in tinnitus. Chapman (1996) summarized that tissue trauma: (1) excites spinoreticular as well as spinothalamic pathways that converge on the dorsal and medial parts of the thalamus (and these overlap with the starting points of the auditory extralemniscal pathways; Graybiel, 1973); (2) trauma generates concomitant affective and sensory processes; (3) trauma activates predominantly noradrenergic limbic structures to produce the affective dimension of pain; (4) the HPA-axis-mediated stress response plays a role in chronic pain (and is also involved in tinnitus; Hébert et al., 2004; Chapter 11).

Neuropathic pain likely arises as a result of changes in the properties of neurons in the central nervous system (CNS): called central sensitization. Several mechanisms that may cause the central sensitization of pain have been described (Milligan and Watkins, 2009). The best-characterized mechanism involves a change in the function of NMDA receptors in the spinal cord dorsal horn neurons. Activation of sensory neurons by painful stimuli leads to activation of pain-projection neurons in the spinal cord. During strong and/or persistent nociceptive stimulation sufficient amounts of substance P and glutamate are released to sustain the depolarization of the spinal cord neurons. When this happens, Mg^{2+} ions that normally block the NMDA channel are removed, allowing Ca^{2+} to flow through the channel into the neuron. The influx of Ca^{2+} causes the production and release of nitric oxide by Ca^{2+} -activated neuronal nitric oxide synthase and of prostaglandins by cyclooxygenase enzymes. These molecules both enhance the excitability of spinal cord neurons in response to incoming pain signals and cause an exaggerated release of neurotransmitters from sensory presynaptic terminals to the spinal cord. Together, these downstream effects of NMDA activation result in the amplification of pain messages being relayed to higher brain centers. Later in this book I will present evidence that similar changes in the CNS are introduced by salicylate and likely also by noise trauma.

Before central sensitization was discovered, there were two major models of pain (Latremoliere and Woolf, 2009). The first was that of a labeled-line system, in which specific "pain pathways" were activated only by particular peripheral "pain stimuli" and that the amplitude and duration of pain was determined solely by the intensity and timing of these inputs. The second model evoked "gate controls" in the CNS, which, by opening or closing, enabled or prevented pain. It is now appreciated that there are specific nociceptive pathways and that these are subject to complex facilitatory and inhibitory controls suggesting that both models were reflecting separate parts of a more comprehensive model. It is also known that changes in the functional properties of the neurons in these pathways are sufficient to reduce the pain threshold, to increase the magnitude and duration of responses to nociceptive input, and to permit normally innocuous inputs to generate pain sensations. Pain is thus not simply a reflection of peripheral inputs or pathology but is also of central neuronal plasticity, in which deafferentation or prior experience leads persisting changes in neuron response properties that affect perception and behavior. Plasticity profoundly alters sensitivity to an extent that it is a major contributor to many clinical pain syndromes (Latremoliere and Woolf, 2009). Central auditory system plasticity is similarly invoked as a major factor in severe tinnitus (Salvi et al., 2000; Eggermont and Roberts, 2004; Møller, 2007) as is "gate control" (Rauschecker et al., 2010; Chapter 11).

Phantom pain belongs to the complex group of phantom phenomena that often develop after amputations. Milder phantom phenomena involve feeling the presence of the previously amputated extremity. Pain in a non-existing body part develops in 50–80% of all amputees (Flor et al., 2006). Similarly, partial deafferentation of the auditory system gives rise to tinnitus with a pitch reflecting the missing inputs, and may therefore be termed a phantom sound (Jastreboff, 1990).

The presence of phantom sensations (e.g. tactile sensations allocated to the amputated arm upon stimulation of the face, which neighbors the arm in the somatotopic map in the primary sensorimotor cortex, S1) gave reason to investigate the somatotopic arrangement of S1 in amputated patients. Using magnetoencephalography (MEG; Chapter 4) it was inferred that the mouth area of S1 also activated that of the former hand. Interestingly, the extent of this effect was highly correlated with the intensity of phantom limb pain (Flor et al., 1995). Cortical reorganization has also been inferred from MEG recordings in tinnitus patients (Wienbruch et al., 2006). Surprisingly, these neuroplastic changes in S1 reorganization are reversible by training (Flor et al., 2001) but this has so far not been demonstrated in chronic tinnitus patients.

1.5 Is tinnitus aberrant spontaneous activity?

The brain is spontaneously active and the auditory parts thereof, specifically the inferior colliculus (IC) and the auditory cortex, are among the most active in the brain as reflected by their metabolic level (Sokoloff, 1979). For instance (Figure 1.1) glucose metabolism in the IC is about double that of the nearby superior colliculus, and the activity in auditory cortex is about one-third more than that in visual cortex (Kennedy et al., 1978).

Even the most peripheral neurons of the auditory system, the auditory nerve fibers (ANFs), are firing spikes in the absence of sound. In the absence of external sound some ANFs fire as much as 100 spikes/s whereas others <1 sp/s. The synapses of the inner hair cells (IHCs) with the ANFs vary in properties, such as the amount of spontaneous transmitter release and the concomitant threshold level of activation for the nerve fibers (Liberman and Kiang, 1978). Higher spontaneous transmitter release corresponds to lower threshold of firing and higher spontaneous firing rates (SFRs). Thus the activity of a single IHC is distributed in some staggered fashion onto 10 or so ANFs. These fibers will

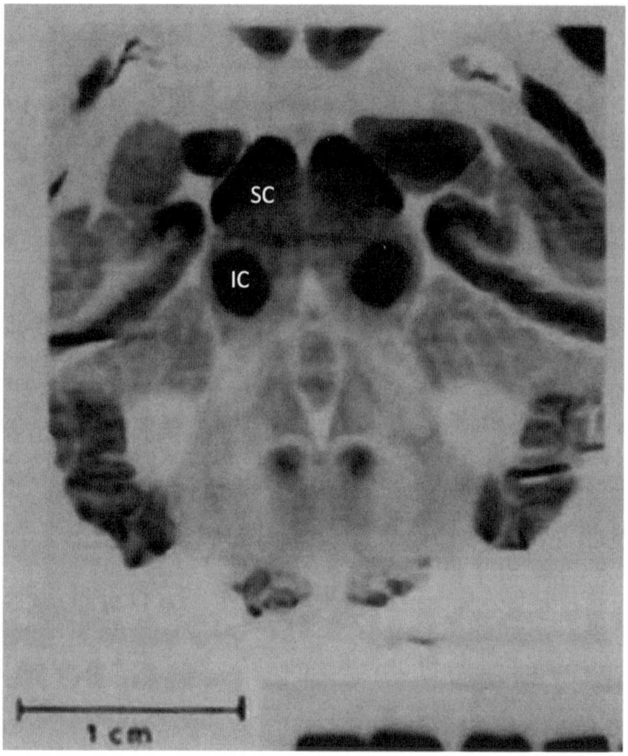

Figure 1.1 Coronal section through the midbrain of a macaque monkey. The inferior colliculi (IC, darkest areas) have the highest metabolic rate of all structures in the brain, and this rate is relatively uniform throughout the structure. This is in contrast to the superior colliculus (SC) where only the most superficial portion, the visual input part, has a high metabolic rate. Note that metabolic rate is not synonymous with spontaneous firing rate. It reflects mostly synaptic activity and does not distinguish between excitatory and inhibitory synapses. Reproduced from C. Kennedy, O. Sakurada, M. Shinohara, J. Jehle, and L. Sokoloff (1978). Local cerebral glucose utilization in the normal conscious macaque monkey. *Annals of Neurology*, **4**, 293–301, Figure 1, © John Wiley and Sons, 1978 with permission.

all be tuned to the frequency that best activates the IHC and they will be most sensitive (respond at the lowest sound levels) to that frequency, the characteristic frequency (CF). This distinction between high and low SFR fibers, as we have seen, represents properties of the intervening synapses because destruction of the IHCs destroys spontaneous activity in all ANFs innervating these hair cells. Destruction of outer hair cells (OHCs) has no such effect; it only increases the response threshold of the IHCs and thus of the ANFs through the loss of mechanical amplification of the basilar membrane movement.

In the auditory nerve of the cat or guinea pig, nearly all fibers show spontaneous activity with interspike intervals that are exponentially distributed and independent of each other (Kiang et al., 1965). This means that the duration of the interval between two action potentials is independent of the length of the previous interval. The exponential nature of the distribution indicates that there are many short intervals between spikes and few long ones. It has been suggested that fibers with the same CF, and thus possibly innervating the same IHC, have uncorrelated firing patterns (Johnson and Kiang, 1976). This is the same as saying that the spontaneous transmitter release times are uncorrelated at neighboring synapses. In contrast, the stimulus-induced release will be highly synchronous and thus ANFs innervating the same hair cell will then likely show highly correlated firing patterns. It may be that this correlation is one of the determinants of sound perception (Eggermont, 1990a). On the other hand it is known that the threshold of hearing correlates better with a significant increase in firing rate above the SFR than with the degree of stimulus locking of the firings, i.e. of correlation in their firing patterns (Javel et al., 1988). This suggests that increased neural synchrony cannot be the only aspect that determines a sound sensation.

In the IC and auditory cortex, spontaneous firing patterns are not as simple as in ANFs and burst firing is common (Eggermont et al., 1993). The spontaneous firing is likely not dependent on that in ANFs and may also exist in the absence of spontaneous activity in the auditory nerve as could be the case when all hair cells are destroyed, or the cochleas are ablated. This effect can already be seen in the ventral cochlear nucleus (VCN), where spontaneous activity disappeared after auditory nerve section whereas that in the dorsal cochlear nucleus (DCN) was not affected (Koerber et al., 1966).

Spontaneous activity in the auditory system is typically defined as neural activity that occurs in the absence of an external auditory stimulus. Spontaneous activity is reflected for instance in the patterns of spontaneous electroencephalography (EEG; Chapter 4). Some authors considered spontaneous activity as neuronal noise that adds to the stimulus-induced activity (Siebert, 1965). The presence of this internal noise results in variable stimulus-related activity and thereby sets a limit to the detection capabilities of the CNS and makes perceptual decisions probabilistic. Another viewpoint is that spontaneous activity reflects the main processes in the nervous system and that stimuli merely modulate the spontaneous activity (Rodieck et al., 1962; Luczak et al., 2009). In this view spontaneous activity acts as an information carrier. The first point of view potentially applies more to the sensory periphery, whereas the second conforms more to the central nervous system. In the auditory periphery, ANFs with high SFR tend to have lower thresholds than those with lower SFR (Liberman, 1978), suggesting that there are benefits of high spontaneous activity. There is also evidence that amplitude modulated stimuli already modulate the SFR of ANFs at levels 10–15 dB lower than those that produce appreciable changes in average firing rate, which correlates with hearing threshold (Javel et al., 1988).

These two distinct views of spontaneous activity, either as unwanted noise or as information carrier, may also determine how one views the neural mechanisms of tinnitus. If one considers spontaneous activity in the auditory system as unwanted noise, the favored concept about tinnitus is that it is likely to result from too much noise. The suggested neural substrate will then be increased SFR in the auditory system. In that case, low intensity stimulus-evoked activity will be additive to the SFR but may be perceived differently. The saturation value, i.e. the upper boundary of the neuron's firing rate, will, however, be reached at lower stimulus levels compared to normal, and sound-induced activity in the brain will be less than in normal cases (Melcher et al., 2000). On the other hand, if one considers spontaneous activity as the information carrier of the brain, sound modulates this firing rate and reorganizes it. In this model external sound can even suppress tinnitus, potentially by decreasing an underlying neural synchrony. Tinnitus in this model results from increased neural synchrony, i.e. the pathology also reorganizes the spontaneous firing times either in the form of serial correlations, i.e. burst firing, or as parallel correlations, i.e. as synchronous firing among neurons (Eggermont 1984, 2000a; Chapter 9).

As a consequence, if a person does not have a hearing loss and somatic tinnitus can be excluded, it is then possible (say, after long use of a low aspirin dose) that the emerging central gain increase (Sun et al., 2009) amplifies environmental sound in such a way that it becomes audible and the person may potentially interpret this as tinnitus or may still externalize it. In other words, can increases in central gain, i.e. hyperacusis, cause the tinnitus complaint? We will provide a partial answer in Chapter 11. As tinnitus can also occur without hyperacusis, the mechanism may be partially different from that of tinnitus with hyperacusis.

An important question now arises: given that spontaneous activity is so diverse and widespread in the auditory system why do we not perceive this spontaneous activity as some sort of sound? We might potentially perceive it as such if there is no external sound to mask it. Normal-hearing persons placed in a soundproof room nearly invariably experience some "ear noise" (Heller and Bergman, 1953) and this may well be that ubiquitous spontaneous activity. The first to describe this was Zwaardemaker (1905), a physiologist in Utrecht, the Netherlands. He built a soundproof room (details in Zwaardemaker, 1910) that can easily compete in specifications with current commercial ones and used this to measure absolute hearing thresholds. He writes (Zwaardemaker, 1905) in German:

> In dieser nur gelegentlich unterbrochenen Stille saust jedes Ohr. Nur in den ersten Augenblicken, nachdem man hineingetreten ist, spürt man nichts, aber sobald man die Aufmerksamkeit auf was überhaupt zu hören ware, gelenkt hat, nehmen jüngere und ätltere Beobachter das physiologische Ohrensausen wahr. Es ist ein eigentümliches sanftes Raüschen, wie der Wind im Walde, doch viel schwächer, eher hoch als niedrig, mit einem fast unmerkbaren, leichten, langsamen Ab- und Anschwellen ohne feste Periodik. Daneben hört man oft ein hohes Zirpen, ungefär aus der 6. Oktave. Für kurze Momente kann man auch das erste, typische Sausen verlieren; es kehrt aber recht bald

> zurfick. Dieses normale Ohrensausen, dass man nur bei grösster, absichtlich hergestellter Stille wahrnimmt, kann durch Stimmgabeltöne zum Verschwinden gebracht werden (das Tick-Tack der Uhr ist hierzu nicht imstande). Eine in der Mitte des Kabinets aufgestellte elektrisch getriebene c^1-Stimmgabel machte das physiologische Ohrensausen verschwinden,.... Man darf annehmen, dass eine Schallenergie von ungefahr 68×10^{-3} Erg pro cm^2 und pro Sekunde ausreicht, um das physiologische Ohrensausen zum Verschwinden zu bringen.

I translate: "In this only occasionally disturbed silence every ear rings. Only during the first moments after one has entered [the room], one detects nothing, but immediately after one directs his attention to anything that is audible, and both young and old observers perceive this physiological ringing. It is a particularly soft sound resembling wind in a forest, but much softer, more likely high [pitched] than low, with a nearly unperceivable, weak, slowly rising and falling amplitude without a clear periodicity. Besides, one also can hear a high [pitched] chirping approximately in the 6th octave (8192 Hz). It can happen that one loses this ringing for short moments but it comes right back. This normal ringing in the ears that one only hears in this specially created silence can be abolished by the sound of a tuning fork (tuned at $c^1 = 261.6$ Hz) but not by the 'tick-tack' of a clock. . . . I have established that a sound energy of 68×10^{-3} erg cm^{-2} s^{-1} is sufficient to make the ringing disappear."

This estimated sound energy is about 38 dB above the normal hearing threshold and may thus be the first measurement of "normal" tinnitus loudness by a masking procedure. From this one would get the impression that nearly everyone has tinnitus, but that it becomes audible only after a certain period in complete silence. This could be the result of a rapid increase in the gain of the auditory system, which takes likely more time (Formby et al., 2003), or more likely by the loss of masking by environmental sound.

It may be that normal spontaneous activity does not violate our expectations, i.e. there is no information in it and our brain actively ignores it, i.e. is habituated to it. This is likely determined by a higher-order brain function, i.e. more central than primary auditory cortex (Kral and Eggermont, 2007). There is recent evidence that auditory or visual attention affects the perception of "spontaneous" sound in a silent soundproof room (Knobel and Sanchez, 2008).

It is also possible that we do not hear spontaneous activity because it is not sufficiently well synchronized between ANFs. In auditory cortex, however, there is normally a spontaneous correlation of spike firing between pairs of cells (Eggermont, 1992a). This correlation generally increases during sound stimulation (Eggermont, 1994) and especially the correlation between spike firing in different cortical areas increases (Eggermont, 2000b). This intercortical synchrony may be crucial for perception to occur. Synchronized activity is also considerably more able to activate downstream neurons than asynchronous activity (Abeles, 1991) and it may well be that only under pathological hypersynchronization spontaneous activity may become audible as tinnitus.

Responses to an external sound adapt as a function of time since sound onset, typically with a time constant of the order of 30 ms in auditory nerve fibers (Eggermont, 1985), but in addition also at a much slower rate (time constants of the order of 1–10 s) in auditory cortex (Gourévitch and Eggermont, 2008). In auditory nerve fibers, the adapted firing rate is much less dependent on sound pressure level than the onset-firing rate (Geisler, 1998). One could equate the difference between onset-firing rate and steady-state firing rate as the information gained about the stimulus and the steady-state response as representing continuing new information being transmitted. One would expect that pure tone-induced activity would then adapt differently, i.e. more strongly, than that induced by narrow-band noise. Of course the frequency-filter properties of the neuron interact with that. Spontaneous firing rates do not adapt, or one could say that they are adapted to the level that reflects the remaining important information. Tinnitus does not adapt either, so the underlying spontaneous activity albeit increased, still behaves as normal spontaneous activity, i.e. does not adapt because it may have reached a new set point.

1.6 **Mechanisms that may be involved in SFR changes**

The main problem is to find the mechanisms that cause the increase in SFR in the CNS that potentially gives rise to tinnitus. There are several candidates—the simplest one is to assume that hearing loss causes an overall reduction in central tonic inhibition in the frequency region of the loss. There is evidence for downregulation of the inhibitory neurotransmitters, glycine and gamma-aminobutyric acid (GABA), following insults that cause hearing loss, either partial or complete, and as a result of aging. This will be described extensively in Chapters 6–10. Since SFRs in the CNS are likely determined by the balance of excitation and inhibition to which these neurons are subjected, and this balance may be subject to plastic changes in the brain as well, it could give rise to new set points for SFR.

Homeostatic mechanisms stabilize the mean firing activity of a neuron over timescales on the order of days, and typically do so by scaling the efficacy of the neuron's synapses (Turrigiano, 1999). An important aspect of synaptic scaling is that the direction of change of synaptic strength depends on both the nature of the synapse and the nature of the postsynaptic neuron. Cortical pyramidal neurons are embedded in complex networks with extensive recurrent excitatory and inhibitory feedback (Figure 1.2). Pyramidal-neuron firing rates reflect not only their excitatory drive, but also the balance between excitatory inputs from other pyramidal neurons and inhibitory inputs from GABAergic interneurons. Activity in these interneurons is, in turn, driven by excitation from the pyramidal neurons. This suggests that to increase pyramidal-neuron firing rates, excitatory input to pyramidal neurons should be increased but excitatory input to interneurons should not. The neurotrophin, brain-derived neurotrophic factor (BDNF), has these properties: it is produced by cortical pyramidal neurons and its high-affinity receptor, tyrosine kinase B (trkB), is present on both pyramidal cells and interneurons. Whereas BDNF reduces pyramidal-neuron synaptic strengths, it increases synaptic strengths of interneurons. As a consequence, manipulations that decrease BDNF levels reconfigure cortical networks to recruit more excitation onto pyramidal neurons (Figure 1.2A), and manipulations that increase BDNF levels act to recruit more inhibitory input to pyramidal neurons (Figure 1.2B).

In the healthy auditory system, homeostatic plasticity could help to ensure that auditory neurons are active within the right range of firing rates independent of the

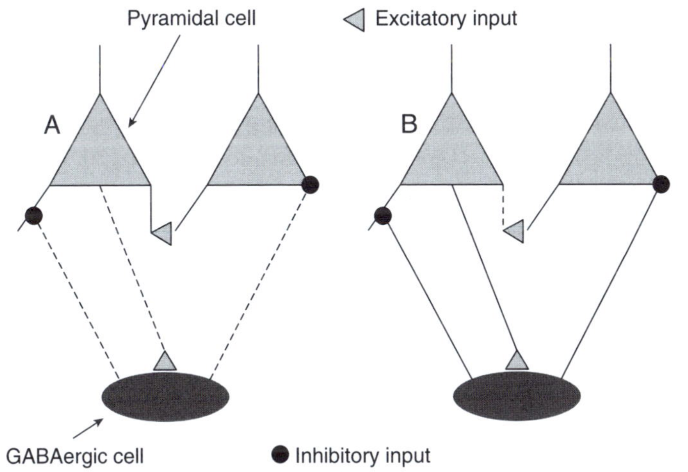

Figure 1.2 Homeostatic regulation of pyramidal-neuron firing rates by the activity-dependent release of brain-derived neurotrophic factor (BDNF). A) When pyramidal neuron firing rates fall, BDNF release from pyramidal neurons decreases. This causes excitatory connections between pyramidal neurons to increase in strength (thick line), and reduces inhibition from the interneurons (broken line). These effects combine to increase the net excitatory drive to pyramidal neurons, and to raise pyramidal neuron firing rates. B) When activity rises, BDNF release increases. This strengthens excitatory connections from pyramidal neurons onto interneurons (thick line), thus increasing the feedback inhibition onto pyramidal neurons and lowering pyramidal neuron firing rates. High activity also decreases the strength of excitatory connections between pyramidal neurons, but this effect is not mediated by BDNF, but by some other unidentified factor. Adapted from G.G. Turrigiano (1999). Homeostatic plasticity in neuronal networks: the more things change, the more they stay the same. *Trends in Neurosciences*, **22**(5), 221–27. Copyright (1999), with permission from Elsevier.

prevailing acoustic environment. Homeostatic plasticity in auditory neurons might also prevent us from perceiving spontaneous neuronal activity as sound. Schaette and Kempter (2006) modeled the effects of homeostatic plasticity by a change in a gain factor proportional to the deviation of the mean activity from a certain target rate. In their model, homeostatic plasticity restores the mean firing rate of a neuron in the DCN after hearing loss. This mean firing rate includes both driven and spontaneous parts, and both are scaled upwards to the target level. This applies to all neurons along the auditory pathway. Restoring the mean rate therefore likely increases the spontaneous rate throughout the auditory system. We will see in Chapter 8 that increased efficacy of somatosensory inputs to DCN neurons is potentially part of this upregulation of SFR.

A third proposed mechanism, superficially similar to a homeostatic one, is that the neuron acts like a hedonistic, selfish agent that strives to acquire activity (Klopf, 1982). In other words neurons "seek" excitation and "avoid" inhibition. Whereas homeostasis refers to a steady-state, the hedonistic neurons strive to optimize a particular internal variable. Klopf called this state heterostasis, and noted that learning does not occur by maintaining a steady state (homeostasis) but by heterostasis, by maximizing particular variables. In an engineering sense the hedonistic neuron acts as a self-interested adaptive element that works toward causing its input to be as high as possible according to its own measure of preference (Barto, 1985). There is a potential parallel with the selfish gene competing for survival (Dawkins, 1976), whereby a synapse is considered to compete for (excitatory) neural connections (De Ridder and Van den Heyning, 2007). This is possible if synapses in turn are "hedonistic," and responding to a global reward signal by increasing their probabilities of vesicle release or failure, depending on which action immediately preceded reward. Hedonistic synapses might learn based on a stochastic approximation to the gradient of the average reward (Seung, 2003).

In case of hearing loss, there are parts of the central auditory system that receive no or only a small signal from the auditory nerve and in that silence the brain starts to produce its own activity through any of the above mechanisms, and this results in tinnitus. With a variant of Aristotle's "Nature abhors a vacuum" one could say that the "Brain abhors silence." The phenomenology and neural substrates of this phantom sound will be the topic of this book.

1.7 Summary

Tinnitus is a perceived sound that is variable in pitch and loudness and is without an attributable external source. Even people who do not experience it in normal circumstances experience tinnitus when they are in a soundproof room for a few minutes. This suggests that normal environmental sounds are capable of masking this normal spontaneous activity. Objective forms of tinnitus can typically be traced to body sounds. If these can be excluded the remaining tinnitus is called subjective tinnitus, which from now will simply be called tinnitus. Tinnitus is person specific and its description varies widely with "ringing" or "hissing" among the most used descriptions. It is a disorder that has probably always been with us, and has been described and suffered from by poets and composers in Roman times, the Middle Ages, and Renaissance up to the present. Its current prevalence is increased by exposure to occupational and recreational noise, but not necessarily by aging. Hearing loss is the most common underlying substrate of tinnitus. It is likely a result of abnormal spontaneous neural activity and in its characteristics resembles "phantom pain" where the brain substitutes neural activity in the absence of sensory-induced ones. The accompanying increased sensitivity to sound (hyperacusis) has its similarities to the increased sensitivity to somatic stimuli (hyperalgesia) found in people with neuropathic pain.

How people get tinnitus and at what time in life is the topic of the next chapter.

Chapter 2

Epidemiology and etiology

Epidemiology is the study of factors that affect the health and illness of populations, and serves as the foundation for interventions made in public health and preventive medicine. Etiology is the study of causation or origination of an affliction, in this case, tinnitus. The first four sections deal largely with epidemiology aspects of tinnitus, and the remaining seven sections with its causes.

2.1 What is significant tinnitus?

The biggest problem in the epidemiology of tinnitus appears to be the definition of "significant" tinnitus. This restricted definition aims to eliminate cases with short transient ringing in the ear, always pure-tone like, that often are accompanied by fullness in the ear and elevated thresholds. This transient problem seldom lasts >30 s. So in order to exclude this as a type of tinnitus, the National Study of Hearing (NSH) in the UK (Davis, 1989) proposed the now widely adopted operational definition of "prolonged spontaneous tinnitus." Its key criteria are that the respondent's tinnitus must last for 5 min or longer and not be occurring only immediately after exposure to loud noise. Most significant epidemiological studies (data to follow) adhered to the ">5-min" criterion. This criterion certainly overestimates the prevalence of tinnitus whereas excluding only the few-hours duration of transient tinnitus after a loud noise exposure reduces the prevalence. For instance, Fabijanska et al. (1999) found in a sample of 10,349 participants, aged >17 years, tinnitus lasting >5 min in 20.1% of the population whereas 4.8% of the participants reported a constant tinnitus.

Nondahl et al. (2002) showed that in a group of 3753 people aged 48–92 years, those with significant tinnitus were more likely to have a history of occupational noise exposure. Davis and El Refaie (2000) reported tinnitus prevalence in the NSH of 7.5% in people with little or no noise exposure compared to 20.7% for people with a high lifetime-level of noise exposure; however, the contribution of noise exposure was non significant once hearing loss (likely caused by the noise exposure) was taken into account. So the determining etiological factor would be the presence of hearing loss, regardless of how it was acquired. This puts gradual life-long noise exposure in the same league with a single traumatic noise exposure and with ototoxic drugs. It is presently far from clear if gradual acquired hearing loss results in tinnitus in the same percentage of people as that resulting from a single noise trauma or short-duration application of high levels of ototoxic drugs such as an aminoglycoside or cisplatin (Dille et al., 2010).

2.2 **Tinnitus prevalence across the life span**

Tinnitus prevalence is defined as the percentage of people with tinnitus in a certain age group (usually with a range of 10 years). I have compiled these data based on three reviews and the original publications contributing to those overviews as well as on more recent papers not included in those three reviews. One review provides an in-depth reanalysis of a few large epidemiology studies (Hoffman and Reed, 2004), one study also covered some older epidemiology where different criteria for inclusion of tinnitus were used (Davis and El Refaie, 2000), and the third one presented a more general (but without a prevalence by age group) overview of a larger number of epidemiology studies (Sanchez, 2004). All in all they covered 14 studies that illustrate an upward trend of tinnitus prevalence with age that is generally the same for all studies but where the absolute levels depend on the questions asked and the type of tinnitus included.

The prevalence of tinnitus by age group in some large studies covering two Scandinavian countries, the USA, and the UK is shown in Figure 2.1. I added a recent study covering 14,178 participants in the 1999–2004 National Health and Nutrition Examination Surveys (Shargorodsky et al., 2010). In this particular study a distinction was made between any, i.e. occasional, tinnitus and frequent tinnitus. I included only the last category. Note that the early studies (Hinchcliffe, 1961; Leske, 1981) did not use the ">5 min" criterion and did not exclude tinnitus occurring only after noise exposure or colds. Therefore they found approximately twice the values obtained in the newer studies. The newer studies generally adopted the ">5 min" criterion, but did not always exclude the noise criterion, which introduced the larger prevalence in the Swedish and Norwegian studies. The lower

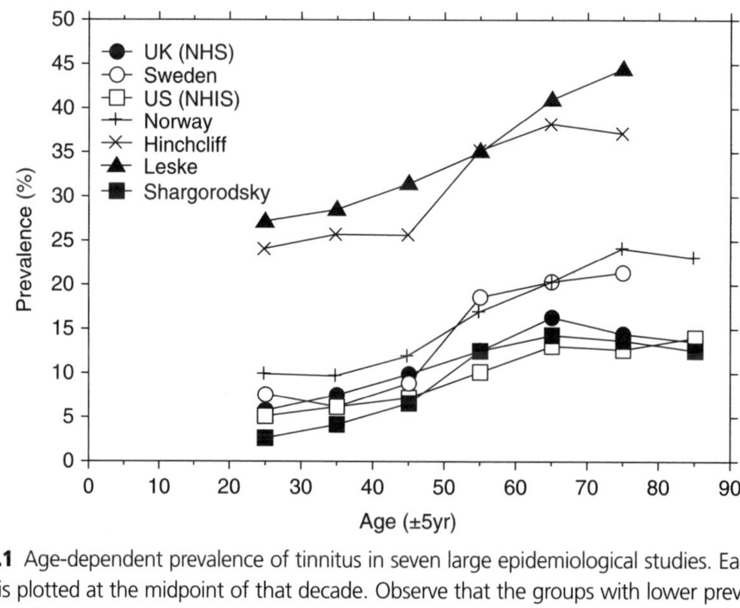

Figure 2.1 Age-dependent prevalence of tinnitus in seven large epidemiological studies. Each decade average is plotted at the midpoint of that decade. Observe that the groups with lower prevalence conform to the ">5-min" criterions, whereas the top two also cover occasional tinnitus.

value for the US (NHIS) study resulted from the requirement that the tinnitus be bothersome. The overall prevalence of the tinnitus in the sample groups were: US 8.4%, UK 10.1%, Sweden 14.2%, and Norway 15.1%. One observes a tendency for the prevalence to level off in the seventh decade of life.

Sanchez (2004) surveyed representative studies covering a wide range of European countries, the US and Australia, of which only two are also incorporated in Figure 2.1. Most studies included the >5 min criterion for prolonged tinnitus or an acceptable equivalent. The overall prevalence is shown in Table 2.1.

A few studies need some comment. The Beaver Dam study in the US reported troublesome tinnitus in 8.2%; this value is by definition much lower than for the less restricted >5 min criterion. The Blue Mountain study in Australia found a staggering 30.3% (>5 min) in an older group; a reason for the deviation from other studies was not presented. Another Australian study covering people older than 70 years of age found a prevalence of "only" 17.8%, suggesting again that the prevalence may go down after 70. Finally, a UK study covering "working" people found prolonged tinnitus in 27% (25%) of the male (female) population, of which 6% (3%) was classified as continuous.

This overview of epidemiological surveys gives the impression that the prevalence of occasional tinnitus may be around 35%, whereas prolonged tinnitus (>5 min) averages at about 18% (range 10.1–30.3%), and continuous tinnitus occurs only in approximately 5% of the population. Bothersome tinnitus had a prevalence of about 8% (the two US studies), a number close to that of continuous tinnitus.

In a first report suggesting that ethnicity affects tinnitus, Shargorodsky et al. (2010) showed that non-Hispanic blacks and Hispanics had lower prevalence of frequent tinnitus than non-Hispanic whites. Although decreased prevalence in hearing loss has been reported previously in non-Hispanic blacks and Hispanics compared with non-Hispanic whites, the fact that significant associations between ethnicity and tinnitus were maintained in participants without hearing impairment suggests a mechanism contributing to tinnitus that is independent of hearing impairment, i.e. potentially a genetic one.

Table 2.1 Prevalence of prolonged spontaneous tinnitus

Population	Prevalence	Authors
Sweden, 20–80 yrs	14.2%	Axelsson and Ringdahl (1989)
Denmark, 53–75 yrs	17%	Parving et al. (1993)
Italy, >18 yrs	14.5%	Quaranta et al. (1996)
Poland, >17 yrs	20.1%	Fabijanska et al. (1999)
Germany, >10 yrs	13%	Pilgramm et al. (1999)
Australia, >70 yrs	17.8%	Sanchez et al. (1999)
USA, 48–92 yrs	8.2%	Nondahl et al. (2002)
UK, 16–64 yrs	27%	Palmer et al. (2002)
Australia, 49–97 yrs	30.3%	Sindhusake et al. (2003)

2.3 Tinnitus in children

Tinnitus is very prevalent in school-age children (Table 2.2) and about twofold higher than for the 20–30-year-olds and thus more in line with the "occasional" (<5 min duration) tinnitus (Figure 2.1). This suggests that no distinction between <5 min and >5 min duration was made either in the questions or in the answers. Graham and Butler (1984) reported on a group of 158 children with sensorineural hearing loss in which 49% had tinnitus. Children with a slight to moderate hearing loss more often experienced tinnitus (66%) than children with a severe hearing loss (29%). Mills et al. (1986) questioned 93 healthy children attending routine school and community medical examinations about tinnitus, 27 of them (29%) described tinnitus. Nine children (9.6%) were bothered by their tinnitus whereas the others were not. The incidence of tinnitus in this group was compared with a group of 109 children with ear disease. Forty-two children (38.5%) in this group reported tinnitus. This difference was not statistically significant. The reported frequency of occurrence varied between twice a day and three to four times a year. Holgers (2003) reported a tinnitus prevalence of 12% in 7-year-old school children (n = 964). Audiometric data did not correlate with the prevalence of tinnitus and no gender differences were found. Tinnitus immediately following noise exposure was reported in 2.5% of the children. Holgers and Juul (2006) studied 274 students between 9–16 years of age from a mainstream school. Children younger than 9 years of age were not included due to their problems in estimating time aspects of tinnitus. In this group 53% of the children had experienced tinnitus following noise exposure at least once, and 46% had experienced tinnitus without noise exposure. Age was not significant with respect to how often the children perceived tinnitus. 22.6% of the children reported tinnitus to be annoying sometimes or more often.

None of the categories in Table 2.2 ("Normal hearing," "Hearing impaired," or "Ear disease") alone or combined show a trend with year of publication, suggesting that the prevalence basically has stayed the same over more than three decades.

In a large study of 1100 children, normal hearing as well as hearing impaired, between 6–16 years (mean age 11.9 years) 374 reported tinnitus when asked and 71 spontaneously complained about it (Savastano, 2007; Savastano et al., 2009). The reported sounds covered a wide range and suggested in many cases objective tinnitus (banging, clicking, like a machine, beep-beep) but in a majority of cases they were named similar to those in adults. Of these, ringing occurred in 22.7% comparable with the average percentage of 27.7% in adults (Chapter 1, Table 1.1). However, buzzing was mentioned much more frequently in children (18.9% vs. 3.5%), as were humming (8.8% vs. 4.6%), whistling (7.5% vs. 2.2%), and sea noise (6.4% vs. 1.7%). No cricket sounds were reported. Brunnberg et al. (2008) found tinnitus always or often present in 39% of 15–16-year-olds with hearing impairment, and in 6% of those with normal hearing from mainstream schools in the county of Örebro, Sweden.

Distributions of the age at which complaints about tinnitus peak in two large studies are age 15 (Holgers and Juul, 2006) and age 12 (Savastano, 2007). This could point to changes in the hormonal regulatory mechanisms in the brain such as those changing stress responsiveness (Ernst et al., 2009), a known factor in tinnitus generation.

Table 2.2 Prevalence of prolonged spontaneous tinnitus in children

Population	Prevalence	Authors
NH	13.3%	Nodar (1972)
HI	64%	Graham (1979)
HI	66%	Graham (1981)
Ear disease	22%	Graham (1981)
HI	29.5%	Mills and Cherry (1984)
Ear disease	43.9%	Mills and Cherry (1984)
HI	55%	Nodar and LeZak (1984)
HI, N = 158	49%	Graham and Butler (1984)
NH, N = 93	29%	Mills et al. (1986)
Ear disease, N = 109	38.5%	Mills et al. (1986)
HI	23%	Viani (1989)
HI	50%	Martin and Snashall (1994)
NH, 7-yr-olds, N = 964	12%	Holgers (2003)
NH, 9–16 yrs, N = 274	46%	Holgers and Juul (2006)
NH, HI, 6–16 yrs, N = 1100	34%	Savastano (2007)
NH, 15–16 yrs, N = 2730	6%	Brunnberg et al. (2008)
HI, 15–16 yrs, N = 148	39%	Brunnberg et al. (2008)

NH, normal hearing; HI, hearing impaired.

2.4 Tinnitus in the elderly

As the earlier-mentioned prevalence studies show, tinnitus is about twice as frequent in the elderly as in young adults. This may be related to the prevalence of hearing loss and other age-related diseases. There are many changes in life that occur with aging (Chapter 10), such as illness, retirement, and reduced social activity, and these may result in changes in stress levels. These life events have the potential to increase the perceived loudness of tinnitus or to exacerbate the response to tinnitus (Henry and Wilson, 2001). Some patients report stress as the precipitating factor for their tinnitus (Meikle and Griest, 1989). The increase in tinnitus prevalence in older patients does not necessarily mean that tinnitus as a separate and distinct symptom will increase with age (Hoffman and Reed, 2004). However, McFadden (1982) has argued that age-related tinnitus presents a distinct pathology and is related to degeneration at all levels of the auditory system. As we will see (Chapter 10) it is not clear if that is really the case.

Among the epidemiological studies that included only older adults, the study by Nondahl et al. (2002) of 3753 people aged 48–92 years, reported no statistically significant difference in the prevalence of tinnitus by gender or by age group. The Blue Mountain study of people aged 55 years and over (Sindhusake et al., 2003; Gopinath et al., 2010a) also did not find a significant age-related change in the prevalence of tinnitus. Sanchez et al. (1999) reported tinnitus data from a study of people >70 years of age. As mentioned earlier, participants were grouped according to the presence or absence of tinnitus at baseline and at 2-year follow-up. The proportion of female and male participants and their ages in each of the four groups were also considered. There was no relationship between tinnitus status and either gender or age, and no interaction between tinnitus status, gender, and age.

As we have seen in Chapter 1, tinnitus comes in two types—objective and subjective tinnitus— and these have different causes. Objective tinnitus can be heard both by the patient and the examiner and is the result of body sounds. Subjective tinnitus can only be perceived by the patient, and is defined here as a sound experienced in the absence of any external or internal physical sound source.

2.5 **Body sounds (objective tinnitus)**

Clicking tinnitus may indicate a mechanical cause for the tinnitus, often resulting from repetitive contractions of muscles in the middle ear or nasopharynx. An open Eustachian tube can also result in pulsatile tinnitus, secondary to inward and outward movement of the tympanic membrane in association with respiratory movements (Liyanage et al., 2006). Objective pulsatile tinnitus has numerous causes, including benign intracranial hypertension, glomus tumors, and atherosclerotic carotid arteries. Pulsatile tinnitus is almost always the result of the sound of turbulent blood flow that is transmitted to the inner ear. This can occur in systemic diseases causing a general alteration of the hemodynamics or in local disorders that are anatomically close to or within the petrous bone (Waldvogel et al., 1998). Hearing one's own heart murmur has been reported in adults; Anderson et al. (2009) reported the case of a young child who clearly heard her own venous hum, causing her to complain of pulsatile tinnitus. Venous hums are innocent heart murmurs that are caused by turbulence in the jugular veins. These veins often narrow as they pass underneath the collarbone and the narrowing causes the sound often described as "whooshing." Typically, the intensity of venous hums changes as the patient turns his or her head from side to side, thus helping the clinician confirm the diagnosis. Undetected venous hums, especially when due to jugular flow that is dominant at night, may potentially be underlying some cases of subjective tinnitus in the absence of hearing loss (Bektas and Caylan, 2008).

Sounds generated by the ear itself have also been suggested as causes of objective tinnitus. These spontaneous otoacoustic emissions (SOAEs) are low-level sounds of very precise and constant frequency emitted by the healthy normal ear, and are recordable with sensitive microphones inserted in the ear canal. Otoacoustic emissions are generated by the OHCs and they diminish in size with increasing OHC loss. In about 6–12% of tinnitus sufferers with normal hearing, SOAEs are considered at least partially responsible for the tinnitus (Lonsbury-Martin and Martin, 2004). This is somewhat surprising as one expects the brain to habituate to continuous and only mildly changing sounds (Pienkowski and Eggermont, 2009). In most cases, however, SOAEs and tinnitus are independent phenomena. In this respect it is interesting to note that salicylate, which in itself can cause tinnitus, abolishes SOAEs in the acute usage phase, and this could function as a simple test for the co-occurrence of emissions and tinnitus (for more details see Chapter 4).

2.6 Subjective tinnitus of peripheral origin

In 1831, J. Fosbroke, M.D. stated in *The Lancet* that:

> Deafness varies from a diminution of hearing, to an almost extinction of the sense, A *noise in the ears*, resembling either the roar of the see, the ebullition of boiling water, or the rustling of the wind among trees, accompanied sometimes with *noise in the head*, exists in *almost every* case of deafness, to whatever cause the deafness may be owing. (Emphasis in the original.)

Subjective tinnitus most often accompanies noise-induced hearing loss (NIHL), but also co-occurs in patients with age-related hearing loss (presbycusis), Ménière's disease, and sudden deafness. Impacted cerumen in the ear canal, middle ear problems such as otosclerosis and otitis media, and other forms of hearing loss can also cause tinnitus. Removing the cerumen relieves the tinnitus (Roesser and Ballachandra, 1997). Such mild conductive hearing losses likely result in the perception of tinnitus by reducing the loudness of environmental sounds that normally mask tinnitus. The presence of a small air–bone gap (>5 dB) appears to be strongly correlated with unilateral tinnitus lateralized to the air–bone gap ear (Davis, 1995). The conductive hearing loss could unmask the tinnitus that is potentially present in every normal hearing subject, where it only becomes audible in very quiet environments (Mills et al., 1986; Del Bo et al., 2008). Infections such as otitis media, and other infectious or inflammatory processes that affect hearing thresholds, comprise about 8% of tinnitus cases.

In a sample of 1037 children that were part of the Dunedin (New Zealand) Multidisciplinary Health and Development Study, otitis media was assessed at 5.7 and 9 years of age (Dawes and Welch, 2010). At 32 years of age, 970 of the 1015 living study members (96%) answered questions about tinnitus. Childhood otitis media with an associated hearing loss in the low and high frequencies was associated with a greater probability of experiencing tinnitus in adulthood. Neither childhood otitis media alone nor elevated thresholds alone predicted adult tinnitus.

In a study of >1000 patients with otosclerosis, the mean hearing threshold levels for air conduction and bone conduction at the speech frequencies did show a significant association with tinnitus, but not in accordance with what was expected (Gristwood and Venables, 2003). Namely, the probability of finding a patient with tinnitus actually fell with an increase in the bone conduction and air conduction mean hearing levels.

2.6.1 Noise-induced hearing loss

Occupational noise exposure Waterfowl hunters are a striking example of a population at risk for tinnitus. This group of recreational firearm users typically does not utilize sound barriers or earmuffs to protect against loud sound, and high incidences of hearing loss and tinnitus are found (Stewart et al., 2009). Waterfowl hunters typically shoot using large-bore semiautomatic (12-gauge) shotguns, which generate peak sounds in excess of 160 dB sound pressure level (SPL), often position themselves in reverberant enclosures (blinds of hides) during hunting activities, frequently hunt in groups, and are often exposed to multiple shots in a single outing from both their own firearm and those of companion hunters. Consequently they nearly all have NIHL and tinnitus.

Loud noise exposure is considered an important risk factor for developing tinnitus. Consequently, a history of recreational, occupational, and firearm noise exposure may all be associated with increased likelihood of acquiring tinnitus. The relation between noise exposure and frequent tinnitus, however, differed depending on the presence or absence of hearing impairment. Occupational noise exposure was associated with increased odds of frequent tinnitus in participants with hearing impairment, while leisure-time noise exposure was associated with increased odds of frequent tinnitus in participants without hearing impairment (Shargorodsky et al., 2010).

Patients with traumatic brain injury form a new particular group with tinnitus complaints (Lew et al., 2007). Traumatic brain injury often results from blast-related injury caused by explosives that emit over-pressurization shock waves or "blast waves." Because blast waves affect both gas- and fluid-filled structures (such as the middle and inner ear), they tend to be destructive to the auditory system.

About 20% of tinnitus occurrence is related to NIHL. Regulations of the Occupational Safety and Health Administration (OSHA), or equivalent, limit the level of daily noise exposure in the workplace. However, regulations about admissible sound levels do not generally apply (with the UK being an exception) to recreational areas such as bars, sport, and concert venues, and if they do apply then typically only to what is audible outside the venue. Ear protection inside these establishments is advisable but not considered a social grace, and generally not complied with. For instance, the average noise levels in a hockey arena can be more than 100 dB(A) during the 3-hour game duration (Hodgetts and Liu, 2006; see also Saunders and Griest, 2009 for more examples). Taking 85 dB(A) and 8-hour exposure as the industrial norm (e.g. in the USA and most of Canada) and allowing an exchange of 5 dB for time doubling or halving, would then result in an allowable exposure time at 100 dB(A) of about 20 min per day. Often these games are watched in addition to the 8-hour allowable level of workplace noise and the cumulative effect of these exposures will result in hair cell damage.

Music-induced hearing loss The average sound level at rock and pop concerts has been shown to be about 95 dB(A), with a range of 73–109 dB(A). Classical musicians are also often exposed to sound levels >85 dB(A) for long periods of time, both during practice and performance, resulting in a high prevalence of tinnitus and temporary threshold shifts in this group (Jansen et al., 2009). Personal listening devices such as MP3 players that can produce outputs of 91–121 dB(A) at maximum settings (Saunders and Griest, 2009) are new and potentially harmful sources of recreational noise and could cause or exacerbate hearing loss, especially when used on a daily basis and for extended time periods. Yet, in a study that surveyed audiometric hearing loss in a group (N = 2526) of young workers entering the workforce of a large company, with workplaces widely distributed across the USA, over a 20-year period ending in 2004, Rabinowitz et al. (2006) found no difference in the prevalence of hearing loss over this 20-year period. One might argue that the personal listening device fashion was not adequately sampled, as this had barely started around 2000, but these data suggest that the presumption of increased hearing loss in young adults does not apply to all groups. However, more recently Shargorodsky et al. (2010) found that the prevalence of any hearing loss in 12–19-year-olds increased significantly from 14.9% in 1988–1994 (N = 2928) to 19.5% in 2005–2006 (N = 1771; $P = 0.02$). In 2005–2006, hearing loss was more commonly unilateral and involved the high frequencies. This is somewhat unexpected since personal listening devices typically have bilateral insert phones and one would expect a more symmetrical hearing loss.

Using transient-evoked otoacoustic emissions (TEOAEs) Rosanowski et al. (2006) investigated whether young adults who frequently visit discos would acquire OHC damage. Eighty-eight young adults (47 women, 41 men; average age about 23 years) were examined. None suffered from permanent tinnitus. Sixteen percent reported transient tinnitus after every visit to a discotheque and 58% after nearly every visit. Eight percent suffered from transient hearing loss after every visit to a disco and 37% after nearly every visit. Three percent reported tinnitus every morning after visiting a disco and 4% nearly every morning after. TEOAE levels and reproducibility decreased significantly with an increased number of visits to discos, suggesting OHC damage without measurable pure-tone hearing loss. It has been recently shown in young mice that noise exposure that causes only temporary threshold shifts may cause delayed damage in spiral ganglion cells that shows up in adulthood (Kujawa and Liberman, 2009).

Disc jockeys' (DJs') hearing loss or tinnitus may be related to their exposure to music and length of time in the profession. Potier et al. (2009) surveyed a group with an average age of ~26 years (SD = 6 years) that were on average 6.6 years in that profession and were on average exposed for ~22 h (SD = 13 h) weekly. Seventy-six percent of them had tinnitus of which in 89% it was perceived as inside the head. Most frequently the tinnitus pitch was high (64%). The DJs' audiograms (Figure 2.2) showed the expected hearing loss at 6 kHz, but also low-frequency losses at 125–500 Hz. Three-quarters of them had tinnitus with a pitch corresponding to the frequencies of the hearing loss. These results are similar to those of Bray et al. (2004) who described 74% of DJs as having tinnitus, albeit that only 3/23 had NIHL. Potier et al. reported that the tinnitus was described as a whistling sound in 64% of cases, as a buzzing in 27%, and as both types in 9%. Bogoch et al. (2005) found that tinnitus and other hearing symptoms were experienced by 85% and 38% of attendees in nightclubs, respectively. Tinnitus pitch matching confirms these descriptions with two-thirds of the tinnitus matching frequencies between 4–8 kHz, and the other third, described as buzzing, characterized between 125–500 Hz (Figure 2.2 bottom part). These results are in accordance with studies showing that the pitch of the tinnitus generally corresponds to the predominant hearing loss frequencies (Noreña et al., 2002; Chapter 3).

Schmuziger et al. (2006) evaluated both ears in 42 non-professional pop/rock musicians and included pure-tone audiometry in the conventional and extended high-frequency range. They also assessed uncomfortable loudness levels, tinnitus, and hyperacusis. Exclusion criteria were, among others, the presence of acoustic trauma and excessive noise exposure during occupational activities not including their music activity. Thus, the observed effects could potentially be attributed as music-induced hearing loss. Eleven of the musicians (26%) were found to be hypersensitive to sound, and seven (17%) had tinnitus. There was no clinically significant psychological distress in these musicians.

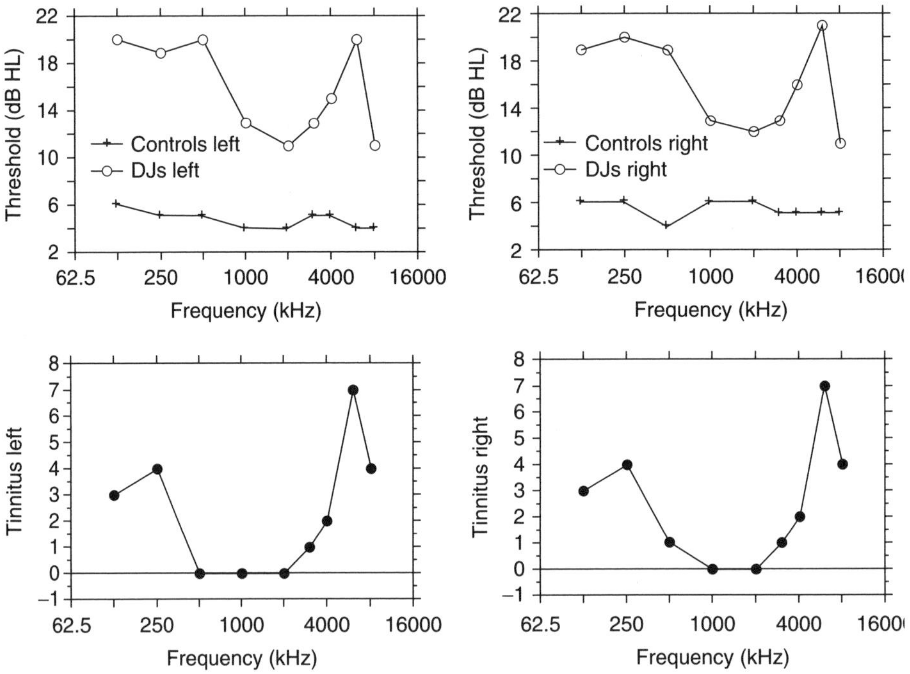

Figure 2.2 Audiogram and tinnitus pitch matching in DJs. Control audiograms are also shown. Note the hearing losses at 6 kHz and in the low frequencies. The pitch of the tinnitus, here plotted as the number of ears vs. tinnitus pitch, is correlated with the hearing loss frequencies. Data from Potier et al. (2009).

Hearing loss as primary cause of tinnitus Is noise or music exposure a primary cause of tinnitus? Rubak et al. (2008) investigated the relationship between noise exposure and tinnitus among workers with normal hearing and hearing loss, respectively. They conducted a cross-sectional survey of 752 workers employed at 91 workplaces in Aarhus County, Denmark. Tinnitus was not associated with the workers' present occupational noise level, with the duration of occupational noise exposure, or with the cumulative occupational noise exposure if participants had normal hearing. As expected, tinnitus was correlated with these noise exposure aspects if participants had a hearing handicap. Based on these data, one has to be cautious in ascribing tinnitus to noise exposure in the workplace if the person has a normal audiogram. These data indicate no risk of noise-induced tinnitus at exposure levels where no hearing loss would be expected, e.g. as usually encountered in non-industrial workplaces. In animals, however, noise exposure without permanent thresholds shift (Kujawa and Liberman, 2009) may over time cause patchy degeneration of ganglion cells and this could be causal to a later-onset tinnitus in the presence of a normal audiogram.

2.6.2 Ototoxic drugs

Tinnitus is a frequent side effect of many commonly used or prescribed drugs, including salicylates (aspirin), non-steroidal anti-inflammatory drugs, quinine, aminoglycoside antibiotics, loop diuretics such as furosemide, chemotherapeutic agents such as cisplatin and carboplatin, and commonly-accepted recreational drugs such as nicotine, caffeine, and alcohol. However, since most of drug-induced tinnitus is reversible when the drug use is stopped (except tinnitus accompanying a permanent hearing loss caused by aminoglycoside antibiotics, cisplatin, and carboplatin, or long-duration high-dose salicylate use), it contributes only 2% to chronic tinnitus cases. Drugs that are sometimes blamed for causing tinnitus include lidocaine, anticonvulsants, antidepressants, cannabinoids, antihypertensives, beta-adrenergic blocking agents, opioids (buprenorphine), and antihistamines (Enrico and Goodey, 2011). I will describe the effects of the most common tinnitus-causing drugs.

Cisplatin, a drug commonly used in the treatment of recurrent pediatric brain tumors, produces a high-frequency hearing loss not unlike that caused by aminoglycosides (Coupland et al., 1991) by initially affecting predominantly the OHCs. In contrast, carboplatin at moderate dose largely affects the IHCs (so far only definitively shown in chinchillas), thereby reducing the output of auditory nerve fibers in specific frequency regions accompanied by a mild audiometric hearing loss (Takeno et al., 1994). Despite the very different effect upon the cochlea, both carbo- and cisplatin produce tinnitus-like behavior (Chapter 5) in chinchillas (Bauer et al., 2008).

In a prospective study evaluating the occurrence of tinnitus in 488 veterans receiving chemotherapeutic agents (cisplatin, carboplatin), ototoxic antibiotics (primarily aminoglycoside), or non-ototoxic drugs (control medications), Dille et al. (2010) found that subjects with exposure to ototoxic medications had significantly increased risk for developing tinnitus. Those on chemotherapeutic agents were found to have the greatest risk. Cisplatin elevated the risk by 5.53 times while carboplatin increased the risk by 3.75 over non-ototoxic control medications. Ototoxic antibiotics resulted in borderline risk (2.81) for new tinnitus.

Ototoxicity is a common side effect of high-dose aspirin (salicylate) treatment in patients with rheumatoid arthritis that often results in hearing loss. Low-dose daily aspirin administration in animal models presumably causes tinnitus without hearing loss (Bauer et al., 2000; Cazals, 2000), so it is possible that it could occur in humans who use a maintenance dose for cardiovascular or other reasons and show oversensitivity to the drug. Hearing loss and tinnitus intensity generally increase progressively with the aspirin dosage and increasing plasma salicylate concentrations (Day et al., 1989).

Quinine, previously used to prevent night cramps (e.g. in long-distance runners) and as an anti-malaria drug, is a potent potassium channel blocker. Quinine works on the cochlea by partially blocking the sensory gates at the tips of the hair cell stereocilia, thereby producing a hearing loss. By its interference with K^+ channels in peripheral as well as central axons it also prolongs the duration of action potentials (Lin et al., 1998), which could result in a stronger depolarizing effect on postsynaptic neurons, hence resulting in increased firing rates and thus could contribute to tinnitus.

Combined noise and aminoglycoside exposure can lead to auditory threshold shifts greater than simple summation of the effects of the two insults. The synergistic toxicity of acoustic exposure and aminoglycoside antibiotics is not limited to simultaneous exposures.

A prior acoustic insult, which does not result in permanent threshold shifts, potentiates aminoglycoside ototoxicity. In addition, exposure to sub-damaging doses of aminoglycosides aggravates noise-induced cochlear damage (Fausti et al., 2005; Li and Steyger, 2009).

2.6.3 Caffeine, alcohol, and smoking

Caffeine, by activation of the ryanodine receptors, increases the tendency of the intracellular calcium release channels in the endoplasmic reticulum to open and thus increases intracellular free calcium (Berridge, 1997). This could lead to increased spontaneous transmitter release in IHCs and thus increases in SFRs. At the synapse level, various synaptic receptors and channel activities are modulated by caffeine via this mobilization of intracellular calcium, inhibition of phosphodiesterase, antagonism of adenosine receptors and GABA receptors. These actions of caffeine enable neurons to induce plastic changes in the properties of synaptic activities, such as synaptic transmission efficiency and morphology. At the network level, caffeine has the ability to activate cortical neural oscillators that deliver repetitive NMDA receptor-dependent signals to surrounding areas, causing strengthening of long-range intercortical communications. Caffeine might thus allow reorganization of cortical network functions via synaptic mobilizations (Yoshimura, 2005) that may be important for tinnitus (Chapters 4 and 11).

Sininger et al. (1992) reported tinnitus in an individual that could be induced by drinking a cup of good strong coffee, and presented some objective indicators for its presence. However, no general correlation appears to exist between enjoying or abstaining from coffee and changes in the nature of existing tinnitus (Kemp and George, 1992; Juliano and Griffiths, 2004).

The data on a relationship between alcohol and tinnitus are also insufficient to draw any firm conclusions. McFadden (1982), Goodey (1981), and Quaranta (1996) found a positive correlation between alcohol consumption and tinnitus. On the other hand, Kemp and George (1992) and Pugh et al. (1995) found no significant effects of alcohol consumption on tinnitus. However, moderate alcohol consumption (odds ratio, OR = 0.63 for ≥141 g/week vs. <15 g/week) was found to be associated with decreased risk (Nondahl et al., 2010).

In the Blue Mountain Hearing Study (Gopinath et al., 2010b) of 2956 participants (aged 50+ years) alcohol consumption and smoking status were measured by using an interviewer-administered questionnaire. Logistic regression was used to obtain ORs with 95% confidence intervals that compared the chances of having hearing loss in participants who did or did not smoke or consume alcohol, after adjusting for other factors previously reported to be associated with hearing loss. Cross-sectional analysis demonstrated, in agreement with Nondahl et al. (2010), a significant protective association between the moderate consumption of alcohol (>1 but ≤2 drinks/day) and hearing function in older adults (compared with non-drinkers), OR = 0.75 (95% confidence interval, 0.57–0.98).

Nicotine affects the cochlea (Maffei and Miani, 1962) and smoking is accompanied by a higher incidence of high-frequency hearing loss. Cruikshanks et al. (1998) found that current smokers were 1.69 times as likely to have a hearing loss as non-smokers (95% confidence interval, 1.31–2.17). This relationship remained for those without a history of occupational noise exposure and in analyses excluding those with non-age-related hearing loss. There was weak evidence of a dose–response effect. Non-smoking participants who lived with a smoker were more likely to have a hearing loss than those who were not exposed to a household member who smoked (OR = 1.94; 95% confidence interval, 1.01–3.74). The inclusion of leisure-time noise exposure or the number of medications used did not alter the results (data not shown). Cigarette smoking may affect hearing through its effects on antioxidative mechanisms or on the vasculature supplying the auditory system. The Blue Mountain Hearing Study (Gopinath et al., 2010b) found that smokers that were not exposed to occupational noise had a significantly higher likelihood of hearing loss after adjusting for multiple variables, OR = 1.63 (95% confidence interval, 1.01–2.64). The interaction between smoking and noise exposure was not significant.

Follow-up studies in the "Established Populations for Epidemiologic Studies of the Elderly" sample (Ferrucci et al., 1998) found among the 10,118 participants, 1406 (12.4%) that reported hearing problems at baseline; of those with no baseline hearing problems and complete follow-up information (n = 8495), 1120 (13.2%) developed new hearing problems. Smoking was associated with higher prevalence and incidence rates of hearing impairment. In both cases the association was weak although statistically significant. Compared with participants with no history of smoking, those who had ever smoked were more likely to report hearing problems at baseline (OR = 1.2; 95% confidence interval, 1.0–1.3) and more likely to develop new hearing problems over the follow-up period (OR = 1.6; 95% confidence interval, 1.4–1.8).

The significant associations between tinnitus and smoking and hypertension in the study by Shargorodsky et al. (2010) also suggested that vascular disease might have a greater contribution to the etiology of tinnitus than previously reported. They showed that current and past smoking confer increased odds of experiencing tinnitus. This was corroborated in the study by Nondahl et al. (2010) showing that the risk of developing tinnitus in a 10-year period was significantly associated with a history of ever smoking (OR = 1.40).

2.7 **Tinnitus of somatic origin**

The characteristic features of somatic tinnitus (Levine, 1999) are: (1) tinnitus is closely associated with factors related to the head or upper neck; (2) tinnitus is always described as coming from the ear ipsilateral to the somatic event, and comprises a high-pitched constant ringing; (3) there are no other associated hearing or vestibular complaints and no abnormalities on the neurological examination; (4) pure tone and speech audiometry of both ears is symmetric and often within normal limits. It is generally assumed that somatic (craniocervical) tinnitus arises from somatic–auditory interactions occurring within the central nervous system (see Chapter 8). Specific origins of somatic tinnitus are temporomandibular joint (TMJ) disorder, dental problems, and head and neck injuries.

2.7.1 Temporomandibular joint disorder and dental problems

TMJ dysfunction and other dental disorders frequently co-occur with tinnitus, suggesting a role of the trigeminal nerve in tinnitus. Bernhardt et al. (2004) studied 4228 subjects of a population on the south shore of the Baltic Sea, screened for cofactors of tinnitus. Increased ORs were found for tenderness of the masticatory muscles (OR = 1.6 for one to three painful muscles and OR = 2.53 for four or more painful muscles), TMJ tenderness to dorsal cranial compression (OR = 2.99), and frequent headache (OR = 1.84) A relationship between tinnitus and TMJ was established in both examinations. Tinnitus patients seem to suffer especially from myofascial and TMJ pain. Tuz et al. (2003) surveyed 200 TMJ patients in Ankara, Turkey and found between 44–59% of the patients complaining about tinnitus. This was much higher than in an age-matched control group where the percentage was 26%.

2.7.2 Head and neck injuries

Head and neck injuries, including whiplash, cause about 10% of tinnitus cases (Folmer and Griest, 2003). Head and neck injuries may affect the auditory system in various ways as the potentially injured cranial and spinal nerves affect both the vascular supply to the cochlea and also innervate the granule cells in the DCN (Shore and Zhou, 2006).

In addition, hearing dysfunction can occur as a consequence of head injuries because of damage to the auditory nerve, the cochlea, or middle ear including the tympanic membrane and ossicular chain. Whiplash injuries can cause abnormal firing activity in the medullary somatosensory nuclei (MSN) located in the lower medulla and upper cervical spinal cord. The MSN normally provides inhibitory innervation to the DCN. During some whiplash accidents, MSN to DCN pathways are damaged. Subsequent disinhibition of the DCN and associated auditory structures could result in the generation and perception of tinnitus (Levine, 1999).

Increased transient otoacoustic emission amplitudes were found in subjects with normal audiograms and tinnitus resulting from head trauma. This included a doubled incidence of spontaneous acoustic emissions, and reduced contralateral tinnitus suppression. These findings point to a potential deficit in the medial olivocochlear bundle activity to the OHCs (Ceranic et al., 1998). Thus, stimulated otoacoustic emissions are useful in the delineation of some mechanisms involved in cochlear functioning that might accompany tinnitus. As we have seen in section 2.5, the increased SOAEs do not generally relate to tinnitus itself.

2.8 Genetics of tinnitus

Tinnitus appears to have a low heritability according to a large study (Kvestad et al., 2010) comprising a sample of 12,940 spouses, 27,607 parent–offspring, and 11,498 siblings from Nord-Trøndelag County in Norway. The study was based on self-report questionnaires and structural equation modeling of the data. They found an upper limit for heritability of 0.11, suggesting that genetic effects explain only a small proportion of the variance of tinnitus in the population. The authors recognize that tinnitus is a symptom described in a heterogeneous group of diseases, and thus the heritability could differ substantially, depending on the biological nature of the underlying disease. The results represent average values across different types of tinnitus rather than particular types of tinnitus. The findings of a link between ethnicity and tinnitus prevalence (Shargorodsky et al., 2010) suggests that the survey group in Norway was likely of low ethnic composition.

About 50% of infant deafness (occurring at 1:1000) is hereditary. By age 65, one in three people suffer from hearing impairments that interfere with speech perception and again 50% is attributed to a genetic origin (Morton, 1991). The occurrence of tinnitus in the over-65 group comprises 15–20% of that population. This is much lower than the one in three occurrence of significant hearing loss in the same age group, but about equal to the fraction attributed to a genetic origin of that hearing loss. It may thus be worthwhile to look into the genetics of progressive hearing loss as this may be accompanied by tinnitus.

2.8.1 Genetics of hearing loss

Peripheral hearing loss Genetic deafness may occur in isolation (non-syndromic) or along with additional clinical abnormalities, such as blindness (syndromic). More than 400 syndromes associated with hearing loss have been identified, and account for 30% of the cases of hereditary hearing loss (Nie, 2008). However, the vast majority (>70%) of inherited hearing disorders are non-syndromic. The hearing loss disorders are classified according to their mode of inheritance: DFNA (autosomal dominant), DFNB (autosomal recessive), and DFN (X chromosome-linked). In general, the autosomal recessive families have a congenital and severe-to-profound hearing impairment, whereas the autosomal dominant ones have a later-onset progressive hearing impairment. In X-linked families, both types of hearing impairment have been reported (Fransen and Van Camp, 1999). More than 100 gene loci for non-syndromic deafness have been mapped to the human genome and >50 genes have been identified (http://www.ncbi.nlm.nih.gov/books/NBK1272/).

Despite the extreme heterogeneity of genes involved in non-syndromic hearing loss, mutations in a single gene (*GJB2*) are responsible for as many as half of cases of autosomal recessive hearing loss. By contrast, most genes responsible for autosomal dominant hearing loss have been detected in only a single or a few families so far; the only exception being DFNA2 based on a *KCNQ4* gene mutation, which has been found in families around the world (Nie, 2008). The protein encoded by this gene forms a K^+ channel that is thought to play a critical role in the regulation of neuronal excitability, particularly in sensory cells of the cochlea. The current generated by this channel is inhibited by muscarinic M1 acetylcholine receptors and activated by retigabine, a novel anti-convulsant drug.

Another candidate gene for tinnitus is *KCNE1*, a K^+ channel subunit gene that has been implicated in maturation defects of central vestibular neurons, in Ménière's disease, and in noise-induced hearing loss. Sand et al. (2010) screened 201 Caucasian outpatients with a diagnosis of chronic tinnitus for mutations in the *KCNE1* open reading frame and in the adjacent sequence by direct sequencing. Allele frequencies were determined for 46 known variants, plus two novel *KCNE1* mutations. An allele is one of two or more forms of the

DNA sequence of a particular gene. When genotypes were grouped assuming dominance of the minor alleles, no significant genotype or compound genotype effects were observed on tinnitus severity. In relation to allele frequencies in healthy control populations from earlier studies, the more common *KCNE1* variants are thus unlikely to play a major role in chronic tinnitus.

Tinnitus has only been associated with the more rare forms of autosomal dominant non-syndromic deafness, i.e. DFNA20/26 based on a mutation of the gamma actin gene *ACTG1* (de Heer et al., 2009) and DFNA9 where a mutation of the *COCH* gene is responsible (Fransen et al., 1999; Verhagen et al., 2000). The gamma actin coexists in most cell types as components of the cytoskeleton, and as mediators of internal cell motility. Actin gamma 1, encoded by *ACTG1*, is a cytoplasmic actin found in non-muscle cells such as hair cells. Hybridization to this gene was detected in spindle-shaped cells located along nerve fibers between the auditory ganglion and sensory epithelium. These cells accompany neurites at the habenula perforata, the opening through which neurites extend to innervate hair cells.

Central causes of tinnitus The serotonergic system has been proposed to play a prominent role in auditory perception and in the etiology of tinnitus (Simpson and Davies, 2000). However, in a study involving 4400 individuals genotyped for a functional length polymorphism in the regulatory region of the 5-HT transporter gene (*5-HTTLPR*), no significant effect was noted on susceptibility to chronic tinnitus (Sand et al., 2007). Similarly, generation of tinnitus was not associated with another serotonin transporter gene (*SLC6A4*) polymorphism and possibly with serotonergic mechanisms. The "*ll*" genotype variant of the *SLC6A4* polymorphic promoter region seems associated with the limbic and autonomic nervous system symptoms of the patients with tinnitus (Deniz et al., 2010).

Neurotrophins, through their role in neuron survival and regeneration, are involved in the tonotopic organization in the central auditory pathway (Reser and Van de Water, 1997). BDNF-induced changes in glutamatergic signaling also suggest modulatory effects on SFRs in the auditory cortex (Lessmann, 1998) and thus potentially on tinnitus (Kaltenbach, 2000). A significantly lower risk of developing chronic tinnitus with hearing impairment was observed among carriers of a BDNF Val66Met missense variant (Kleinjung et al., 2006). Similarly, glial cell-derived neurotophic factor (GDNF) likely plays a modulatory role in chronic tinnitus (Kleinjung et al., 2009).

2.8.2 Genetics of depression, anxiety, and migraine

Many individuals with tinnitus experience impaired emotional health and deficits in hearing, sleep, or concentration (Mrena et al., 2002). Assuming that this comorbidity involves a common genetic background, this information may be used in investigations of the genetic susceptibility to tinnitus. Genome-wide scans of susceptibility to major depression have identified candidate regions on chromosomes 15q, 17p, and 8p, among others (Holmans et al., 2007).

Tinnitus has also been described in people suffering from various forms of headache (Erlandsson et al., 1992). In one case series, 63% of individuals suffering from primary headache were diagnosed with concomitant tinnitus (Farri et al., 1999), and tinnitus is a known risk factor for developing headache. In individuals with basilar-type migraine, a prevalence of 26% has suggested aberrant neural activity along the auditory pathway and a central nervous etiology likely different from vasospasm (Volcy et al., 2005). Clustering of these phenotypes warrants more detailed research into the genetic background of dually affected individuals in the light of emerging loci for familial subtypes of migraine on 19p, 1q, and 2q (Colson et al., 2007).

2.9 Vestibular schwannoma and gaze-evoked tinnitus

Vestibular schwannoma (VS) and cerebellopontine angle tumors can cause tinnitus by putting pressure on the high-frequency ANFs that lie on the outside of the auditory nerve, thereby causing a partial conduction block that limits output to the brain. In addition, VS may disrupt cochlear blood flow and cause a sensory hearing loss, which in turn could cause tinnitus. In a series of 941 patients with unilateral VS (Baguley et al., 2006a), statistically significant associations were found between tinnitus presence/absence and tumor size and type of hearing loss (progressive, sudden, fluctuant, or none) with a tendency for patients without hearing loss to be less likely to experience tinnitus.

Translabyrinthine removal of unilateral VS (Baguley et al., 1992, 2005a) showed that if there was no tinnitus preoperatively it was unlikely to develop. If it did (in 50% of cases according to Berliner et al., 1992), it would not be severe enough to significantly affect patients' quality of life (Levo et al., 2000). If patients had mild or moderate tinnitus it was most likely to stay the same, or become less intense. If a patient had severe tinnitus it was very likely to improve following the surgery and not affect his/her future quality of life (Fahy et al., 2002; Kameda et al., 2010).

Complete and acute unilateral deafferentation of the auditory periphery (auditory and vestibular afferents) can induce changes in the central nervous system that may result in unique forms of tinnitus (Coad et al., 2001; Cacace, 2003). These tinnitus percepts can often be controlled (turned on and off) or modulated (changed in pitch or loudness) by certain overt behaviors in other sensory/motor systems. In a cohort of 359 patients who had undergone translabyrinthine removal of a VS, gaze modulation of tinnitus was identified in 19% of patients in this series (Baguley et al., 2006b). Explanations for this form of tinnitus include crossmodal plasticity, which has been hypothesized to involve neural sprouting between the para-abducens nucleus, or medial longitudinal fasciculus, and the auditory pathway, perhaps at the level of the cochlear nucleus. Yang et al. (2010) suggested a dysfunction in vergence eye movements involving their control by cortical–brainstem–cerebellar circuits. They hypothesized that central auditory dysfunction related to tinnitus percept could trigger mild cerebellar–brainstem dysfunction or that tinnitus and vergence dysfunction could both be manifestations of mild cortical–brainstem–cerebellar syndrome reflecting abnormal cross-modality interactions between vergence eye movements and auditory signals.

Alternatively, the development of gaze-evoked tinnitus may involve the unmasking of a previously-inhibited pathway linking auditory gain and eye movement (likely via the ophthalmic division of the trigeminal ganglion; Chapter 8). The deafferentation associated with VS removal would potentially disinhibit such a pathway (Cacace et al., 1999a,b).

2.10 Stress and tinnitus

Emotional or physical stress initiates the release of corticotrophin-releasing factor, which in turn leads to the release of cortisol and adrenalin from the adrenal gland and the activation of the sympathetic nervous system. Hébert et al. (2004) measured cortisol levels in saliva in tinnitus patients and controls. The high tinnitus-related distress group had chronic cortisol levels that were greater than both the low tinnitus-related distress and control groups, and also displayed greater intolerance to external sounds (hyperacusis). This suggests a link between intolerance to both internal (tinnitus) and external sounds in people with tinnitus, which is compatible with the clinical observation that severe tinnitus is associated with high stress levels (see also Chapter 11). Predictive factors for tinnitus severity include anxiety disorders and poor well-being at tinnitus onset (Holgers et al., 2000, 2005). Psychosocial stress and social problems are also more common in patients with severe tinnitus than in patients with less severe tinnitus, where severity is defined as absence from work for >1 month and more than three visits to the therapist or medical audiologist (Holgers et al., 2005).

2.11 Tinnitus of unknown etiology

The etiology of tinnitus is unknown in about 40% of patients. Henry et al. (2005) wrote: "in many patients, the emergence of tinnitus as a problem occurs long after the underlying medical condition (most commonly hearing loss). The trigger for the adverse or intrusive effects of tinnitus is sometimes unrelated to the associated condition."

A considerable number of patients (about 8% in a study by Barnea et al., 1990; 18% in Stouffer and Tyler, 1990a) complain about tinnitus in the absence of hearing loss or obvious somatic problems. Tinnitus without hearing loss and no obvious signs of somatic tinnitus remains something of a mystery. One possibility already mentioned could be undetectable venous hum, which would put it in the objective tinnitus category. Another one could be that there are narrow dead regions in the cochlea that cannot be detected by standard audiometry. A test has been developed (Moore et al., 2000) to detect these dead regions, and Weisz et al. (2006) used this technique to demonstrate that most high-pitched tinnitus in the absence of hearing loss could be the result of the presence of such dead regions. These dead regions may cause mini reorganizations and disinhibition in the central auditory system (see later). Delayed damage to the spiral ganglion following temporary threshold shifts induced by mild noise trauma (Kujawa and Liberman 2009) could underlie these dead regions.

One also wonders whether some people with tinnitus in the absence of hearing loss could have suffered from prolonged conductive hearing loss in childhood. This causes tinnitus with a pitch related to the shape of the audiogram (Savastano, 2007), and after recovering from this childhood hearing disorder the tinnitus might have retreated temporarily, only to show up again in adulthood. This would be due to the same mechanisms that induce tinnitus following mild NIHL (described earlier). However, mild NIHL is here to stay and the conductive loss usually disappears, although there are persistent changes in the central auditory system following long term conductive hearing loss (Chapter 7). One would expect brain plasticity to return aberrant neural activity back to normal after the hearing loss disappears, so the assumption of dormant tinnitus following conductive hearing loss remains speculative (but see section 2.6).

Tinnitus in normal hearing subjects and in patients with hearing loss may differ. Based on 520 consecutive tinnitus sufferers of which 223 had normal hearing and 297 had a SNHL, Savastano (2008) found no differences in residual inhibition (RI) between the two groups. However, the characteristics of tinnitus in normal hearing subjects, except for the subjective judgment of tinnitus loudness, the pitch and the RI, were significantly different for those observed in subjects with hearing loss. The different quality of tinnitus in hearing loss seems to be based on the increase of the perceived severity of the symptom, potentially as a result of increased likelihood of hyperacusis (Chapter 11).

One has to realize that some of the more detailed etiology reports only include people with tinnitus that seek help from a clinic. Although there are wide-ranging investigations based on questionnaires using surface mail (reviewed in Davis and El Rafaie, 2000) or Internet (Andersson et al., 2004), these mostly cover the incidence and prevalence of tinnitus and its relation to hearing loss, smoking etc. It is somewhat strange that the "unknown" etiology for tinnitus is found in about half of the cases. This could reflect the inability to relate a current problem to a cause that may be years or decades in the past, as tinnitus becomes prevalent with age. An example is found again in delayed ganglion cell degeneration after temporary threshold shifts (Kujawa and Liberman, 2009). Age appears to exaggerate existing conditions that may have been induced by hearing loss, because of the general reduction in central inhibition with age. Animal studies suggest that aging in itself, i.e. without hearing loss, does not induce neural signs of tinnitus (Chapter 10). On the other side of the age spectrum, the occurrence of childhood tinnitus may need to be correlated with (previous or existing) otitis media.

This all serves to show that the range of etiologies of tinnitus is vast and that tinnitus perception is likely a phenomenon with common properties regardless of the cause, but with a biological substrate that depends on its etiology. We will specifically address three etiologies that have been used as model systems in animal research—salicylate, noise trauma, and somatic insults—in Chapters 6–8. A model for a potentially common substrate for these forms of tinnitus, the hypersynchrony model, will be presented in Chapter 9. Chapter 10 will look in depth at the effects of aging on the peripheral and central auditory system in animals. However, before we focus on the animal studies, we will present the results of the effects of tinnitus on the human brain by asking patients to listen to their tinnitus (and so extracting the psychoacoustical and psychological aspects of tinnitus; Chapter 3), and by using the various structural and functional imaging methods currently available (Chapter 4). Chapter 5 provides a bridge between Chapter 3 and the animal studies and describes in depth the potential (and the problems) of behavioral animal models for tinnitus.

2.12 **Summary**

The prevalence of tinnitus is high in children with normal hearing as well as in those with hearing loss or ear disease, and reaches levels in the 20–40% range. These levels have been constant over the last four decades. It may be that the strict criteria about duration (">5 min"), adhered to in most adult epidemiology, have not been enforced in these studies. The prevalence of tinnitus (">5 min") is much lower in young adults (~8%) but taking any tinnitus into account it also reaches about 25%. Frequent tinnitus (">5 min") increases in the population from 8% slowly up to 65–70 years of age (~15%) after which it stabilizes. The incidence of tinnitus in age groups >50 years of age, defined as the percentage of people who did not have tinnitus at the onset of the period but acquired it over a 5–10-year span, varies between 14–18% (depending on the original age). Occupational and recreational noise, smoking, and hypertension increase the risk of acquiring tinnitus, whereas moderate alcohol consumption appears to lower that risk. Hearing loss that originates from noise exposure or from the use of ototoxic drugs appears to be the single-most cause of tinnitus. Somatic insults or activity can both induce and modulate existing tinnitus. Tinnitus overall has no or very low heritability, whereas non-syndromic hearing loss that may result in tinnitus clearly does. Factors accompanying tinnitus such as depression also seem to have heritable aspects.

The quantification of tinnitus quality and its impact on life is the topic of the next chapter.

Chapter 3

Listening to tinnitus

We have seen (Chapter 1) that there are no unique subjective descriptors of the tinnitus sound. Variability may arise from the phantom sounds that tinnitus sufferers hear, and from variability in the words they choose, which depend on personal history and context. Nevertheless tinnitus patients do hear sounds that others do not, and most of the descriptors they choose fall into the broad category of persistent tonal, ringing and hissing sounds, or noise. Psychoacoustic methods can be used to evaluate the sounds they hear. Psychoacoustics is the study of quantifying subjective human perception of sounds, and investigates the psychological correlates of the physical parameters of sound. In this chapter the focus will be on pitch and loudness of tinnitus, its maskability, and estimating the degree of annoyance tinnitus causes.

3.1 Pitch

The spectral characteristics of tinnitus have been recognized as a fundamental aspect for many years. Fowler (1944) argued that it was important to match the loudness and pitch of tinnitus by using tones presented to the contralateral ear. However, since about half of tinnitus cases are bilateral (Chapter 1) this is not universally practical. For pitch matching, the importance of presenting matching tones at levels equal to the perceived tinnitus loudness was stressed. Pitch matching to a pure tone indicates the most prominent pitch of tinnitus, and in a large study this pitch matching could be obtained in 92% of 1033 patients (Meikle, 1995). The reliability of pitch matching, however, varies widely across people with tinnitus. Pitch can vary from day-to-day or even within a day (Stouffer and Tyler, 1990a), and circadian rhythms may have something to do with it (Chapter 11). Nevertheless, Nageris et al. (2010) found that testing of pitch and loudness of tinnitus resulting from NIHL gave similar group means when contrasted between sessions separated by 1 month or more, suggesting that some degree of reliability is present. Methodological aspects of the measurement procedure are likely to be important.

Pitch matches for tinnitus occur often in the frequency region of maximum hearing loss or at the edge frequency of the audiogram (Penner, 1980), and the most common pitch for NIHL tinnitus appears to match that of a 3-kHz tone. Tinnitus pitch increased over time in 20% of patients; the reason for that is not clear, but with aging the frequencies with the highest thresholds tend to shift upward more so than thresholds at lower frequencies, which may also change. The prediction of over-representation of edge frequencies in tonotopic maps after noise trauma (Chapter 7), implying that tinnitus pitch would match the edge frequencies (Rauschecker, 1999), could not be confirmed

(Pan et al., 2009). However, this failure may be related to octave errors in pitch matching. Following the training of participants to avoid these errors, the mean pitch matches were close to the values of the edge frequency, with a correlation of 0.94 (Moore and Vinay, 2010). Still, there is strong evidence that tinnitus pitch is generally located away from the audiometric edge (see later in chapter).

In his prescient study, Fowler (1944) stated that even tinnitus that sounds like a pure tone is in fact always composed of a narrow band of frequencies. Tinnitus-matching sounds can be synthesized by combining pure tones (Penner, 1993) in potentially the following way (based on Noreña et al., 2002). After threshold determination, subjects adjusted the intensity of tones within the hearing loss range (one randomly selected tone at a time) to match the loudness of their tinnitus. The person then stated whether the frequency corresponded to one of the components of their tinnitus spectrum and, if it did, gave a rating on a 10-point scale (0 = no match, 10 = tinnitus) of the extent to which the frequency was part of their tinnitus sensation. Frequencies were selected randomly from the tested range and repeated until a total of three measurements had been obtained for each frequency. Tones were presented monaurally either to the tinnitus ear or to the ear where tinnitus was most pronounced. In each of the 10 cases tested by Noreña et al. (2002), the rated tinnitus spectrum spanned the region of hearing loss without a dominance of edge-frequency ratings. The tones were then combined according to their rated importance into a spectrum, and the sound corresponding to that spectrum was presented to the person for feedback about the degree of similarity to his/her tinnitus. In Figure 3.1 it is shown schematically that the tinnitus spectrum is often mirroring the audiogram shape with highest thresholds frequencies contributing mostly to the tinnitus spectrum. It is likely that a tinnitus spectrum as shown on the top would represent a "wooshing noise" percept whereas for the one shown at the bottom it would be more tonal or ringing. This, however, may be an oversimplification as some subjects also indicated tonal tinnitus for broad tinnitus spectra, the pitch then corresponding to a peak in the spectrum (Noreña et al., 2002). The correspondence between the hearing loss range and the tinnitus spectrum was corroborated by Roberts et al. (2006, 2008) using a highly similar method for measuring tinnitus percepts, in which subjects rated each of 11 sounds of constant bandwidth but differing in center frequency for the degree of "likeness" to their tinnitus. Kay et al. (2008) compared tinnitus ratings for the same patients when tinnitus frequencies were assessed by the "likeness" method and by a two-alternative forced-choice procedure designed to specify a tinnitus "pitch." The peak of likeness ratings tended to correspond to the tinnitus pitch, but the subjects preferred the likeness method. This may have been because likeness judgments avoided difficult choices when neither sound resembled the tinnitus or when both did.

3.2 **Loudness**

Loudness relates to the number of neurons activated, their rate of firing, and potentially also to the degree of synchrony of their firings. It is likely a centrally determined percept as it can be influenced by habituation (Formby et al., 2003). Tinnitus loudness is usually

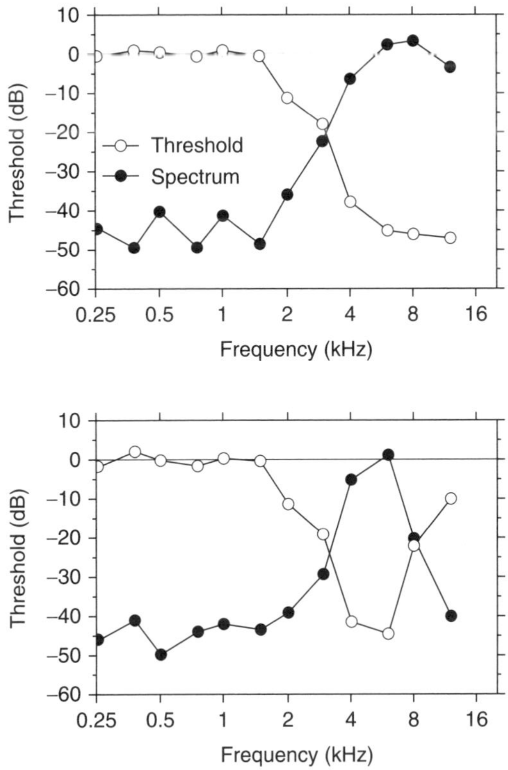

Figure 3.1 The tinnitus spectrum, that is, the range of frequencies making up the tinnitus, is closely related to the inverse of the hearing loss. This is shown here schematically. The scale for the correspondence of a given tone with the tinnitus percept typically is scaled from 0–10, but here a scaling is used with 0 being the highest value. The top example with a relatively broad frequency loss likely would represent a hissing tinnitus, whereas the bottom example with a limited frequency range of loss could represent a more tonal tinnitus. A strict one-to-one correspondence of hearing loss and tinnitus pitch does not hold however.

measured by rating on a visual analog scale (VAS) from 1–10, or by matching to the level of external sounds. Loudness matching is typically done by adjusting the level of a pure tone presented to the same or the other ear to equal that of the tinnitus. Fowler (1944) commented that patients often described their tinnitus as very loud, yet the tinnitus could usually be matched to external tones with levels of only 5–10 dB sensation level (SL), i.e. 5–10 dB above their threshold at that frequency. This is understandable if one realizes that SLs represent a sound level (physical) in dB above the person's threshold at a particular frequency and are not a measure of loudness (psychological). It is likely that these measurements are affected by the loudness-recruitment phenomenon that is present in sensorineural hearing loss (SNHL). Loudness recruitment is the abnormal increase in the loudness for a given increase in sound level (for more detail see Section 3.3). Tyler and Conrad-Armes (1983) converted sensation level into conventional HL, which is the dB level with respect to the average threshold in a large group of normal hearing persons.

Their conversion first takes the hearing loss into account and secondly converts this (the tinnitus loudness in dB HL) into sones (a measure of perceived loudness). This resulted in much higher loudness values than the SL indicated. The sone was proposed as a unit of perceived loudness by Stevens (1955). According to Stevens' definition, one sone equals the loudness level of a 1-kHz tone at 40 dB SPL. The number of sones was chosen so that a doubling of the number of sones is perceived by the human ear as a doubling of the loudness, which also corresponds to increasing the sound pressure level by approximately 10 dB. The relationship between SPL and loudness in sones for sounds >1 s can be approximated by a power function with an exponent of 0.6 when plotted against sound pressure, or 0.3 when plotted against sound intensity (Stevens' power law). More precise recent measures show that loudness grows more rapidly (with a higher exponent) at low and high levels and less rapidly (with a lower exponent) at moderate levels (Moore and Glasberg, 2004). Further evidence that recruitment may give spuriously low estimates of tinnitus loudness is that loudness matches to control sound frequencies presented below the hearing loss region (where recruitment may be reduced) are typically about twice that for sound frequencies in the hearing loss region where recruitment may be stronger (Henry and Meikle, 2000; Roberts et al., 2008).

Cope et al. (2011) also found that, although tinnitus is generally matched by a sound with a low SL, this does not necessarily imply that the tinnitus is perceived as very soft. When matches were expressed as loudness levels in phons, using the loudness model of Moore and Glasberg, the majority (14/18) of the loudness matches for the comparison group fell between 20 and 50 phons, 4/18 matches fell above 70 phons. By definition, 1 phon is equal to 1 dB SPL at a frequency of 1 kHz. Equal-loudness contours map the dB SPL of pure tones of different frequencies to the perceived loudness level in phons. The dependence of tinnitus loudness on level of hearing loss, i.e. assuming that tinnitus loudness is ~10–15 dB SL, may be a reason why the tinnitus spectrum has a shape that is roughly the inverse of the hearing loss (Figure 3.1).

Just as tinnitus pitch fluctuates, so does its loudness (Stouffer and Tyler, 1990b). This fluctuation likely represents the cumulative effects of test–retest variability, actual fluctuation of the tinnitus loudness, and changes in tinnitus pitch or loudness produced by the measurement stimulus if presented to the tinnitus ear. Presenting a matching stimulus to the contralateral ear might also influence certain aspects of the tinnitus because of its potentially central origin. One way to avoid this sound–tinnitus interaction is to use cross-modal loudness matching. Cross-modal matching compares tinnitus loudness to either the brightness of a visual stimulus (visual analog scale, VAS), the length of a bar, etc. Somatosensory stimuli as comparators should be avoided because of their potential modulation of tinnitus properties (Chapter 8).

A relatively new method called constrained scaling may be useful in obtaining reliable loudness matches in tinnitus subjects (West et al., 2000). The procedure consists of training subjects in loudness estimation on a standard response scale that closely resembles the sone scale. Following a training phase, the tinnitus subjects judged either silence (i.e. tinnitus) or pure tones at 500 Hz and 65 Hz (to compare their use of the standard response scale with that of normal-hearing listeners). Subsequently, they judged the loudness of sounds at their tinnitus frequency (pure tones at either their pure-tone tinnitus frequency match or at the center frequency of the tinnitus-matching band of noise). In each of the later phases, subjects judged 17 non-standard stimuli interleaved with the same 17 standard stimuli used in training—the latter to maintain calibration on the standard response scale. Intensities in both sets were presented in different randomly shuffled orders for each subject and for each phase. All 17 stimuli at each frequency consisted of levels from 40–88 dB in 3-dB steps, except of course for the silence stimuli, which consisted of endogenously generated tinnitus. The entire constrained scaling procedure could be completed in approximately 1 hour. A much-improved consistency of loudness matching in tinnitus patients was found (Ward and Baumann, 2009). They found on average loudness levels of 52.5 (mild hearing loss group) and 58.7 phons (moderate hearing loss group).

3.3 **Loudness recruitment**

Loudness recruitment is an abnormal relationship between sound intensity and perceived loudness that may accompany NIHL (Moore, 2000). In this condition, the threshold of hearing is elevated, but a given increase in intensity above threshold causes significantly greater than normal increase in loudness. The rate of loudness growth in the impaired ear is often so steep that, at high intensities (60–100 dB SPL; e.g. Stillman et al., 1993), the loudness in the impaired ear matches that in the normal one. Loudness recruitment has been primarily studied in humans and has also been demonstrated behaviorally in animals with NIHL (Pugh et al., 1979). The physiological mechanisms of loudness recruitment are not well understood. It is commonly thought that the loudness of a sound reflects some aspect of the overall activation pattern of peripheral structures such as the auditory nerve, e.g. the total number of action potentials. From this point of view enhanced output from the cochlea would be expected based on the loss of the cochlear compressive non-linearity consequent to damage to OHCs (Robles and Ruggero, 2001). In agreement, psychophysical results are consistent with an explanation of recruitment in terms of a loss of compression of the amplitude of basilar membrane vibration (Oxenham and Bacon, 2003). Steeper than normal rate-level functions (RLFs) of ANFs have also been observed in SNHL following ototoxic damage to OHCs in cats (e.g. Harrison, 1981). Eggermont (1977) recorded compound action potentials (CAPs) from the cochlea promontory in humans with normal hearing and in patients with SNHL (Figure 3.2). The stimuli were 2-kHz tone pips presented at a rate of 7/s. The median CAP amplitude-level function for 20 normal ears typically continues down to levels of 0 dB HL (by definition). The individual patient amplitude-level functions shown in this figure reflect their hearing loss of 50–75 dB at 2 kHz. Note that at high stimulus level these functions approach or exceed the median curve for normal hearing people. If loudness can be equated with CAP amplitude it clearly indicates a peripheral marker for the recruitment of loudness that these patients showed behaviorally. This hypothesis is consistent with the observation that the velocity–intensity relationship of the basilar membrane becomes steeper after the loss of OHC amplification (Ruggero and Rich, 1991).

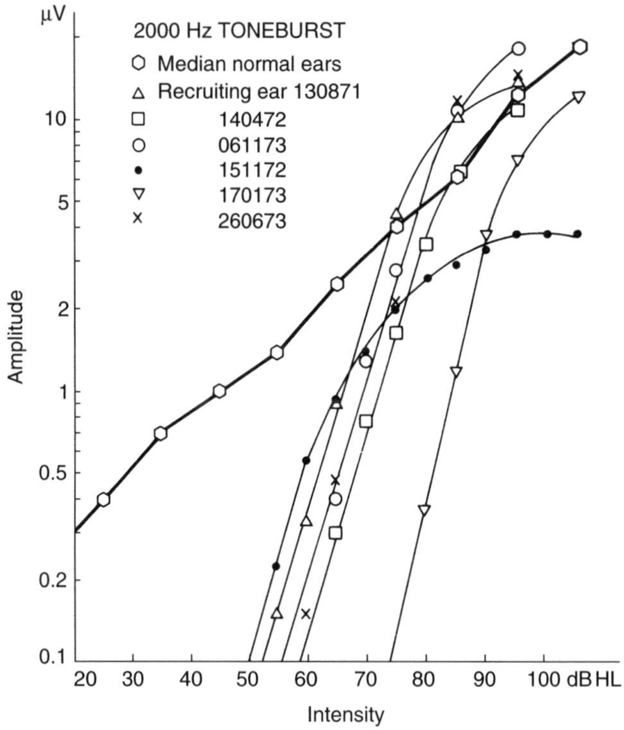

Figure 3.2 Compound action potential (CAP) amplitude-level functions for a group of 20 normal hearing people (median shown with fat line) and for five people with SNHL. Data were recorded from the promontorium of the cochlea by inserting a thin needle through the tympanic membrane under local anesthesia. Reproduced from Eggermont, J.J. (1977). Electrocochleography and recruitment. *Annals of Otology Rhinology and Laryngology*, **86**, 138–49 © with permission.

However, the overall story appears to be more complex. Following acoustic trauma, OHC loss can be accompanied by IHC damage (Liberman and Dodds 1984), which compromises cochlear transduction and *lowers* the firing rates of ANFs (Liberman and Kiang 1984). As a result, the slopes of their RLFs become *shallower* on average (Salvi et al., 1983; Heinz and Young 2004). In cats following acoustic trauma, the summed RLFs of ANFs over either a narrow or wide range of CFs surrounding the stimulus frequency did not show abnormally steep slopes, which is a result that was deemed inconsistent with loudness recruitment (Heinz et al., 2005). Two considerations may contribute to resolving the conflicting results. First, the RLFs of auditory neurons reflect both IHC and OHC function. When OHC damage is present, steeper RLFs arising from loss of compression may mask reduced output from the cochlea arising from IHC damage. The CAPs of Figure 3.2 reflect OHC damage and loss of compression (these changes signaled threshold shifts), but IHC damage may nonetheless also have been present. Second, the role of compensatory processes in central auditory structures needs to be considered. Increased neural synchrony of ANF firings after noise trauma will lead to larger CAPs and potentially more powerful activation in cochlear nucleus neurons (e.g. stellate cells, octopus cells, fusiform cells) that rely on synchronous activation in order to fire. Hyperactivity is seen in central auditory neurons after SNHL and is characterized by (among several other changes) an abnormally rapid increase of response with stimulus intensity or by a heightened maximum response, or both (Salvi et al., 2000). This central hyperactivity is a potential contributor to loudness growth.

To explain recruitment, central hyperactivity must affect patterns of neural activity that code for perceived loudness. What might the central correlate of loudness be? Coincidence-detecting cells that receive ANF inputs with different CFs can potentially decode the spatiotemporal cue for sound level (the synchronous firing of a neural population). Such spatiotemporal processing could be performed at any level of the auditory brainstem where cells receive converging inputs from lower levels. For example, neurons recorded in VCN are typically classified based on temporal discharge patterns in response to tones (four types are shown in Figure 3.3). There is physiological evidence that cells in the anteroventral cochlear nucleus (AVCN) that receive convergent ANF inputs are sensitive to the temporal response pattern across the inputs (Carney, 1990). In particular, onset chopper

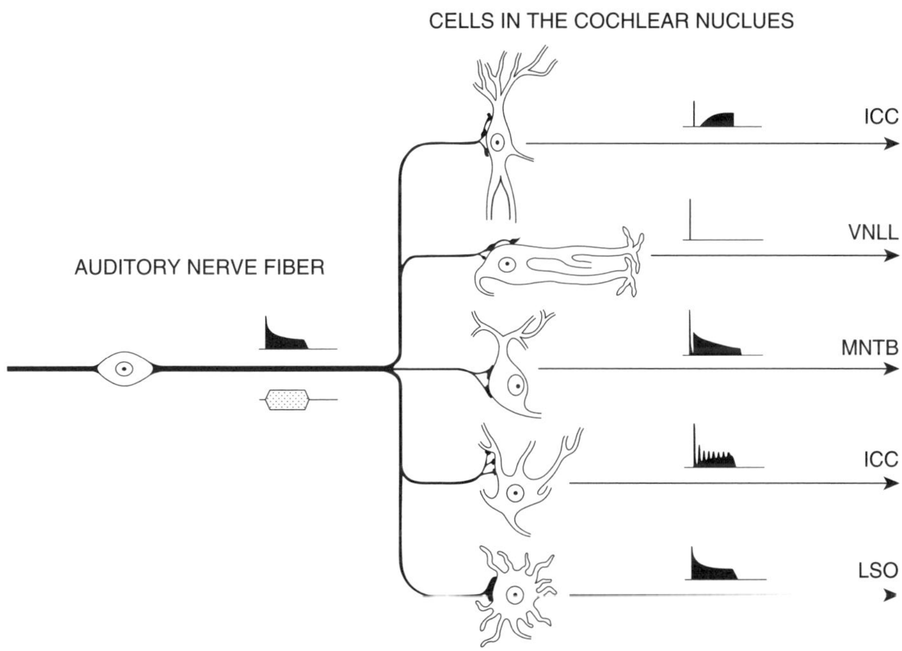

Figure 3.3 Response types and cell types in the cochlear nucleus. The histograms (cell types) for the CN cell responses are from top to bottom: pauser (fusiform cells in DCN); on-type (octopus cells in PVCN); primary-like with notch (globular bushy cells in AVCN); chopper (T-stelate cells in AVCN); and primary-like (spherical bushy cells in AVCN). ICC, central nucleus of the inferior colliculus; LSO, lateral superior olive; MNTB, medial nucleus of the trapezoid body; VNLL, ventral nucleus of the lateral lemniscus. Reproduced from Kiang, N.Y. (1975). Stimulus representation in the discharge patterns of auditory neurons. In: D.B. Towers (ed.), *The Nervous System*, volume 3. © Lippincott Williams, and Wilkins, 1975, with permission.

response types (T-stellate cells) in the ventral cochlear nucleus, which are known to receive convergent inputs from a wide tonotopic region of the auditory nerve root and to have wide dynamic ranges (Smith and Rhode, 1989), may act as coincidence detectors coding for sound intensity. Some major projection sites are indicated for each cell type. The source of input to these cells is also important. As discussed in Chapter 1, synapses on IHCs span a range of different thresholds such that synchronous activity detected from among a subset of them could give rise to a population response related to sound intensity.

One mechanism that may maintain or enhance the neural response to a sound in the face of IHC damage is homeostatic plasticity (Turriganio, 1999), which enhances synaptic efficacy to maintain a relatively constant level of neural activity when input from the periphery is reduced (Schaette and Kempter 2006). However, there are a number of unanswered questions, including at which station(s) of the central pathway the plastic changes occur, in which populations of neurons, and how the changes interact with peripheral changes to produce enhanced central responses. Most studies have used evoked potential recordings (e.g. Salvi et al., 1990) that did not allow identification of the neuronal sources of changes. Cai et al. (2009) analyzed RLFs summed over various populations of VCN neurons following acoustic trauma and showed that the responses of non-primary-like types of neurons, especially chopper neurons, showed changes in rate responses consistent with loudness recruitment (as suggested by Carney, 1990), whereas primary-like neurons did not, except in cases where spread of excitation in the population was included. Cai et al. (2009) suggest that the VCN is the most peripheral stage of the auditory pathway at which overexcitability of neurons appears following acoustic trauma. Moreover, they point to the synaptic processing in the non-primary-like, especially the stellate, cells as the initial locus of central compensation for reduced rate responses in the ANFs.

As said before, loudness is assumed to be proportional to the summed activity of some group of auditory neurons (Moore and Glasberg, 2004), but the nature of that summation and the relevant populations of neurons are unclear (Heinz et al., 2005). In considering recruitment, one assumes that changes in the loudness function are mirrored by similar changes in neural response growth functions, steeper neural growth functions corresponding to steeper loudness growth, for example. Evoked potentials and multineuron recording show recruitment-like overexcitability of central auditory neurons following cochlear damage in a variety of animal models (Saunders et al., 1972; Salvi et al., 1990; Wang et al., 2002). This behavior has also been reported in the human auditory system through functional imaging (Langers et al., 2007). These neural data match the expectation of recruitment qualitatively in that the neural responses of populations are stronger after cochlear damage. Neural hyperexcitability is commonly reported in cortex and IC, but the results in cochlear nucleus have been mixed, with both positive (Saunders et al., 1972) and negative (Salvi et al., 1990; Qiu et al., 2000) findings.

A more quantitative comparison of perceptual and neural response growth can be obtained by comparing loudness balance functions to rate-equality functions. Equating the perceptual loudness between a normal and a damaged ear should behave similarly to equating the summed discharge rates of corresponding groups of neurons as long as loudness is in some way a monotonic function of summed neural activity (but that may be not generally valid as the chopper cells require coincidence detection). This approach avoids the difficulties posed by non-linearities in the loudness growth functions themselves (Moore and Glasberg, 2004). If the neural basis of perceived loudness is in some way related to the discharge rates of central neurons (and/or the number of discharging neurons), then the loudness recruitment seen in normal listeners under conditions of noise masking, where compression is intact, may have its origins more in central neural processes than in cochlear ones (Phillips, 1987). This is not inconsistent with my own results (Eggermont, 1977) from CAP recordings in humans, which point to a peripheral synchrony correlate as well when OHC damage is present.

3.4 Hyperacusis

The terms "hyperacusis" or "phonophobia" are interchangeably used to describe an unusual hypersensitivity or discomfort induced by exposure to sound. Dauman and Bouscau-Faure (2005) found that the prevalence of hyperacusis in tinnitus patients was high (79%). Sleep complaints in the elderly (mean age 67 years) were mainly explained by hyperacusis rather than by hearing loss (Hébert and Carrier, 2007).

As described earlier, CAPs of the auditory nerve (Eggermont, 1977) and chopper neuron responses in the AVCN (Cai et al., 2009) show sharply elevated amplitudes and discharge rates respectively at high sound levels (recruitment) following acoustic trauma. This behavior could provide an explanation for auditory hyperacusis, sometimes named over-recruitment, in which intense sounds become too loud for comfortable listening (Nelson and Chen, 2004; Sherlock and Formby, 2005). Of course, this hypothesis depends on the nature of the additional adjustments in rate responses at later stages of the auditory system. In reality, hyperacusis patients often cannot stand relatively soft sounds and complain that they are too loud. We thus have to look for a potential neural correlate that shows strongly increased responses starting just above threshold.

3.4.1 Peripheral aspects of hyperacusis

There are several documented auditory peripheral hearing impairments that give rise to oversensitivity for loud sounds (Baguley, 2003). Hyperacusis has been described following abolition of the stapedial reflex as seen in facial nerve palsy, Ramsey Hunt syndrome, and myasthenia gravis. Hyperacusis can also occur with Ménière's disease. Here, changes in perilymph pressure may cause acoustic reflex changes, and endolymphatic fluid pressure variation may modify the dynamics of the cochlea and causing changes in the excitation pattern of the OHCs. Many people with hyperacusis have "normal" audiograms, thereby excluding hyperacute thresholds as well as hearing impairment (Anari et al., 1999). Hyperacusis is accompanied by increased amplitude of distortion product otoacoustic emissions (DPOAEs) in tinnitus patients without hearing loss (Sztuka et al., 2010).

3.4.2 Central mechanisms of hyperacusis

There is increasing evidence that tinnitus frequently co-occurs with hyperacusis (Hazell and Sheldrake, 1992; Dauman and Bouscau-Faure, 2005). Jastreboff and Hazell (1993) described hyperacusis as a "manifestation of increased central gain" (Formby et al., 2003; Noreña and Chery-Croze, 2007), which may cause enhanced perception of peripheral signals, and considered hyperacusis as a pre-tinnitus state.

Clinical conditions that are unrelated to peripheral lesions may co-occur with hyperacusis (Marriage and Barnes, 1995; Katzenell and Segal, 2001). Hyperacusis occurs among others: (1) in migraine, where the prevalence is reported to be between 70–83% during attacks and 76% between attacks. It co-occurs with unusual hypersensitivity or discomfort induced by exposure to light (Phillips and Hunter, 1982); (2) in depression and post-traumatic stress disorder; (3) in benzodiazepine dependence; (4) in Williams's syndrome, a rare genetic disorder occurring in one in 50,000 persons caused by a deletion of about 26 genes from the long arm of chromosome 7. Patients have a developmental delay coupled with unusually strong (i.e. for persons who are diagnosed as developmentally delayed) language skills, and cardiovascular problems. Nearly 95% of patients with Williams's syndrome demonstrate hyperacusis (Nigam and Samuel, 1984); (5) hyperacusis is also common among patients with complex regional pain syndrome type 1 related dystonia (de Klaver et al., 2007); (6) people with fibromyalgia displayed significantly greater sensitivity to all levels of auditory stimulation (Geisser et al., 2008); (7) in autism spectrum disorder (Khalfa et al., 2004; Orekhova et al., 2008); (8) in multiple sclerosis (Weber et al., 2002).

Although there are a number of different neurotransmitter abnormalities in these clinical conditions it is likely not a coincidence that in each of these conditions there is a disturbance in serotonin (5-HT) function (Marriage and Barnes, 1995; Simpson and Davies, 2000; Al-Mana et al., 2008). Serotonin (5-hydroxytryptamine or 5-HT) is formed from the amino acid tryptophan. In the brain, the highest levels of serotonin are found in the dorsal and medial raphe nuclear complex. As a CNS neurotransmitter or neuromodulator, serotonin plays an important role in many behavioral and mood functions. Because 5-HT is altered in depression, this may also explain the comorbidity of severe tinnitus/hyperacusis and psychological disorders (Zöger et al., 2001). 5-HT also plays a role in anxiety control and is implicated in the organization of sleep. Headache and migraine are associated with specific receptor subtypes of 5-HT. The serotonergic system modulates neuronal habituation responses to repetitive stimulation and may provide a "gain-control" of excitatory and inhibitory mechanisms (Hegerl and Juckel, 1993). For instance, 5-HT produces inhibition of nociception and responses to painful stimulation. Wutzler et al. (2008) found that the sound level dependence of auditory evoked potentials (AEPs) and loudness of sound is inversely related to serotonergic neuronal activity.

Hyperacusis may be at the basis of observations (Ison et al., 2007) that mice with presbycusis also show exaggerated acoustic startle responses (see also Turner and Parrish, 2008; Sun et al., 2009). This is a potential secondary mechanism for sustaining chronic, debilitating hyperacusis and underlies the importance of behavior modification in parallel to the specific desensitization techniques for management of hyperacusis (Noreña and Chery-Croze, 2007).

3.5 **Masking**

Masking is based on two mechanisms: a so-called "line-busy" effect where the masking sound activates the neurons and prevents them from firing to a probe sound (e.g. another tone or tinnitus), and a suppression effect where the masker interferes with the cochlear mechanical activity pattern of the probe sound (Delgutte, 1990). This suppression causes the basilar membrane to vibrate less than normal to the probe tone in the masker frequency region and thus produces less firing to the probe tone in the ANFs innervating the IHCs in that region.

Although pure tones were reported to mask tinnitus completely in 91% of patients, masking of tinnitus does not reflect the standard effects that a masker has on an external probe sound. Tyler and Conrad-Armes (1984) found that the psychophysical masking curve for a real tone (matched in pitch and loudness to the tinnitus) was different from masking their tinnitus (see also Burns, 1984). If the tinnitus was generated at the cochlear level, the masking thereof would include suppression. Consequently, the masking profile of an external pure tone would be similar to the masking profile of tinnitus. Penner (1980) found that suppression mechanisms were impaired in tinnitus patients as evidenced by broadening of the psychoacoustic tuning curves. Thus the differences in the tuning curve and the tinnitus masking profile points to a central generation site for tinnitus.

Masking effects on tinnitus were categorized (Feldman, 1971) into three main categories according to whether auditory thresholds and tinnitus masking curves converged: (1) only at some frequencies ("convergence," 34% of patients with chronic tinnitus), (2) at most frequencies ("congruence" type, 32% of patients), or (3) showed only a weak trend ("distance" type, 22% of patients). These pattern types were confirmed by subsequent studies using pure tones or narrow band noise as masking stimuli (Mitchell, 1983; Tyler and Conrad-Ames, 1984), with convergence tending to be the most common form (53% in the study of Mitchell). When masker intensity is calculated as sensation level (SL) in each of these studies, it appears that for each pattern the sound intensity needed to mask tinnitus is lowest when the masker frequency is in the region of the hearing loss (Roberts et al., 2008). Almost all participants in these studies report masking of their tinnitus when presented with sounds in this hearing loss region (94% in the study of Mitchell, 1983). Noreña et al. (2002) elucidated this by showing that the tinnitus spectrum resembles that of the hearing loss frequencies.

Maskers are effective against tinnitus only when presented above individual hearing thresholds (Vernon and Meikle, 2003). When a masker of sufficient intensity is presented in the region of hearing impairment, excitation arrives at the cortex via thalamocortical pathways into the affected frequency region followed by comparatively stronger feedforward inhibition after one synaptic delay (Cruikshank et al., 2007). This inhibition would be expected to disrupt the abnormal synchronous neural activity that is believed to underlie tinnitus, and thereby diminishing its perceptual salience (Eggermont and Roberts, 2004). By comparison, masking sounds presented at other frequencies (i.e. at frequencies outside of the region affected by hearing loss) may leave the tinnitus relatively intact, because tinnitus is not generated in these frequency regions. However, at high masking levels the upward spread of masking may affect tinnitus with a pitch above the masker frequencies. Masking appears to be most efficient for sounds whose frequency is just below the dominant tinnitus frequency (Terry et al., 1983). The feedforward lateral inhibition may be stronger at these frequencies where hearing may be relatively better preserved. Alternatively, given the challenge of tinnitus pitch measurement, studies of the frequency specificity of masker efficiency may be prone to uncertainties.

3.6 **Residual inhibition**

Following the offset of an appropriate masking stimulus, tinnitus may remain suppressed for typically less than a minute (Feldman, 1971; Tyler and Conrad-Armes, 1984). This phenomenon is known as "residual inhibition" (RI; Roberts, 2007). As RI often lasts for a relatively long time (usually seconds, but can last from minutes to hours), it likely is of central origin. If the origin were cochlear, then the standard timescale of forward masking would determine the duration of the inhibition, and this is rarely >200 ms (Eggermont, 1985). It is most likely that residual inhibition is based on habituation induced by the masking stimulus (Condon and Weinberger, 1991; Sakai, 2007) and shows a recovery time generally of the same order as the stimulation duration. Whereas the original studies used periodically presented tones, it may be essentially the same for continuous multifrequency sound. So in that case the masker habituates the neurons involved in tinnitus and the associated decrease in gain may affect both stimulus-driven and spontaneous firing, hence a temporary reduction or abolishing of tinnitus. This is not supported by findings that the auditory steady-state response (ASSR) amplitude following masking is increased compared to the pre-masking amplitude in tinnitus patients, but not in controls (Roberts, 2011). As we will see in reviewing the animal experiments the gain change typically only affects stimulus-driven activity and not spontaneous firing (Chapter 6), so RI may potentially produce a gain increase combined with a decrease in spontaneous activity.

Roberts et al. (2008) found that psychoacoustic functions relating the strength and duration of tinnitus suppression ("residual inhibition") to the center frequency of band-passed noise masking sounds appear to span the region of hearing loss, as do psychoacoustic measurements of the tinnitus spectrum (Figure 3.4). The RI was generally largest for sounds in the hearing loss range and for sounds that resembled the tinnitus spectrum. The results suggested that cortical map reorganization induced by hearing loss, which results in an over-representation of the edge frequency in the audiogram, is not the principal source of the tinnitus sensation. Were that the case, one would expect the tinnitus pitch to match the edge frequency (which appears to be dependent on avoiding octave errors, see section 3.1) and that these edge frequencies would result in the largest RI.

In a group of 59 subjects with bilateral tinnitus (Roberts et al., 2008), 70% reported at least some degree of RI averaged over three trials during RI testing. Of the total sample, 22% reported near elimination of tinnitus to at least one masker averaged over three trials, and 19% reported RI duration persisting between 35–45 s (the latter the highest duration that could be measured with the system). RI duration measured at its peak was on the order of 25 s for most subjects. The prevalence of RI in Roberts et al. (2008) (some degree of RI reported by ~70% of bilateral cases) is less than in the reports by Henry and Meikle (2000) and Vernon and Meikle (2003) where RI could be demonstrated in ~88% of tinnitus cases. A likely explanation for the difference that Roberts et al. (2008) presented masking sounds for 30s (to allow measurement of RI for 12 different masking

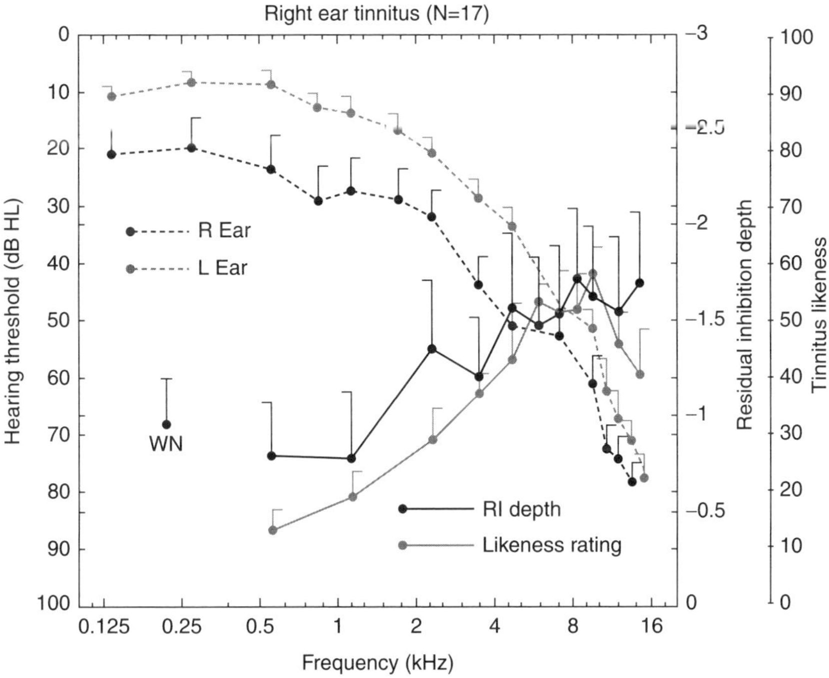

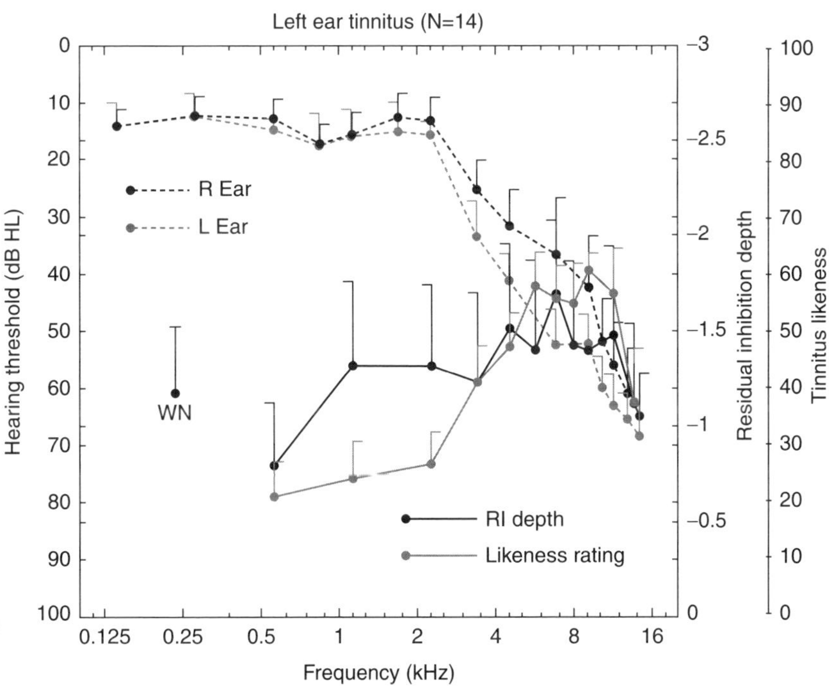

Figure 3.4 Residual inhibition functions and tinnitus spectra (Likeness rating) inversely track the region of threshold shift in unilateral tinnitus. The group averaged audiogram is also shown. Error bars denote 1 standard error. With kind permission from Springer Science + Business Media: Roberts, L.E., Moffat, G., Baumann, M., Ward, L.M., and Bosnyak, D.J. (2008) Residual inhibition functions overlap tinnitus spectra and the region of auditory threshold shift. *Journal of the Association for Research in Otolaryngology*, **9**(4), 417–35. Figure 6.

stimuli over an hour) whereas Terry et al. (1983) found that masking for 1 minute gives better RI. Terry et al. (1983) further investigated in *unilateral* tinnitus cases whether maskers presented ipsilateral or contralateral to the tinnitus ear differed in their effectiveness. Only masking in the ipsilateral ear induced RI, implying suppression of tinnitus related activity at some level in lemniscal projection pathways (75% of lemniscal fibers cross over above the level of the inferior colliculus). This points to a subcollicular locus for residual inhibition.

In agreement with deafferentation models of tinnitus (Rauschecker, 1999; Eggermont and Roberts, 2004; Llinás et al., 2005), Roberts et al. (2008)'s findings implicated hearing impairment as a factor. In particular: (1) tinnitus spectra and RI functions increased incrementally at the edge of normal hearing and proportionately with the extent of threshold shift, (2) the tinnitus spectrum shifted to higher frequencies with increases in the audiometric edge, and (3) the ear with poorer hearing was a predictor of the tinnitus ear in unilateral cases.

In a positron emission tomography (PET) study, Osaki et al. (2005) found that the right anterior middle and superior temporal gyri (Brodmann areas 21 and 38) were activated during residual inhibition, while the right cerebellum was more activated during tinnitus perception in the tinnitus patients. Thus during RI there was a suppression of regional blood flow in the cerebellum, whereas there was more blood flow in BA21 and BA38. The cerebellum has been implicated in habituation (Timmann et al., 1998), but also in focused auditory attention (Benedict et al., 1998). Manganese-enhanced magnetic resonance imaging (MRI) indicated that in rats with evidence of tinnitus, activity was generally significantly elevated in the cerebellar paraflocculus as well as in auditory brainstem structures (Brozoski et al., 2007). Manganese, an activity-dependent paramagnetic contrast agent, accumulates in active neurons through voltage-gated calcium channels, primarily at synapses, and serves as both a structural and functional indicator. The parafloccular lobule receives input from the auditory cortex and ultimately provides feedback to the IC (Sens and Alameida, 2007). It has been proposed that the cerebellum is involved in the development of disabling tinnitus (Shulman and Strashun, 1999).

3.7 **Psychological aspects**

Tinnitus has both psychoacoustic aspects (e.g. pitch, loudness) and psychological aspects (e.g. distress, anxiety, sleep problems). Tinnitus, just like pain, may cause emotional and psychological distress that is often out of proportion with the magnitude of the injury. In this regard, tinnitus is similar to other phantom sensations such as phantom pain (Jastreboff, 1990). The strong emotional implications of sound in general may be the basis for psychological distress in some tinnitus patients (Hallam et al., 1984; Jastreboff et al., 1996). In the first years of serious study of tinnitus, the 1970s and 1980s, the psychoacoustic aspects and related treatments such as tinnitus masking, were attracting most of the attention whereas currently it is the physiological aspects and its associated behavioral therapies (Eggermont, 2012). These qualitative aspects were recognized early on as well but a shift from the audiological treatment to the psychological ones required a shift in perspective, especially from the patients (McKenna, 1998). The term qualitative does not mean that these aspects cannot be quantified; that is the very aim of using tinnitus questionnaires. However, the personal nature of the complaints makes it difficult to generalize them to a large population.

The following section draws heavily on the insightful reviews by Noble (2000), Kennedy et al. (2004), and Newman and Sandridge (2004), which describe a large number of tests in great detail. The aim here is to provide an overview with emphasis on cross validation of the various questionnaires, and so to stress both the individual nature of tinnitus and its commonalities. It is this last aspect that can be addressed in the context of neuroscience.

3.7.1 **Tinnitus questionnaires**

The objective for any tinnitus questionnaire is to accurately identify and quantify the patient's tinnitus-associated problems. Assessing the effects of tinnitus on an individual's quality of life is a complex issue that results from a variety of factors. First of all, tinnitus is an entirely subjective percept of the patient. Psychological factors may play a critical role in determining an individual's reactions to tinnitus (e.g. some people can solve their own problems whereas others need help; Erlandsson, 2000). Women appear more likely to report emotional reactions to their tinnitus than men (Meikle and Griest, 1989; Dineen et al., 1997), and the incidence of personality disturbances is greater for male than for female tinnitus patients. Differences in lifestyle and in their occupational acoustic environment may make some patients more likely to experience problematic tinnitus. The quality of life is likely reduced for anyone with chronic tinnitus. Sleep disturbance is reported by about half of those individuals who complain of tinnitus (Erlandsson, 2000).

The use of written and self-administered questionnaires is important for any large-scale tinnitus assessment, but should be used with care. Because of the self-administration, and particularly the time of filling out the questionnaire, this reflects only this instance of the tinnitus experience, and potentially also the general emotional health of the patient. Probably the earliest systematic study relying on the self-report method in relation to tinnitus is that by Tyler and Baker (1983). They applied the open-ended question technique introduced by Barcham and Stephens (1980), and asked self-identified tinnitus sufferers to list the difficulties they experienced and attributed to the disorder. Noble (1998) suggested that interference with hearing functions, such as speech understanding and sound localization, i.e. the psychoacoustic aspects, represents the *disability* component of the tinnitus, whereas emotional, health, and sleep problems represent the *handicap* component.

3.7.2 **Comparative aspects of questionnaires**

There are at least a dozen published outcome instruments (see Table 3.1) that are used to obtain tinnitus severity ratings, and there appears to be little consensus regarding their use across tinnitus treatment centers (Newman and Sandridge, 2004). The questionnaires

Table 3.1 Tinnitus questionnaires

Tinnitus Questionnaire	Author(s)	Number of items and rating	What is assessed? Subscales	Internal consistency: Cronbach's α	Test–retest correlation r
ITI—International Tinnitus Inventory	Kennedy et al. (2005)	8 items 5-point scale	Tinnitus related complaints		
STSS—Subjective Tinnitus Severity Scale	Halford and Anderson, (1991)	16 items	General distress	0.84	
TCQ—Tinnitus Cognitions Questionnaire	Wilson and Henry (1998)	26 items 3-point scale score 0–104	Thoughts about tinnitus Benefit from CBT: 1. Negative cognitions 2. Positive cognitions	0.91	
TCSQ—Tinnitus Coping Style Questionnaire	Budd and Pugh (1996)	40 (33) items 7-point scale	Coping strategies Benefit from CBT: 1. Maladaptive coping 2. Effective coping	0.90	
TH/SS—Tinnitus Handicap/ Support Scale	Erlandsson et al. (1992)	28 items 5-point scale	Attitude of others: 1. Perceived negative attitudes 2. Social support 3. Disability/handicap		
THI—Tinnitus Handicap Inventory	Newman et al. (1996; 1998)	25 items yes/some-times/no	Impact on everyday functioning: 1. Functional effects 2. Emotional response 3. Catastrophic response	0.93	0.92
THQ—Tinnitus Handicap Questionnaire	Kuk et al. (1990), Tyler (1993)	27 items, 100-point scale	Perceived handicap: 1. Physical, emotional, social effects of tinnitus 2. Hearing and com-munication ability 3. Individuals percep-tion of tinnitus	0.95	0.89
TPQ—Tinnitus Problems Questionnaire	Tyler and Baker (1983)	Open ended format	Disability vs. handicap: 1. Effects on hearing 2. Effects on lifestyle 3. Effects on general health 4. Emotional problems		

(Continued)

Table 3.1 Tinnitus questionnaires

Tinnitus Questionnaire	Author(s)	Number of items and rating	What is assessed? Subscales	Internal consistency: Cronbach's α	Test–retest correlation r
TQ or TEQ—Tinnitus Questionnaire/ Tinnitus Effects Questionnaire	Hallam et al. (1988)	52 items 3-point scale	Complaints: 1. Emotional/Cognitive Distress 2. Intrusiveness 3. Auditory difficulties 4. Sleep disturbances 5. Somatic complaints	0.91–0.95	0.91–0.94
TRQ—Tinnitus Reaction Questionnaire	Wilson et al. (1991)	26 items 5-point scale	Coping vs. non-coping: 1. Annoyance 2. Interferes with work 3. Severe distress 4. Avoidance of activities	0.96	0.88
TSI—Tinnitus severity index	Meikle et al. (1984)	12 items 5-point scale	Negative impacts upon activities		
TSQ—Tinnitus Severity Questionnaire	Coles et al. (1992)	10 items 5-point scale	Severity and impact: 1. General severity 2. Quality of life 3. Psychological aspects		
TSS—Tinnitus Severity Scale	Sweetow and Levy (1990)	15 items 4-point scale weighted score	Therapeutic progress: 1. Intrusiveness 2. Distress 3. Hearing Loss 4. Sleep disturbance 5. Medication	0.86	

most often used in North America are the Tinnitus Severity Index (TSI), the Tinnitus Handicap Inventory (THI), and the Tinnitus Handicap Questionnaire (THQ). The baseline index scores for THI, THQ, and TSI show comparable intersubject variability. However, an examiner cannot rely solely on one index score to make a clinical severity judgment (Newman and Sandridge, 2004). One wonders, however, how disparate results from different questionnaires are merged into one clinical judgment? Does one retain only the ones that are highly correlated or does one pay more attention to the items that result in different scores? I feel that the clinical experience of the audiologist or psychologist is the ultimate factor, backed up by written tests. This will be different when the questionnaire instrument will be used to assess a treatment; in that case the investigator has to refrain from any personal judgment. The Tinnitus Questionnaire (TQ also known as the Tinnitus Effect Questionnaire, TEQ) is most commonly used in the UK and Germany (Hiller and Goebel, 1992; Hiller et al., 1994), and has been cross-validated with the THI and THQ (Baguley et al., 2000). The TQ was one of the first tinnitus questionnaires (Hallam et al., 1988), still has the largest number of items (52) and illuminates the point that the perceived "loudness and unpleasantness" of tinnitus noise is related to reported emotional distress and not to reported interference with auditory perception.

The most important aspect of questionnaires is their internal consistency, which is a measure based on the correlations between different item scores on the same test (or the same subscale on a larger test). It measures whether several items that propose to measure the same general construct produce similar scores. Internal consistency is commonly assessed using Cronbach's α, a statistic calculated from the pair-wise correlations between items. Internal consistency ranges between zero and one. A rule of thumb is that an α of 0.6–0.7 indicates acceptable reliability, and 0.8 or higher indicates good reliability. High reliabilities (0.95 or higher) are not necessarily desirable, as this indicates that the various items may be largely redundant (Everitt, 1998). Table 3.1 gives the values for those questionnaires that have been evaluated in this respect. It is noteworthy that the α's all exceed 0.8, indicating good reliability. It is also important to realize that both the THQ and TRQ have scores of 0.95 and 0.96 respectively that brings them in the redundant territory. This should be further evaluated.

Questionnaires should have high test–retest reliability, which is important if the instrument is used to assess the effect of a particular treatment on qualitative aspects of tinnitus. For most tests for which this is estimated (see Table 3.1 last column for correlation coefficient) it may only apply to relatively short test–retest intervals since no long intervals were included in the studies. Test–retest correlation coefficients are only available for four questionnaires and are between 0.88–0.94, which suggest that the test results could explain between 77–88% of the variance in the data, e.g. of the effectiveness of the treatment.

The matter of cross-validating different questionnaires also has received some attention. It is important since very high values could remove the need for administering several questionnaires to the same patient, which is time consuming and may not provide new information. Baguley et al. (2000) compared the TQ (of psychological origin) and the THI (of audiological design), Newman and Sandridge (2004) compared the THI, THQ, and TSI. Baguley et al. (2000) found that the THI scores, total and subscale, were very similar to the standardization data (Newman et al., 1996), indicating that the data obtained using the THI are comparable in the UK and US. This similarity to the standardization data is also the case for the TQ. Both instruments thus appear robust when used in a clinical setting (audiological vs. psychological) that is different from that in which they originated. The correlations between the total scores for the two instruments were high at 0.88. Thus there is a high convergent validity between the TQ and THI, and therefore, between the hypothetical constructs of "tinnitus handicap" which underlie each instrument. This finding has two important implications. The first is that studies using either total THI or total TQ scores as outcome measures have been shown to be comparable. The second is that the constructs of "tinnitus handicap" as developed by audiologists (e.g. the authors of the THI), and by psychologists (the authors of the TQ) have been demonstrated to be convergent. The outcome of the two comparison studies is that the instruments are all comparable but still one questionnaire may be more sensitive than the other on specific subfields of questions, e.g. relating to sensitivity to sleep disturbance in the TQ and THI, thereby still suggesting that more than one questionnaire be used.

Another important issue emerged with the advent of cognitive behavioral therapy (CBT) for treatment of tinnitus patients. This occurred partly in reaction to the purely audiological and medical forms of treatment, using masking techniques or drug regimes that were not very effective in answering the needs of tinnitus patients. In addition to the use of the test to measure the effect of the treatment itself, is a prior indication of the potential benefit that the patient may receive from the CBT. This has led to two independently developed questionnaires that explore differences in coping style and cognitive attitudes towards tinnitus, aimed at predicting treatment success. These are the TCSQ and the TCQ (Table 3.1) and they are very similar in their aim and subscales (Noble, 2000), and the TCQ outcomes correlate with those of commonly used instruments such as TQ, THQ, and TRQ. An overview of the questionnaires subjected to cross-validation is shown in Table 3.2, a + sign indicates which pairs have been cross-validated.

In the first part of this chapter we have evaluated the quantitative aspects of tinnitus as resulting from psychoacoustic methods. These aspects are important to probe for tinnitus mechanisms and to some extent for amelioration of tinnitus by external sounds. However, it was realized very early on (Meikle et al., 1984) that these are not the indicators that assess the success of a particular treatment, and that this can only be done by tinnitus questionnaires. The large-scale study by Meikle et al. (1984) included loudness rating and loudness matching techniques, as well as a 10-point rating scale of tinnitus "severity" (TSI). The definition of severity was not explicit but, in conformity with previous clinical observations, rated severity of tinnitus did not correlate with matching loudness level. Severity was weakly related to whether or not the tinnitus sound could be successfully masked. So, using the definitions of Noble (2000) we can say that there is some interaction between the *handicap* and *disability* aspects of tinnitus.

Table 3.2 Cross-validation of questionnaires

	TQ/TEQ	THQ	TRQ	THI	TCQ	TSQ	TH/SS
TQ/TEQ		+	+	+	+		
THQ	+			+	+		
TRQ	+				+		
THI	+	+					
TCQ	+	+	+				
TSQ							+
TH/SS						+	

3.8 Summary

Tinnitus has psychoacoustic (pitch and loudness, masking ability) as well as psychological (annoyance, depression) properties. The psychoacoustic properties can be well quantified with standard procedures and are often replicable and accurate, whereas the psychological ones vary greatly over time and context. The psychological aspects that are typically extracted from questionnaires, of which there are a good dozen each with slightly different emphasis, appear also to be well validated. Tinnitus has a spectrum that matches that of the hearing loss. Assigning a distinct pitch may have the problem of octave errors; however, when those errors are eliminated by training the pitch can correspond more closely with the edge frequency of the audiogram. How this fits with the concept of the tinnitus spectrum is not clear. Tinnitus loudness is typically only a few to 10 dB above the threshold level, but the presence of loudness recruitment may make this correspond to a much higher perceived loudness. The masking properties of tinnitus are different from those of external sounds, likely missing the effect of lateral suppression in the cochlea, and suggesting a central origin of tinnitus. Residual inhibition lasts from seconds to minutes and also fits better to a central than peripheral origin. The annoyance of tinnitus appears to have no relation to its loudness, but appears to be determined more on the level of hearing loss, the presence of hyperacusis and signs of depression.

The next chapter will investigate where in the brain tinnitus resides using imaging and neurophysiological techniques in humans.

Chapter 4

Objective assessment of tinnitus

Objective measurements of human auditory system activity can be obtained from nearly all levels from cochlea to cortex, i.e. using techniques as diverse as OAE recordings that probe the function of the OHCs, with evoked potentials or magnetic fields registering synchronous brain activity, and with various imaging techniques. Measurements of human cortical activity are non-invasive with few exceptions, e.g. only in preparation for surgery to provide relief from epilepsy, and are generally confined to scalp recorded activity using EEG/MEG or functional imaging with PET, single-photon emission computed tomography (SPECT), and functional magnetic resonance imaging (fMRI).

4.1 Otoacoustic emissions

Spontaneous otoacoustic emissions (SOAEs) are low-level sounds emitted by the healthy normal ear, and are recordable with sensitive microphones inserted in the ear canal. In at least 4% of normal-hearing persons, SOAEs are considered partially responsible for the tinnitus (Penner and Jastreboff, 1996). In the vast majority of cases, however, SOAEs and tinnitus are independent phenomena (Penner and Burns, 1987; Penner, 1992). An important point here may be that it is difficult to record SOAEs at frequencies above 6 kHz, which correspond with a large fraction of reported tinnitus pitches. Thus, the incidence of the correlation of tinnitus and SOAEs could be higher. Although spontaneous emissions could theoretically represent increased "spontaneous firing" in neurons innervating the basilar membrane at the emission site, via the potentially increased local amplifying action of the OHCs, CNS habituation may preclude their audibility. Occasionally, some subjects hear intermittent SOAEs as intermittent tinnitus (Burns and Keefe, 1992). Aspirin administration was reported to abolish OAE-related tinnitus, likely by interfering with the OHC amplifier (Penner and Coles, 1992). This could potentially be used as a test to eliminate SOAEs as the source of tinnitus. However, as presented in Chapter 6, chronic salicylate application increases SOAEs.

Stimulated OAEs are sensitive indicators of OHC loss and the influence of the efferent system on the cochlea. Distortion product emissions (DPOAE) originate from the non-linear transformations in the normal cochlea that result from the OHC amplification of the basilar membrane movement. DPOAEs usually reflect the 2f1–f2 cubic distortion product, where f1 is the lower and f2 the higher frequency (with f2/f1 ~1.25) that are simultaneously presented to the ear. Typically normal-hearing tinnitus subjects have lower amplitude transient OAEs and DPOAEs than normal-hearing controls (Granjeiro et al., 2008). However, Sztuka et al. (2010) found higher DPOAE amplitudes in tinnitus patients with normal hearing who also had hyperacusis. Some patients with normal hearing acuity who have acute tinnitus seem to have a less effective functioning of the cochlear efferent system, because the application of contralateral noise either enhanced the DPOAEs or suppressed them less intensely than it did in a control group (Attias et al., 1996; Fávero et al., 2006; Riga et al., 2007; Hesse et al., 2008). Nottet et al. (2006) found in military personnel, 24 h after an acute acoustic trauma, that OAEs appeared to be a better predictor of the persistence of tinnitus than hearing thresholds alone. After head injury, significantly higher spontaneous and stimulated OAE amplitudes, and reduced medial olivocochlear suppression, were reported in patients with tinnitus, as compared to subjects without tinnitus (Ceranic et al., 1998). Thus, stimulated and spontaneous OAEs are sensitive indicators of subthreshold hearing loss and changes in efferent activity at the cochlear level, but SOAEs only rarely are the cause of tinnitus.

4.2 Metabolism- and blood-flow-based non-invasive measures of brain function

4.2.1 Positron emission tomography

Imaging techniques used to assess human brain function, are based on the relationship between cerebral blood flow (CBF), energy demand, and neural activity (Heeger and Ress, 2002). The energy requirement of the brain can be expressed in terms of oxygen consumption, because ~90% of the glucose is aerobically metabolized, and therefore parallels oxygen consumption. Oxygen consumption is proportional to neural activity and is four times greater in gray than in white matter. PET images are spatial maps of the radioactivity distribution, i.e. positron emitting sites, within tissues (Johnsrude et al., 2002; Ruytjens et al., 2006). It is the annihilation of these emitted positrons by encountering an electron that produces gamma rays emitted in opposite directions and the basis for estimating the annihilation site (very close to the generation site) of the positrons. The most common radioactive tracer for quantifying metabolic activity is [^{18}F]fluorodeoxyglucose (FDG). This compound is phosphorylated by the brain metabolism in the same way as glucose. The deoxyglucose measure thus reflects energy use by neurons and glial cells. The gamma emissions produced by annihilation of the positron emitting FDG are used to construct a picture of the distribution of enhanced metabolism. Alternatively, the gamma emissions from the ^{15}O isotope incorporated in water and injected into the blood supply relate to local changes in vascular parameters deduced from the local concentration of the tracer in cortical microvessels, and reflect blood flow. The time resolution of PET is poor (of the order of tens of seconds) and so is the spatial resolution (of the order of a cm^3). Single subject studies are typically not possible and most studies rely on group averages. The big advantage of PET for auditory and tinnitus research is that it is a silent technique so that spontaneous brain activity can be measured (Lanting et al., 2009).

4.2.2 Single-photon emission computed tomography

SPECT is an imaging technique using gamma rays. The basic technique requires injection of a gamma-emitting radioisotope, into the bloodstream of the patient. SPECT is similar to PET in its use of radioactive tracer material and detection of gamma rays. In contrast with PET, however, the tracer used in SPECT emits gamma radiation that is measured directly, whereas PET tracers emit positrons which annihilate with electrons up to a few millimeters away, causing two gamma photons to be emitted in opposite directions. A PET scanner detects these emissions "coincident" in time, which provides more radiation event localization information and thus higher resolution images than SPECT.

4.2.3 Functional magnetic resonance imaging

The most studied signal in fMRI is known as the blood oxygenation level-dependent (BOLD) signal (Fox and Raichle, 2007). The temporal resolution of the BOLD signal (~1 s) is much higher than that of the ^{15}O-PET response, which can only be computed after integration of gamma activity over a time period of about 30 s. The BOLD signal is based on the measurement of changes in magnetic susceptibility of hemoglobin, depending on whether it carries an O_2 molecule or not. The BOLD response reflects a rather small difference signal from a control or rest state, generally not more than 2–3% of baseline. This implies that for standard studies one has to average across groups of about 10 people to get decent statistics, i.e. small SDs. A sudden increase in synaptic metabolism is followed by a transient drop in oxyhemoglobin concentration in vessels neighboring activated neurons and consequently in the BOLD signal (Logothetis, 2002). A major increase in oxyhemoglobin concentration then occurs as a consequence of vessel dilatation, with a peak observed 5–6 s after stimulus onset. Logothetis et al. (2001) found that the BOLD response directly reflects a local increase in neural activity. For the majority of recording sites, the BOLD signal was a linear function of local field potential (LFP) amplitude, and the firing rate of small neural populations. LFPs, however, were a substantially more reliable predictor than spike firing rate, and were covarying with the BOLD signal even when spike firings adapted. This suggests that the BOLD response primarily reflects the input (post-synaptic potentials, PSPs) and local processing in neural circuits rather than the output (spike) signals, which are transmitted to other regions of the brain by the axons of the principal neurons. Thus it is widely believed that increased blood flow follows directly from increased synaptic activity. Estimates of the metabolic costs of brain activity show that most of the energy is used by the neurons (and likely also glial cells) and that energy usage depends strongly on firing rates (Heeger and Ress, 2002).

4.3 Imaging of tinnitus

4.3.1 Structural changes

Structural changes in the auditory system of tinnitus patients appear to be limited to gray matter increases at the thalamic level. Outside the auditory system, gray matter decrease was found in the subcallosal region including the nucleus accumbens (NAc; Mühlau et al., 2006). They speculated on a gating function for the NAc that when impaired would lead to tinnitus (Chapter 11). Landgrebe et al. (2009) found significant gray matter decreases in the tinnitus group in the right IC and in the left hippocampus. However, no changes in the subcallosal area or in the thalamus as described by Mühlau et al. (2006) were observed.

They also found no correlation between gray matter changes and tinnitus duration or severity. Schneider et al. (2009) did find an association of tinnitus with structural changes in the auditory cortex. The medial part of Heschl's gyrus (mHG) was studied in people with and without chronic tinnitus using MRI. Patients exhibited significantly smaller mHG gray matter volumes than controls. In unilateral tinnitus, this effect was almost exclusively seen in the hemisphere ipsilateral to the affected ear. The authors interpreted this as the ipsilateral hemisphere auditory cortex causing less inhibition in the contralateral cortex, thereby facilitating the generation of tinnitus. In bilateral tinnitus, mHG volume was substantially reduced in both hemispheres. The tinnitus-related volume reduction was found across the full extent of mHG, not only in the high-frequency part usually most affected by hearing loss-induced deafferentation. However, there was also evidence for a relationship between volume reduction and hearing loss. Husain et al. (2011) examined neuroanatomical alterations associated with hearing loss and tinnitus in three groups of subjects: those with hearing loss and tinnitus, those with hearing loss without tinnitus, and normal hearing controls without tinnitus. Structural MRI scans and voxel-based morphometry and diffusion tensor imaging (DTI) were used. No significant changes were found in gray or white matter in subjects with tinnitus and hearing loss compared to normal-hearing controls. In agreement with Landgrebe et al. (2009) the finding of thalamic-level gray matter decreases in tinnitus relative to normal hearing (Muhlau et al., 2006) could not be confirmed.

From this limited number of studies one has to conclude that other aspects beyond tinnitus may be confounding the picture. One can think of age, hearing loss, and difference in brain structure that make some people more prone to developing tinnitus. Brain connectivity is also relevant. Lowry et al. (2004) reported a patient (an otolaryngologist) whose chronic tinnitus was permanently abolished by a small stroke that affected thalamocortical pathways projecting to frontocentral cortex and to the regions of the neostriatum (caudate nucleus and putamen) in the left hemisphere. Communication among brain regions appears to be critical for tinnitus and is another feature to be assessed by imaging studies.

DTI can study the microstructure of white matter in the CNS *in vivo* on basis of the anisotropic flow of water parallel and perpendicular to white matter tracts. Connectivity studies using DTI (Crippa et al., 2010) incorporate an a priori bias because of the selection of the seed areas. Several measures can be used, such as fractional anisotropy (FA), the asymmetry in the water diffusion direction in a single voxel, which can be used to estimate probabilities to track fiber directions (Beaulieu, 2002). This can be extended to the mean FA of a path. Crippa et al. (2010) used probabilistic tractography to get a connectivity map depicting the voxel-wise probability to reach any given voxel starting from a user defined region of interest (ROI). This is defined as the percentage of samples leaving from the starting ROI that pass through that voxel. A weighted FA was used in which the calculation of the mean path FA was weighted by the probability of each voxel as defined earlier. The path strength was defined as the percentage of samples leaving the ROI that were able to reach the target ROI. Because of statistics and crossing of fiber paths with other fiber paths that neither start, nor end in the respective ROIs, it matters at what

ROI one starts or ends, i.e. the measures going from IC to auditory cortex (AC) may not be equal as those starting at AC and ending in IC, i.e. the measures are generally not symmetric. Their most important finding is an increased patency of the white matter tracts between the auditory cortex and the amygdala in tinnitus patients as compared to healthy controls. Also using DTI, Lee et al. (2007) found a significant reduction in the arcuate fasciculis in tinnitus patients. This is a fiber tract that connects frontal and temporal cortical centers, and the results suggest reduced connectivity.

4.3.2 Functional changes

Functional brain imaging in humans would ideally provide an accurate representation of changed neural brain activity during silence, which is generally considered the crucial neural substrate linked to tinnitus. This problem cannot readily be solved at the group comparison level between tinnitus patients and normal controls because of high variability in resting metabolic activity levels among people. Spontaneous BOLD activity is not random noise, but is specifically organized in the resting human brain. Spatial patterns of coherent BOLD activity that differed between normal controls and patients with depression has established differences in the spatial correlation structure of resting neural activity that may well be applicable to tinnitus (Fox and Raichle, 2007; Chapter 11).

PET and SPECT Arnold et al. (1996) using FDG in a PET study reported increased resting metabolic activity in primary auditory cortex for a group of 10 out of 11 patients with chronic tinnitus compared to 14 controls. PET images of regional cerebral blood flow distribution using ^{15}O labeled water obtained during masking of the tinnitus were subtracted from PET images obtained during the presence of the tinnitus sensation (Mirz et al., 1999). They found a larger difference in neuronal activity between unmasked and masked conditions in the tinnitus patients compared to the controls. This larger signal in the unmasked condition was found predominantly in the right hemisphere with significant foci in the middle frontal and middle temporal gyri, larger signals during masking were found in the left transverse and left superior temporal gyri. They interpreted the difference signal as due to tinnitus. Note that a larger signal following masking was found using the auditory steady state response for tinnitus patients but not in controls (Roberts, 2011; Chapter 3)

Lidocaine is a local anesthetic that blocks fast voltage-gated Na^+ channels and temporarily reduces neural firing (Chevrier et al., 2004). However, several randomized, controlled studies found that tocainide (an oral analog of lidocaine) had little benefit for tinnitus. Thus, the tinnitus suppressing action of lidocaine may not be due to the Na^+ channel blocking activity common to these two drugs. In patients who suffer from tinnitus, intravenous lidocaine can suppress this phantom sensation for a short time. The percentage of tinnitus patients who experience a benefit from lidocaine is approximately 60% (Simpson and Davies, 1999). Intravenous infusion of lidocaine had a statistically significant inhibitory effect on tinnitus in patients who underwent translabyrinthine removal of a vestibular schwannoma. The site of action of lidocaine in this instance must be in the central auditory pathway, as the cochlear and vestibular nerves are sectioned during surgery (Baguley et al., 2005b). A PET experiment, including a non-tinnitus control group and also a condition in which a placebo injection was given (Reyes et al., 2002), also showed that not all nine patients responded to lidocaine in the same way. In four patients it suppressed tinnitus, in another four it enhanced tinnitus, and in one it had no effect. Group analysis of PET data (Plewnia et al., 2007) showed tinnitus-related increases of regional CBF in the left middle- and inferior-temporal cortex as well as right temporoparietal cortex and posterior cingulum, when compared to activity following intravenous lidocaine that induced a suppression of tinnitus (Figure 4.1). Like prior imaging studies, the group data from Plewnia et al. (2007) showed no significant tinnitus-related hyperactivity, i.e. that blocked by lidocaine, in the primary auditory cortex (PAC, Brodmann area (BA) 41). The regions affected by lidocaine were in the human equivalent of the parabelt areas of auditory cortex. Andersson et al. (2000) found after lidocaine infusion that tinnitus was associated with increased rCBF in the left parietotemporal auditory cortex, including the primary and secondary auditory cortex with a focus in the parietal cortex (BA 39, 41, 42, 21, 22). Activations were also found in right frontal paralimbic areas (BA 47, 49, and 15).

From these lidocaine experiments arises the suggestion that major blood flow changes generally occur in the parabelt and belt areas of auditory cortex. It is not clear if these are the only areas involved in the generation of tinnitus. This "Where is tinnitus in the brain?" question has early on been addressed indirectly by focusing on people who can modulate the strength of their tinnitus by either changing eye gaze direction or by making orofacial movements. The advantage is that in these studies the patients function as their own controls, but one has to keep in mind that only specific cross-modal mechanisms of inducing tinnitus are probed. Tinnitus as a result of hearing loss typically cannot be modulated in this way.

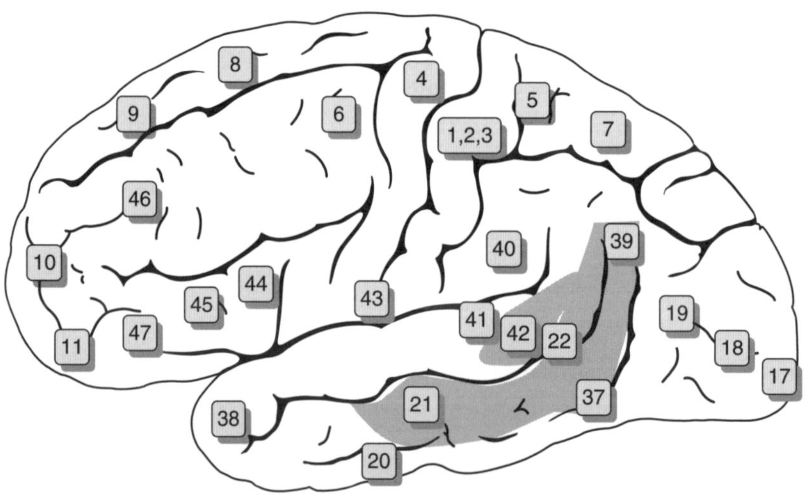

Figure 4.1 Temporal regions where increased activity is found in tinnitus patients using PET (gray areas). Note that the involved regions are not primary auditory areas (BA 41). Based on data from Plewnia et al. (2007), Andersson et al. (2000), Giraud et al. (1999), and Farhadi et al. (2010).

Gaze-evoked tinnitus has most often been reported following complete and acute unilateral deafferentation of the auditory periphery after surgical removal of space occupying lesions from the base of the skull (House, 1982). In its pure form, gaze-evoked tinnitus is absent in certain eye positions (i.e., 0° gaze, from a neutral head-referenced condition) but can be activated when static deviation of eye gaze exceeds a certain displacement (3–10°) in the horizontal or vertical direction (Cacace et al., 1994). In a study of four adults that developed the pure form of gaze-evoked tinnitus which became manifest after unilateral acoustic tumor removal, changes in eye gaze evoked tinnitus. In PET studies, localized sites within temporoparietal association areas, but not in PAC were activated bilaterally. The main peak of activation was located in the secondary auditory cortex (BA 42). This activation extended in the dorsal direction towards the planum parietale (BA 40) and also into BA 21. In the left hemisphere, the activation was more extensive though less significant. Here, the main two peaks of activation were found in the superior temporal gyrus (BA 22) and in the medial temporal gyrus (BA 21). Significant rCBF increases were found in the region between the two peaks and the activation in the region of the former peak (BA 22) extended into the inferior parietal lobule (BA 40). Activation in the superior temporal gyrus (BA 42/22) and the medial temporal gyrus was a pattern common to both hemispheres (Giraud et al., 1999). In a PET study of individuals that could modulate a constant background tinnitus with eye gaze changes of >60°, activation of brainstem (lateral pontine tegmentum, vermis of the cerebellum, cuneus) and auditory cortical areas was observed (Lockwood et al., 2001).

A [^{18}F]-FDG SPECT study carried out on 55 tinnitus patients and eight controls (Farhadi et al., 2010) showed no abnormal uptake in BA 41 and BA 42. The most common sites of abnormal uptake were BA 21, BA 37, and BA 22, corresponding to the areas indicated by the lidocaine experiments (cf. Figure 4.1). There was no significant association between side of tinnitus and side of SPECT coincidence scan abnormality in the brain. Uptake in frontal areas was noted as well.

Summarizing, gaze-induced tinnitus as well as group-based abnormalities appear to localize tinnitus activity dominantly to the parabelt and belt areas of auditory cortex just as lidocaine infusion does. Modulating an existing tinnitus by large gaze changes differentially affects auditory cortex, but also the lateral tegmentum (source of dopamine), the vermis of the cerebellum, and the cuneate nucleus. Orofacial movements produce a change in tinnitus loudness and differentially affect only the auditory thalamus.

BOLD The noise generated by a MRI scanner provides a major limitation faced by fMRI studies of the auditory system. The high-level noise is due to the flexing of the gradient coil loops in the static magnetic field as current passes through the loops during imaging. Even in the absence of this particular scanner noise, the liquid helium circulation pump and the air ventilation systems produce low-level noise that is continuously present in the magnet room (Ravicz et al., 2000). Although MRI scanner noise contains many audible frequencies, the main power of the noise occurs in the frequency range of 1–2 kHz with higher order resonances of lower intensity also present. This scanner noise can reduce the detectable activity within the auditory system through masking thereof, habituation of the nervous system, saturation of the activation, and impairment of cognitive processing. This makes interpreting these functional imaging studies of the auditory system a challenging task. Scanner related acoustic noise induces a BOLD response in the auditory cortex. It has been shown that, similar to other auditory stimuli, scanner noise induces a hemodynamic response within 2–3 s after the onset of acoustic noise that peaks after 3–8 s (hemodynamic delay) and returns to baseline in ~8 s. The hemodynamic responses to magnetic resonance-related acoustic noise and stimulus-induced brain activation do not add up linearly; this implies that a simple subtraction of this activity from that induced by the intended stimulus is not well possible (Moelker et al., 2003).

Typically sound-induced BOLD responses reflect the few percent difference with the non-stimulus (but mind the pump noise) baseline activity. In the case of tinnitus, this baseline activity could be higher than normal, leaving fewer neurons that can be activated with the sound ("line-busy" effect; Chapter 3). In the IC Melcher et al. (2000) found that external sound produced abnormally low activation, additional to baseline, in the IC contralateral to the tinnitus ear. Smits et al (2007) found the same effect in IC, auditory thalamus, and AC. So indirectly this could point to increased spontaneous activity levels in these brain structures. The problem with using sound in any imaging technique in cases of tinnitus is that the sound likely interacts with the resting activity related to tinnitus, such that the estimated spontaneous levels by simple subtraction may be biased. Contradicting the findings of Melcher et al. (2000) and Smits et al. (2007), Lanting et al. (2008) observed increased responses to sound in the IC of tinnitus patients. However, methodological differences, such as continuous sampling versus sparse sampling, and the presence or absence of pumping sounds in the cooling system, may explain this (Melcher et al., 2009). The pump sounds in particular appeared to interfere with the tinnitus group (somewhat expected in the light of the interaction phenomenon postulated earlier) and not with the control group (Figure 4.2). After eliminating these sounds, the results of Melcher et al. (2009) were in agreement with those of Lanting et al. (2008).

It is, however, not clear whether the increased sound-evoked responses in tinnitus patients indicate the presence of tinnitus or are largely due to gain changes in the auditory system as a result of hearing loss (Chapter 3). The inference of spontaneous activity levels underlying tinnitus draws heavily on the linearity assumption; the sound-evoked activity is purely additive (up to a saturation level) to the baseline tinnitus level. In addition, tinnitus often is accompanied by hyperacusis that results from gain changes in the central auditory system. Thus, the BOLD response under stimulus conditions may largely reflect hyperacusis. A recent study (Gu et al., 2010) appears to corroborate this, at least for the IC.

Gu and colleagues (Gu et al., 2010) performed fMRI in subjects with and without tinnitus and/or hyperacusis, all with clinically normal hearing thresholds. They were all tested to assess the presence of hyperacusis. Despite receiving identical sound stimulation levels, subjects with hyperacusis (typically but not always accompanying tinnitus) showed elevated BOLD responses in the IC, thalamus, and primary auditory cortex compared with subjects with normal sound tolerance. Only primary auditory cortex showed increased BOLD responses specifically related to tinnitus, i.e. in the absence of hyperacusis. The results directly link both hyperacusis and tinnitus to gain changes within the central auditory system.

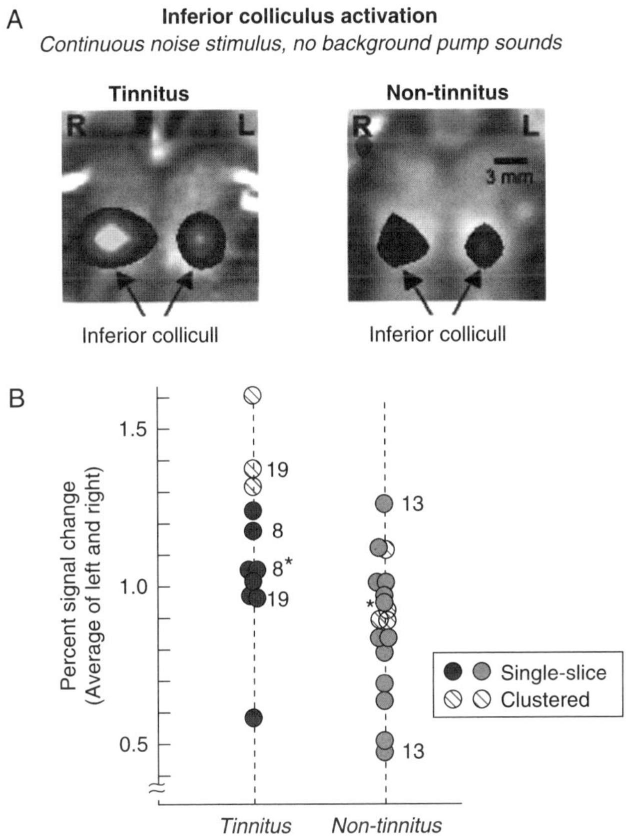

Figure 4.2 (Also see Color plate 1). The inferior colliculi of tinnitus subjects showed abnormally high sound-evoked activation when the scanner coolant pump (and the acoustic noise it produces) was off. The sound stimulus was a continuous, broadband noise (binaural, 50–55 dB SL). (A) Enlarged images of the inferior colliculi in one tinnitus and one non-tinnitus subject (corresponding to circles near asterisks in (B)). A map of activation (color) produced by the continuous noise stimulus is overlaid on a T1- weighted anatomical image (grayscale) obtained in the same imaging session. The color scale in the activation maps indicates the significance of the difference in image signal between stimulus on and off periods according to a t-test (uncorrected for multiple comparisons) (blue: $p = 0.01$; yellow: 2×10^{-9}). (B) Percent signal change in the inferior colliculi of each tinnitus and non-tinnitus subject studied during the "pump off" condition. Each circle indicates% change averaged between the left and right inferior colliculi of a given subject. Reprinted from Melcher, J.R., Levine, R.A., Bergevin. C., and Norris, B. (2009). The auditory midbrain of people with tinnitus: Abnormal sound-evoked activity revisited. *Hearing Research*, **257**, 63–74, © with permission from Elsevier.

It is surprising that an elevated response to sound in the auditory cortex of tinnitus patients without hyperacusis can occur. This elevated response represents clearly a cortical gain change and one would expect it to have the behavioral correlate of hyperacusis as well. There are two ways out of this conundrum: the first is that the spontaneous activity in the cortex of tinnitus patients without hyperacusis is lower than in controls (see Chapters 6 and 7 for animal data that may corroborate this). This would allow a larger

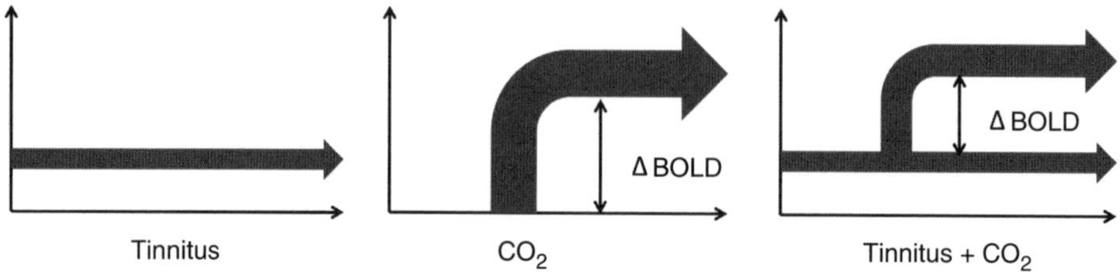

Figure 4.3 The principle of the hypothesized BOLD ceiling fMRI in the context of the actual investigation. Tinnitus should evoke a prolonged BOLD response (tinnitus). The application of only CO_2 should evoke an induced BOLD response, which begins after ~2 min and ends after ~5 min (CO_2). Under the assumption of a BOLD ceiling, i.e., a limitation of the maximum possible BOLD response, we expect that the CO_2-induced ΔBOLD response is reduced in tinnitus-activated areas compared to inactive areas, which are at BOLD baseline state. Tinnitus activated areas can thus be identified based on a reduced CO_2-induced ΔBOLD. After Haller et al. (2006).

increment for stimulus-induced activity. The second is that hyperacusis is a subcortical phenomenon only and auditory cortex is not involved. This would uncouple perception (likely a cortical phenomenon) from the hyperacusis phenomenon, potentially requiring a cortical bypass from the IC to the limbic system.

A potentially promising method of measuring resting levels of activity that does not use sound stimulation and that would be applicable to investigate substrates of tinnitus has been proposed by Haller et al. (2006). They used inhaled CO_2 as a vasodilator to induce a "global" blood oxygenation level-dependent (ΔBOLD) response (Figure 4.3). They implied, albeit indirectly, that spontaneously active brain areas in subjects with tinnitus will exhibit a reduced CO_2-induced change in the ΔBOLD response due to pre-existing tinnitus-induced BOLD response. The capability of cerebral vessel dilation is limited, consequently, there is a ceiling of the cerebral perfusion and thus to the BOLD response. It is hypothesized that inactive brain areas are in the BOLD baseline state and present the full CO_2-induced ΔBOLD amplitude change. In contrast, a reduction of ΔBOLD amplitude change will occur in continuously active brain areas due to a pre-existing pathological, e.g. tinnitus, activity-induced BOLD response. This putative reduction in the change of the BOLD response compared to non-auditory areas might then be exploited for mapping of a continuous neuronal activation that putatively exists in tinnitus. The comparison with the effect in non-auditory areas could provide a within-subject control, which obviously is not possible when using a sound to obtain a BOLD response. This, however, assumes that central manifestations of tinnitus are limited to the auditory nervous system (and it is clear that they are not albeit that there will be cortical regions that are not affected).

4.4 Extracranially recorded neuronal activity of auditory cortex

4.4.1 The electroencephalogram

If large numbers of neurons are activated simultaneously, the changes in their membrane potential occur synchronously. The corresponding currents then add up in phase and become so large that they can be detected at the scalp. Synchronicity, however, is not enough for compound activity to be detectable at the scalp. Currents that flow in opposite spatial directions can partially or totally cancel each other. Thus, for compound voltage changes to be detectable at a distance, i.e. to produce a far-field potential, the majority of the currents of the individual neurons have to flow in the same spatial direction. Only those structures that have a spatial alignment of neurons with the same orientation of their current producing parts (termed open fields) produce far-field potentials (Eggermont, 2007a). In cortex these are the pyramidal cells. The combined electrical activity from a large number of cortical neurons recorded from the scalp is called the electroencephalogram (EEG). The magnetic counterpart of the EEG is called the magnetoencephalogram (MEG) and has the advantage that it records only localized activity, but also is less responsive to activity generated by dipoles oriented radial to the head. Spontaneous EEG or MEG activity relies on neural synchrony, without synchrony the EEG would just be noise. Neural synchrony appears to be confined to several frequency bands. In general, there is a correlation between the distance over which synchronization is observed and the frequency of the synchronized oscillations. Short distance synchronization tends to occur at higher frequencies (γ-band, 30–80 Hz) than long-distance synchronization, which often manifests itself in the β-band (15–30 Hz) but also in the θ-band (4–8 Hz) and α-band (8–12 Hz). The latter activity when occurring over the temporal lobe is typically referred to as tau rhythm.

The EEG changes as a function of the brain state, e.g. drowsy or aroused. A drowsy brain shows large periodic waveforms with frequencies around 10 Hz, a sleeping brain even larger and slower ones with frequencies around 3 Hz (and in addition the slow waves of ~0.1 Hz). In contrast, an awake brain may show low-amplitude high-frequency oscillations in the frequency range above 30 Hz, i.e. the gamma band. The more active the brain is, the fewer low-frequency oscillations and the more high-frequency ones can be observed (Buzsaki and Draguhn, 2004).

4.4.2 Spontaneous EEG and MEG activity in tinnitus

Spontaneous EEG or MEG activity likely relates more to tinnitus than stimulus-evoked activity. The latter may represent hyperacusis as already implied for the sound-evoked BOLD response. We will start with spontaneous activity.

The spontaneous MEG in a group of individuals with tinnitus is characterized by a marked reduction in the tau band (8–12 Hz) power over the temporal cortex, together with an enhancement in delta (1.5–4 Hz) power as compared to normal-hearing controls (Figure 4.4). Correlations with tinnitus-related distress revealed strong associations with this abnormal spontaneous activity pattern, particularly in right temporal and left frontal areas (Weisz et al., 2005a). Reduction of the tau rhythm is a normal cortical reaction to sound presentation (Lehtela et al., 1997) just as visual stimulation (eyes open) reduces the occipital α-rhythm. Thus, the results of Weisz et al. (2005a) suggest that a decrease of tau power in tinnitus patients may be related to the tinnitus percept that plays a similar role to that of an external sound.

The presence of enhanced delta conforms to some extent with propositions made by Llinás et al. (2005), who ascribe this activity in pathological conditions to a thalamic

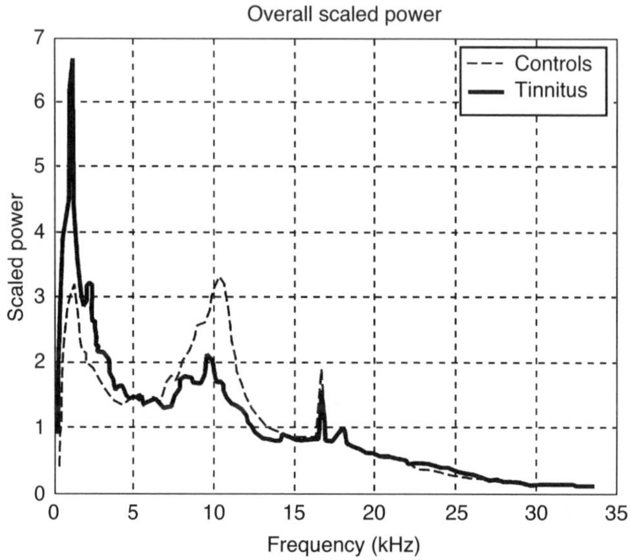

Figure 4.4 Power spectra averaged over all MEG sensors show a reduced alpha peak (~10 Hz) in participants with tinnitus and an enhancement for delta (<2 Hz). The sharp peak centered at ~16 Hz represents an artifact. Reproduced from Weisz, N., Moratti, S., Meinzer, M., Dohrmann, K. and Elbert, T. (2005). Tinnitus perception and distress is related to abnormal spontaneous brain activity as measured by magnetoencephalography. *PLoS Medicine*, **2**(6), e153, © 2005, with permission from PLoS Medicine.

deafferentation followed by reduced inhibition. Tinnitus in their model is part of a group of thalamocortical dysrhythmias (TCD). The power spectra of the TCD patients differ from those of the control subjects in four important respects: (1) the power in the low-frequency band is increased; (2) the peak of this band is shifted towards the theta frequency band (4–8 Hz); (3) the normal alpha-band activity is reduced or absent; and (4) localization of the theta rhythm relates to the type of dysrhythmia generated. The common denominator in TCD seems to be thalamic oscillation at low frequency, associated with abnormal gamma-band activity generated by a disinhibitory "edge effect". Thalamic electrophysiology has shown that brain rhythms have a counterpart in the rhythmic oscillations that occur at single-cell level. It is well established that in thalamic neurons the intrinsic membrane potential oscillations can be modulated by the resting potential of the cell. Membrane hyperpolarization results in deinactivation of low-threshold voltage-gated T-type Ca^{2+} channels. This deinactivation results in spontaneous production of low-threshold spikes crowned by bursts of fast Na^{+} -dependent action potentials with low-frequency rhythmicity. The role of inhibitory interneurons is thus crucial not only in the organization of normal thalamocortical activity at thalamic and cortical levels but also in the generation of pathological conditions leading to neurological and psychiatric symptoms (Llinás et al., 2005).

With that in mind, Weisz et al. (2007) examined gamma-band activity during brief periods of marked enhancement of delta activity (see Weisz et al., 2005a). Results revealed that both control and tinnitus groups showed significant increases in gamma-band activity after the onset of slow waves. However, gamma-band activity was more prominent in tinnitus subjects than in controls. The hemispheric lateralization of the gamma-band activity did correlate with the lateralization of the perceived tinnitus. Ashton et al. (2007) identified discrete localized unilateral foci of high-frequency activity in the gamma range (>40–80 Hz) over the auditory cortex in eight patients experiencing tinnitus during recording. In contrast to the findings by Weisz and colleagues, in five of Ashton et al.'s patients the focus of fast temporal lobe activity was ipsilateral to the perceived tinnitus, while in three it was contralateral. These hotspots were not present in subjects without tinnitus.

Weisz et al. (2007) summarized many of these findings to provide a putative neural code of tinnitus. They note that cochlear damage, or similar types of deafferentation from peripheral input, triggers reorganization in the central auditory system. This produces permanent alterations in the ongoing oscillatory dynamics at the higher layers of the auditory hierarchical stream. The change results in enhanced slow-wave activity reflecting altered corticothalamic and corticolimbic interplay. Such enhancement facilitates and sustains gamma activity as a neural code of phantom perception, in this case auditory. Thus, in tinnitus patients gamma activity during silence appears to play the same role as stimulus induced gamma activity in persons without tinnitus.

A comparison of eight tinnitus patients with eight age-matched controls showed, on average, that the patient group exhibited higher spectral power over the entire frequency range of 2–100 Hz, suggesting increased synchrony on both local and global scales. Using low-resolution electromagnetic tomography (LORETA) source analysis, the generators of delta, theta, alpha, and beta power increases were localized dominantly to left auditory areas (BA 41, 42, 22), and to temporo-parietal, posterior insula, anterior cingulate and parahippocampal cortical areas (Moazami-Goudarzi et al., 2010). This contrasts with the findings of Weisz et al. (2005a), potentially reflecting different tinnitus groups in the two studies.

Van der Loo et al. (2009) showed that in unilateral tinnitus patients (N = 15; 10 right, 5 left) the source strength of resting state EEG gamma band oscillations in the contralateral auditory cortex showed a strong positive correlation with VAS loudness scores (max $r = 0.73$, $P < 0.05$). Auditory phantom percepts thus show similar sound level dependent activation of the contralateral auditory cortex as observed in normal audition. This suggests that unilateral-tinnitus loudness is reflected in gamma band activity in the contralateral auditory cortex.

The same group (Vanneste et al., 2010c) subsequently focused on the cortical and subcortical source differences in resting-state EEG between tinnitus patients with different grades of distress using continuous scalp EEG recordings and LORETA. Results show more synchronized alpha activity in the tinnitus patients with a serious amount of distress with peaks localized to various emotion-related areas. These areas include subcallosal anterior cingulate cortex, the insula, parahippocampal area, and amygdala. In addition, less alpha-synchronized activity was found in the posterior cingulate cortex, precuneus, and dorsal lateral prefrontal cortex (DLPFC).

In addition, a comparison was made between the findings in the tinnitus group with distress and those from a normative database. Here the LORETA current source density in the alpha (8–12 Hz) and beta (21–32.5 Hz) band was higher for tinnitus patients with distress in anterior cingulate cortex (BA 24 and 32). The opposite, namely decreased delta (2–3.5 Hz) and theta (4–7.5 Hz) activity in anterior cingulate cortex (BA 24 and 32) was found for tinnitus patients types with distress in comparison to a normative database. It is interesting that these areas show some overlap with those related to the emotional component of pain. Unpleasant pain also activates the anterior cingulate and prefrontal cortices, amygdala, and insula. As such, it might be that distress is related to alpha and beta activity in the dorsal anterior cingulate cortex, i.e. the amount of distress is related to activity in an alpha network consisting of the amygdale–anterior cingulate cortex–insula–parahippocampal area (Vanneste et al., 2010c). A comparison of the findings by the different groups is shown in Table 4.1. There is a large degree of similarity in the findings. Differences are noted in the area-dependent behavior of the alpha (tau) band, and in the frequency specificity of the changes ranging from only delta, tau and gamma to the entire range of EEG frequencies.

Vanneste et al. (2010d), again using LORETA, localized resting state EEG recordings and found differences in the resting state high-frequency activity (beta and gamma) between unilateral and bilateral tinnitus. This difference is found in the superior prefrontal gurus, right parahippocampus, right angular gyrus, and right auditory cortex. Vanneste et al. (2010f) found by analyzing resting state EEG with LORETA-based source analysis that narrow-band noise tinnitus patients differ from pure tone tinnitus patients in the lateral frontopolar (BA 10), PCC and the parahippocampal area for delta, beta, and gamma frequency bands, respectively. They attributed the activity differences in BA 10 to pitch-specific memory retrieval.

Table 4.1 Findings in spontaneous EEG and MEG

Area/rhythm	Delta	Theta	Tau	Beta	Gamma
AC-contra	⇑ (1,2)		⇓ (1)		⇑ (1,3,5)
AC-ipsi					⇑ (3)
AC-left (bilateral tinnitus)	⇑ (4)	⇑ (4)	⇑ (4)	⇑ (4)	⇑ (4)
Anterior cingulate	⇑ (4)	⇑ (4)	⇑ (4)	⇑ (4)	⇑ (4,6)
Posterior cingulate			⇓ (6)		
(Posterior) insula	⇑ (4)	⇑ (4)	⇑ (4)	⇑ (4)	⇑ (4,6)
Parahippocampal area	⇑ (4)	⇑ (4)	⇑ (4)	⇑ (4)	⇑ (4,6)
Amygdala					⇑ (6)
Precuneus			⇓ (6)		
Dorsolateral prefrontal cortex			⇓ (6)		

(1) Weisz et al. (2005a); (2) Weisz et al. (2007); (3) Ashton et al. (2007); (4) Moazami-Goudarzi et al. (2010); (5) Van der Loo et al. (2009); (6) Vanneste et al. (2010c).

Thus, this research group (De Ridder and colleagues) suggests that tinnitus loudness correlates with the current-source density of gamma band activity in the contralateral auditory cortex. On the other hand they found that the amount of distress in tinnitus patients is related to an alpha network consisting of the amygdala–anterior cingulate cortex–insula–parahippocampal area. In addition, some suggestions for different resting EEG correlates of tonal versus noisy tinnitus and unilateral versus bilateral tinnitus have been proposed. There may be confounds from potentially different etiologies of the tinnitus in these groups, so that the findings either indicate that etiology determines pitch and laterality or that these psychoacoustic properties (loudness, pitch, and laterality) are not limited to auditory nervous system differences.

4.4.3 Auditory evoked potentials

During stimulation with sound, the EEG (and MEG) undergoes changes that are related in time-locked fashion to changes in the sound. These synchronized EEG changes are called auditory evoked potentials (AEPs) or auditory evoked fields (AEFs). They are generally smaller than the peak EEG voltage and can only be recorded after averaging the voltage changes to a series of stimulus trials. The middle latency responses (MLRs) and the steady-state responses, with latencies below 50 ms, may be partially generated by the primary auditory cortex (Lütkenhöner et al., 2003b) and can be considered an average stimulus-evoked gamma oscillation. The long latency components, with latencies above 50 ms, are likely generated in non-primary cortical areas. The best-known and most used long-latency (~100 ms) component is the N_1, which may result from different sources located dominantly in the planum temporale (Näätänen and Picton, 1987).

4.4.4 What do AEPs (AEFs) reflect?

The evoked response amplitude reflects the number of synchronously activated, and spatially aligned, pyramidal cells in auditory cortex. Any changes that occur in the AEP (AEF) amplitude as a result of changes in stimulation, changes in age, or plastic changes as a result of training or hearing loss, can therefore occur as a result of change in each or all of those aspects of pyramidal cell activity. Core cortex in humans is situated on Heschl's gyrus, a concave surface with the potential of partial cancellation when spatial orientations of the individual dipoles change, e.g. through development. This cancellation may also be reflected in the small amplitudes of the MLR. Increasing stimulus level may recruit more neurons over a larger area, whereas peripheral hearing loss may result in increased neural synchrony (Salvi et al., 2000). Thus, whereas fMRI and PET are insensitive to just synchrony changes and require increased synaptic (or glial cell) activity, AEPs (AEFs) are exquisitely sensitive to neural synchrony and thus complement imaging results. The BOLD response appears to be more tightly coupled to the neural activity underlying the transient part of the MEG (and likely also the AEP) than the sustained part of these signals (Gutschalk et al., 2010). This is only partly substantiated by our own findings (Scarff et al., 2004) based on simultaneous high-field (3 T) fMRI and high-density (64-and 128-channel) EEG using a sparse sampling technique to measure auditory cortical activity generated by right ear stimulus presentation. Using dipole source localization, we showed that the anatomical location of the grand mean equivalent dipole of AEPs and the center of gravity of fMRI activity were in good agreement in the horizontal plane. However, the grand mean equivalent dipole was located significantly more superior in the cortex compared to fMRI activity. This would suggest greater contributions to the BOLD signal by a neuronal population that does not produce transient activity or in addition a substantial amount of tonic activity.

A specific class of AEPs is formed by those responses that can only be obtained following an unexpected sound, e.g. as an infrequent (10–15% of the time) tone of 1000 Hz among a series of more frequent (85–90%) tones of 1100 Hz. Under passive listening conditions the difference between the responses to the standard and the unexpected, also called deviant or odd-ball, sound obtained is the mismatch negativity (MMN). It is noted that the MMN is not a potential that can be recorded as such; it is a derived response and depends on the difference of AEPs to frequent and rare stimuli. Various types of auditory stimuli have been used in MMN paradigms, such as harmonic tones, phonemes, and syllables (Picton et al., 2000). If, during the presentation of a series of standard and deviant sounds, the subject is required to press a button or count the deviant tones an additional positive peak with a latency of 300 ms (P300 or P_3) appears. A still later response component, the N400, can be elicited when a word at the end of a sentence is perceived as semantically wrong. Syntax violations generate a late P600 response (Friederici, 2002).

4.4.5 Auditory evoked potentials (fields) in tinnitus

In tinnitus patients, magnetic recording of the stimulus-evoked P_2, which originates largely from secondary and association cortex, was found to be delayed and low in amplitude (or even completely missing), while the amplitude of the N_1 was significantly augmented (Hoke et al., 1989). However, Jacobson et al. (1991) and Colding-Jørgensen et al. (1992) could not confirm these findings. Attias et al. (1993) found that, in tinnitus subjects, the N_1 and P_2 were both reduced in amplitude compared to a group of controls matched for hearing loss and age. The difference in these reports could potentially be related to the presence or absence of hyperacusis in their tinnitus subjects (information not available). This could result from reduced inhibition in particular frequency ranges. Weisz et al. (2005b) compared tinnitus subjects with normal controls and used tonal stimuli at the edge of their hearing loss as well as tonal stimuli with a frequency one-octave below the edge frequency. They found that the N_1 dipole strength for tinnitus subjects and controls was not different for the audiogram edge-frequency tones, but that the N_1 responses were significantly larger for tonal stimuli that were one-octave below the edge-frequency of the audiogram in the right hemisphere of the tinnitus subjects. This points to a reduced inhibition originating from the neurons whose best frequencies are in the hearing-loss range. It also suggest that there is no over representation of edge frequencies. The data from Weisz et al. (2005b) also suggest increased neural synchrony (larger N_1) but in the normal-hearing range, partially in contrast to the interpretation of the Attias et al. (1993) data. In general, the evoked potential data are not very consistent and their interpretation remains elusive if information about the presence of hyperacusis is not obtained or not reported.

Noreña et al. (1999) proposed the different manifestation of the N_1 as a useful typology of tinnitus. They showed that the N_1–P_2 intensity-dependence was higher and N1 latency shorter in tinnitus sufferers compared to controls. The differences also depended on the type of tinnitus. Patients that showed improved tinnitus by masking noise had greater intensity-dependence and longer N_1 latency than patients for whom masking noise aggravated their tinnitus.

Weisz et al. (2004) found an increased MMN in tinnitus sufferers with high-frequency hearing loss that was associated with subjective distress level. The frequent tones were presented at the edge frequency of the audiogram and the deviants at 1–4% lower frequency. The enlarged MMN may be related to the enhanced frequency discrimination that sometimes accompanies the broadened central representation of edge frequencies in steep-sloping hearing loss. But how that would be related to a distress level is unclear. Subjects with hearing loss exhibit slightly reduced detection thresholds (Irvine et al., 2000) and slightly enhanced frequency discrimination (Thai-Van et al., 2003) for sound frequencies at the edge of the affected frequency region. So the findings could just reflect cortical tonotopic map changes as a result of hearing loss.

Tonotopic map changes in tinnitus patients have also been explored and were inspired by finding somatotopic map changes in phantom-pain subjects and tonotopic map changes in animal models of hearing loss. Wienbruch et al. (2006) used the 40-Hz ASSR to compare the tonotopic frequency representations in the PAC between subjects with chronic tinnitus and hearing impairment and normal hearing controls. In normal hearing subjects, frequency gradients were observed in the medial–lateral, anterior–posterior, and inferior–superior axes, which were consistent with the orientation of Heschl's gyrus and with functional organization revealed by fMRI investigations (Formisano et al., 2003). These ASSR frequency gradients were attenuated in both hemispheres in hearing loss subjects with tinnitus. Such degraded frequency representations in tinnitus patients may reflect a loss of intracortical inhibition in deafferented frequency regions of the PAC after hearing injury. Dipole power was also elevated in these patients, and suggested that more neurons were synchronized to the ASSR envelope. The increased dipole power could also reflect an increased central gain in the tinnitus patients.

Tinnitus patients may process sound differently than controls (Schlee et al., 2008). Magnetically recorded ASSRs revealed abnormal connectivity that was widely dispersed over the whole brain. The right parietal source and frontal sites played a prominent role in this network of abnormal coupling. This complements previous evidence for an involvement of both auditory and non-auditory regions in tinnitus patients. Two of these connections revealed a significant and also strong relationship between the strength of phase synchrony and tinnitus annoyance, both exclusively in the tinnitus tone condition applied to the left ear. The coupling between the right parietal source and the anterior cingulate was positively correlated with tinnitus annoyance ($r = 0.75$). The pair between the right frontal source and the anterior cingulate was negatively associated with tinnitus annoyance ($r = -0.65$). Correlations between synchrony and annoyance at the control frequencies were not significant. This lends credibility to the assumption that the interconnectivities are part of a neural network specific to tinnitus (see also Chapter 11).

The ASSR magnetic field was recorded in tinnitus patients and controls, both groups comprised of musicians and non-musicians, all of them with high-frequency hearing loss (Diesch et al., 2010). Stimuli were AM-tones with three carrier frequencies matching the "audiometric edge," the tinnitus frequency or the frequency 1½ octaves above the audiometric edge in controls, and a frequency 1½ octaves below the audiometric edge. In both hemispheres, the dipole-source amplitude of the response was larger for contralateral than ipsilateral input. In non-musicians with tinnitus, this laterality effect was enhanced in the hemisphere contralateral and reduced in the hemisphere ipsilateral to the tinnitus ear, especially for the tinnitus frequency. The hemisphere-by-input laterality dominance effect was smaller in musicians than in non-musicians. In tinnitus patients, dipole-source amplitude was negatively correlated with the MRI-determined volume of the medial partition of Heschl's gyrus (from Schneider et al., 2009 in the same patients; section 4.3.1). Tinnitus patients showed an altered excitatory–inhibitory balance reflecting the downregulation of inhibition and resulting in a steeper dominance hierarchy among simultaneous processes in auditory cortex.

4.5 **Summary**

In patients with tinnitus, functional and structural imaging of brains combined with measurement of the EEG and MEG have suggested fairly specific correlations between aspects of tinnitus and brain functioning. Since tinnitus is a spontaneous activity phenomenon and is masked or otherwise affected by sound, the more easily interpretable results are from measurement of spontaneous brain activity. The best-suited techniques are PET and SPECT, which measure regional blood flow or energy consumption in the brain. By comparing groups of tinnitus patients with control groups, or better, tinnitus in people where the tinnitus can be obliterated by lidocaine infusion or modified by gaze or residual inhibition, one has arrived at participating brain regions in the non-primary auditory cortex, in the limbic system, and in the cerebellum among others. FMRI has the drawback that it is a noisy technique and that it has been used so far mostly to study the response to sound in tinnitus patients. It is likely that the results reflect the presence of hyperacusis more that that of tinnitus. Spontaneous EEG and MEG points to a reduction of alpha-band activity and an increase in delta- and gamma-band activity. The strength of the latter appears to correlate well with tinnitus loudness, whereas that of the alpha- and delta-band relates more to the level of annoyance caused by the tinnitus. The clear changes in spontaneous synchrony in the MEG activity between different neocortical areas in tinnitus patients suggest wide involvement of brain areas in this disorder.

In order to study detailed neural correlates of these changes in tinnitus patients, we first need to establish if animals have tinnitus. That will be the topic of the next chapter.

Chapter 5

Do animals have tinnitus?

The neural substrate of tinnitus can only be studied in depth in animal models that show behavioral evidence of tinnitus under conditions similar to those that cause tinnitus in humans. The question is: do animals have tinnitus and can it be demonstrated? Humans can tell us when they have tinnitus and can describe what it sounds like. Animals cannot provide verbal reports. Thus, other ways to determine if an animal is experiencing tinnitus have been developed; these methods typically require training to consistently respond when tinnitus is present. The methods employ either punishment or reward to achieve that goal. Tinnitus is presumably continuously present in these animals and reward or punishment cannot be timed to its onset as it would be in more standard training procedures where animals have to respond, for instance, to the presence of a sound in a particular time interval. The amended procedures use the distinction between tinnitus and silence to elicit a behavior from the animal. Typically, an animal is trained to respond differently to silence than to a sound presented by the experimenter; the sound preferably would be similar to that of tinnitus. Then the animal receives a tinnitus-inducing agent such as salicylate or noise exposure and is some time later evaluated on its responses to silence and sound as during the previous training procedure. The dominant idea is that tinnitus abolishes the notion of silence, i.e. the absence of an external sound.

Tinnitus in humans shows a wide diversity in descriptions of spectral characteristics: tones, complex sounds, or noises (Chapter 1). Administering salicylate or quinine, is one of the few methods known to elicit tinnitus in humans in a reversible manner (McFadden, 1982), but such induced sensations are long lasting, and their onsets and offsets are predictable only within hours. Extrapolations of the induced tinnitus sensations in humans suggest that the standard behavioral procedures based on fixed time windows following the onsets of conditioned stimuli (CS) and unconditioned stimuli (US) are not well suited for assessing the presence of tinnitus in animals. Accordingly, both the lengthy duration of perceived phantom sound and its unpredictability have pointed to a search for a paradigm in which the background and contextual environment plays a critical role.

How can an animal differentiate between tinnitus, i.e. presumably as a result of increased spontaneous activity, or amplified environmental noise through increased neural gain (as in hyperacusis) that also increases firing rates? If the animal is kept in a soundproof room for the test, environmental noise is kept at a low level but is still present especially at low ambient frequencies. Hyperacusis typically also decreases thresholds in the frequency regions where it occurs and these coincide with the frequency regions where there is increased SFR (Noreña et al., 2006). One can therefore not exclude that the combination of lowered threshold and increased central gain allows a response to this low background noise in cases where there is no or little SNHL. Often this is the case after chronic application of salicylate, which causes gain changes in the CNS (see Chapters 3 and 6) and often tinnitus-like behavior.

The situation is different when there is a sizable permanent hearing loss, as after a noise trauma. In case the tinnitus, as detected behaviorally, occurs in the region of the hearing loss, and the hearing loss does not prevent hearing the test tones, the inference may be correct. However, if the behavior points to tinnitus frequencies outside the hearing loss region, the low thresholds and the presence of hyperacusis in these frequency regions (see Chapter 6) make the attribution of a response to either tinnitus or hyperacusis again nearly impossible. Problems like this also occur when the noise exposure only causes a temporary threshold shift (TTS), often accompanied by hyperacusis, and the testing is done after the hearing returns to near normal.

5.1 Commonly used animal models

The brown rat, or Norway rat (*Rattus norvegicus*) is one of the best-known and most common rats. Selective breeding of *Rattus norvegicus* has produced the laboratory rat. Wistar rats are an outbred strain of albino rats also belonging to the species *Rattus norvegicus*. The Wistar rat is currently one of the most popular rat strains used for laboratory research. It is characterized by its wide head, long ears, and having a tail length that is always less than its body length. The Sprague Dawley rat and Long–Evans rat strains were developed from Wistar rats. Wistar rats are more active than other strains like Sprague Dawley rats. The Sprague Dawley rat is an outbred multipurpose breed of albino rat used extensively in medical research. Its main advantage is its calmness and ease of handling. Long–Evans rats are white with a black hood, or occasionally white with a brown hood. They are utilized frequently as a multipurpose model in behavioral and obesity research.

The Fischer 344 (F344) is an inbred albino rat strain that has been widely used since its introduction in 1920 in cancer research and toxicology. The spontaneous age-related incidence of neoplasms and degenerative diseases is very high in this strain. Numerous reports have accumulated about the toxic influence of several substances on F344 rats (e.g. Sullivan et al., 1987). The first serious attempts to investigate the characteristic features of the aging auditory system in F344 rats appeared around 1990. These pioneering papers were motivated by a recommendation from the National Institute of Aging to use the F344 strain of rats as a model for studying the effects of aging. Strains, lacking melanin, may develop more pronounced degeneration. Following the initial studies of hearing function in F344 rats, the strain was used as an animal model of presbycusis in several labs. Hearing thresholds in F344 rats begin to increase in this strain during the first year of life. Toward the end of the second year, the thresholds are very high. The rapid deterioration of distortion product OAEs, with the majority of OHCs being morphologically intact, is apparently produced by the disruption of prestin. Sound-evoked behavioral reactions are also impaired in old F344 rats (Syka, 2010). Keithley et al. (1992) did not find any essential differences in the signs of degeneration in the cochlea with aging among

Wistar and F344 rats. Their conclusion was that the presence or absence of melanin does not have any effect on cochlear degeneration in the rat with aging.

5.2 Validity criteria for animal models

An appropriate animal model of tinnitus should fulfill the following criteria: strong phenomenological similarities and similar pathophysiology (face validity), comparable etiology (construct validity), and common treatment (predictive validity) as the affected humans. Predictive validity assesses the performance of a model, which is typically accomplished by demonstrating a similarity in treatment efficacy between an animal model and a human clinical condition (e.g. examining them for drug false positives and false negatives and for similarity in pharmacological treatment potencies). A model with high predictive validity should therefore maximize identification of both "true positives" and "true negatives," but should minimize identification of "false positives" and "false negatives" (Willmer and Mitchell, 2002). However, not all of the clinical symptoms of tinnitus can be modeled in animals; symptoms in human conveyed by subjective verbal report are, in principle, excluded.

Construct validity assesses the theoretical rationale of the animal model. Willner (1986) argued that construct validity is established by the demonstration of a similar theoretical base for the paradigm and the clinical condition. However, any evaluation of animal models of tinnitus is intrinsically limited by the rudimentary state of theories of the pathology of tinnitus. Indeed, there is little in the extensive literature describing biochemical markers reportedly associated with tinnitus that can be usefully employed to provide a theoretical standard against which to validate animal models. Even the most basic questions of whether the level of activity in GABAergic or glutamatergic systems is transiently or permanently elevated or decreased in tinnitus remain controversial (Chapters 6–10). Similar problems arise in relation to the evaluation of etiological validity, which is part of construct validity. It is clear that a variety of different factors is implicated in the etiology of tinnitus: "psychological" factors include undesirable life events, chronic mild stress as well as "biological" factors include genetic influences, and a variety of physical insults and ototoxic drugs. More commonly, the pathogenesis of tinnitus may be better understood as the result of an accumulation of a number of these factors. This point has been largely overlooked in the construction of animal models of tinnitus, which in general have assumed a single causal factor.

Again, tinnitus is a heterogeneous disorder and its many symptoms will be hard to produce in laboratory animals. The question therefore remains whether we can know that the animal has tinnitus, including psychoacoustic as well as psychological signs. As we will see, animal models of tinnitus so far only look at psychoacoustic signs.

5.3 Behavioral animal models of tinnitus

5.3.1 The Jastreboff operant conditioning procedure

The classic technique of conditioned suppression (Estes and Skinner, 1941) suggested to Jastreboff and colleagues a promising direction for designing a behavioral animal model of tinnitus. Underlying this technique is the antagonism between ongoing, desire-based behavior and aversively-based responses elicited by Pavlovian training. Jastreboff et al. (1988a) deprived Norway hooded rats of water and had them continuously engaged in licking behavior during each experimental session. A safety signal consisted of a constant 24-h background noise. The CS was the offset of the background noise, which was paired with a mild footshock (US) during Pavlovian training. They established a constant level of thirst that resulted in a stable licking behavior. The occurrence of silence produced a decreased number of licks. The ratio R of licks occurring during 1 minute of silence (B) divided by the sum of licks occurring during 1 minute before the silence (A) and during the silent minute, i.e. $R = B/(A + B)$, was taken as an index of behavioral change in response to the silence. Measures of lick changes indicated that training animals to associate silence with electric foot shock took 4–10 associations corresponding to 1–2 days of training. Passive Pavlovian extinction, in which only the CS without the US was presented, permitted measurement of the strength of learning and the extent of interference from experimental manipulations on acquisition and extinction processes. Extinction of this behavior took 4–5 days.

Jastreboff et al.'s (1988a) procedure was used to detect a continuous background auditory sensation potentially induced by the administration of sodium salicylate to rats. The rationale then is that if tinnitus is induced by salicylate, it will interfere with the perception of silence and will modify the extinction behavior. Using this procedure, Jastreboff and his colleagues showed that rats given salicylate after the training were less likely than control animals to stop drinking when the noise was turned off. This result is taken to indicate that the treated animals still hear a sound when no external sound is present, i.e. they have tinnitus (Jastreboff et al., 1988a,b; Jastreboff and Brennan, 1994). Jastreboff and his colleagues also demonstrated that the effect of salicylate is dose dependent (Jastreboff and Brennan, 1994). In addition, they showed that quinine also produced the behavioral signs of tinnitus and that the behavioral effects of salicylate and quinine can be abolished by administering nimodipine, a calcium blocker, in the drinking water (Jastreboff and Brennan, 1988; Jastreboff et al., 1991).

Heffner and Harrington (2002) critically analyzed the behavioral method of Jastreboff and colleagues. In particular they addressed the various control tests that Jastreboff et al. used to further explore their results. First, they addressed the question of whether animals trained to treat a broadband noise from a loudspeaker as a safe signal would generalize this to a tonal signal that presumably resembled tinnitus. Jastreboff et al. (1988a) showed that presenting control animals with a 7-kHz tone (60 dB SPL) during the silent intervals increased the likelihood that they would continue drinking. Thus, the animals generalized from an external noise to an external tone presented about 60 dB above their threshold (cf. Heffner et al., 1994). However, a later study found that presenting a 10-kHz tone at levels from 32–62 dB above threshold had no effect on the animal's performances and, furthermore, that presenting the 10-kHz tone at higher intensities made the animals less likely to continue drinking (Jastreboff and Brennan, 1994). Thus, the degree to which animals trained with this procedure and broadband noise as a CS generalize to tones is not clear. This makes the use of this method to assess tinnitus pitch not reliable.

The question of the pitch of the animal's tinnitus was addressed by administering salicylate to animals before training so that any tinnitus they developed would be paired with shock (Brennan and Jastreboff, 1991). The animals were then tested by turning off the noise followed by presenting tones ranging from 7–11 kHz. It was expected that the animals would be less likely to continue drinking when presented with tones similar in pitch to their tinnitus. The results showed that the animals were progressively less likely to continue licking as frequency was increased, leading the authors to suggest that the tinnitus was >12 kHz. However, it was later demonstrated that the levels of salicylate used result in a hearing loss that begins at 2 kHz and becomes progressively greater as frequency increases (Brennan et al., 1996). Thus, the reduced drinking may have been due to a hearing loss that made the high-frequency tones less audible.

A second question that Heffner and Harrington (2002) analyzed concerned whether the animal's motivational level or stress could affect the outcome of these tests. The effect of motivational level was addressed by testing animals whose body weights were reduced to 90% ad lib weight, thus making them less thirsty than animals tested at the standard 80% weight (Jastreboff et al., 1988a). The results showed that a significant, albeit smaller, effect of salicylate could still be demonstrated with the less motivated animals. They also demonstrated that salicylate by itself does not affect an animal's water consumption, so that the tendency of salicylate-treated animals to continue drinking during silent intervals does not appear to be due to increased thirst. With regard to stress, however, it has been noted that the introduction of a stressor, such as being handled by an inexperienced technician or being presented with a loud sound, can affect the results. In these situations, the control group may actually be more likely to continue drinking during the silent intervals than the salicylate group (Jastreboff and Brennan, 1994). One wonders if stress also causes tinnitus in these animals or that they don't care about the test-shocks anymore. Thus, the animals must be carefully handled in order to obtain reliable results.

A third question is whether the effects of salicylate are specific to auditory tasks or can also affect non-auditory discriminations. This question was addressed by training animals to stop licking when a light (instead of noise) was turned off (Jastreboff et al., 1988a). The results indicated that there was no effect of salicylate on suppressing to a light cue. Thus, salicylate did not have a general effect on an animal's performance, but, instead, was specific to auditory tasks.

In summary, Heffner and Harrington (2002) concluded that Jastreboff and his colleagues presented evidence that animals respond systematically and reliably following administration of salicylate or quinine. However, it is still possible that the results might have been due to a hearing loss resulting from salicylate. The possibility also remains that the sudden introduction of a hearing loss caused by salicylate may affect performance by initially acting as a stressor. In particular, for salicylate application the also induced hyperacusis may strongly interfere with the interpretation of the behavioral results.

5.3.2 The current state of the classical conditioning procedure

Heffner and Harrington (2002) used a modification of the conditioned suppression technique used by Jastreboff and his colleagues (e.g. Jastreboff et al., 1988a) to test hamsters.

One important difference was that the animals in the present study received extensive training to increase their reliability so that the likelihood of tinnitus in individual animals might be assessed. In addition, both behavioral and auditory brainstem response (ABR) techniques were used to measure the hearing loss resulting from the exposure to loud 10-kHz tones. I describe the relevant details in the set-up used in these experiments.

Behavioral apparatus Testing was conducted in an anechoic, double-walled sound chamber. This is one of the requirements to avoid that the animals start responding to low-level ambient sound, potentially as a result of developing hyperacusis. It also avoids masking effects of the tinnitus by environmental sounds. The equipment for behavioral control and stimulus generation was located outside the chamber and the animals were observed over closed-circuit television (Heffner et al., 2001). A waterspout was present in the front of the cage and was adjusted to a level that permitted an animal to drink comfortably with its head facing forward (Figure 5.1). Water was delivered via a pump with the flow rate adjusted so that an animal could satisfy its daily water needs in a single 15–20-min test session. Because of the small size of hamsters, the movement space had to be restricted by a wire mesh to keep the animal facing the speaker while drinking. Requiring an animal to keep its mouth on the waterspout served to fix its head in the sound field, allowing precise measurement of the sound level at its ears. When an animal made contact with the spout the water was turned on. A power source connected between the spout and the cage floor provided mild shocks.

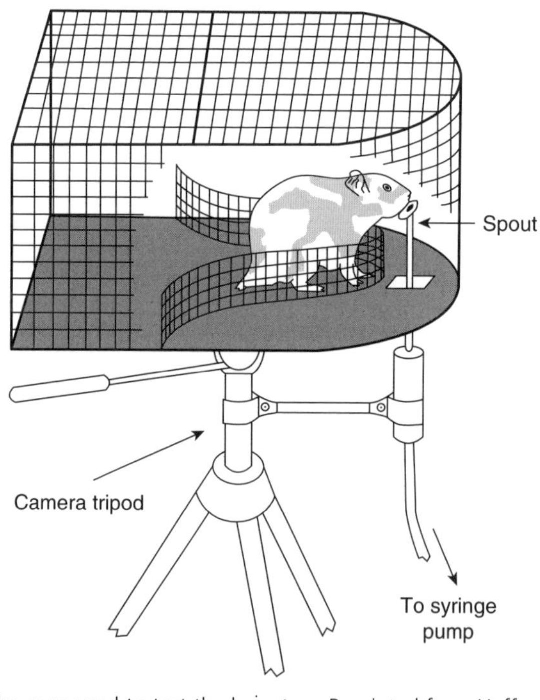

Figure 5.1 The wire cage used to test the hamsters. Reprinted from Heffner, R.S., Koay, G., and Heffner. H.E. (2001). Audiograms of five species of rodents: implications for the evolution of hearing and the perception of pitch. *Hearing Research*, **157**(1–2), 138–52. Copyright (2001), with permission from Elsevier.

Behavioral procedure for assessing tinnitus The animals were trained to drink from the waterspout in the presence of broadband noise and/or tones and to stop drinking in the absence of these sounds (silence) in order to avoid a mild electric shock. They were then tested for tinnitus by determining the percentage of time they drank during noise trials and silent trials when shocks no longer followed the silent trials. The hypothesis was that animals with tinnitus would be more likely to continue drinking during silent trials because they would now hear their tinnitus. Although it was assumed that any tinnitus induced by noise trauma would be tonal, the animals were also trained using noise, making the test sensitive to non-tonal tinnitus as well.

An animal was first accustomed to drinking from the spout in the combined presence of broadband noise (32 dB SPL) and tone (10 kHz, 37 dB SPL) for three to four sessions—the noise and tone were combined to simulate the presence of tonal tinnitus in conjunction with the noise. During training, a 15-s trial was presented in which either the tone alone (tone trial) or no sound (silent trial) was presented. Initially, an animal was shocked if it contacted the spout any time after the first 2 s of a silent trial. Once it had learned to reliably break contact with the spout, it was not shocked unless it contacted the spout during the second half of the trial. Thus, an animal had 7.5 s in which to decide whether or not to break contact. Each trial (tone or silent) was followed by a 15-s interval, in which the noise and tone were presented together after which the next trial began. An equal number of tone and silent trials were presented in a quasi-random sequence. After 10–12 sessions, the animals were trained to generalize to other tones and loudspeaker locations in order to increase the likelihood that the animals would generalize from the training tones to any tonal tinnitus they might develop. The frequency of the tone was changed from one session to the next, but was always the same within a session. The tones ranged from 33 dB (12 kHz) to 51 dB (20 kHz) above the average threshold for hamsters (Figure 5.2). These intensities were chosen because they resulted in performances well above the 0.01 level of chance for all animals. The animals were trained for 32–35 sessions and all performed well above chance levels during the last five training sessions with scores of 70% or better (Mann-Whitney U-test, $P < 0.01$).

Testing began 5 days after the trauma-tone exposure. The reason for waiting 5 days after exposure was because the increase in spontaneous activity in the DCN that follows tone exposure asymptotes at about 5 days post-exposure and did not occur up to 2 days post-exposure (e.g. Kaltenbach et al., 2000; see also Figure 7.4, Chapter 7). Note that spontaneous activity in primary auditory cortex of cats already significantly increases within 2 h after the exposure (Norena and Eggermont, 2003) and remains at that level for months thereafter (Seki and Eggermont, 2003). After exposure to 80 dB SPL, also in DCN a trend of increased SFR is found after 1 h and significant changes after 2 days (Kaltenbach et al., 2005). It would be interesting to see results after short time periods since the animal's response is likely determined by cortical changes in SFR. For tinnitus testing, animals received 15-s trials that alternated between broadband noise and silence. The loudspeakers were located at 90° left, 0°, 90° right, and overhead. No tones were presented and no shock was given. Each animal received 25 noise and 25 silent trials per session for five consecutive sessions—limiting the number of trials meant that testing was completed before satiation could occur. The time that an animal was in contact with the waterspout

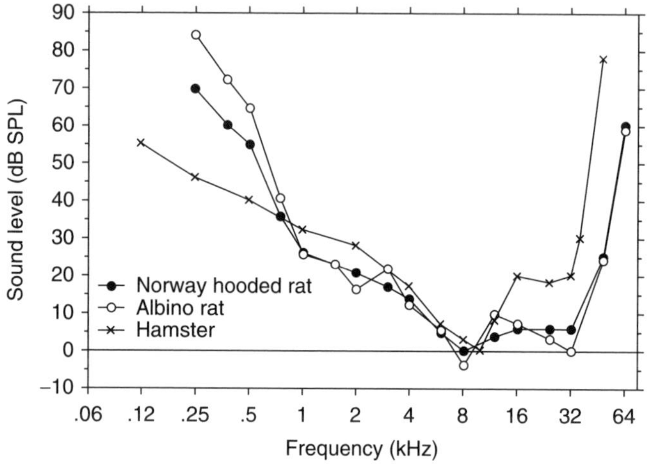

Figure 5.2 Behavioral audiogram of hamsters (x), Norway hooded rats (filled symbols) and albino rats (open symbols). Data from Heffner et al. (1994, 2001).

during the last half of each trial (7.5 s) was automatically recorded. Performance was calculated as the average percentage time that an animal was in contact with the spout during noise trials and was not in contact during silent trials (indicated as percent correct in Figure 5.3). Scores could range from 50% (random performance) to 100% (perfect performance). The results demonstrated that hamsters trained to stop drinking during silence are more likely to continue drinking following exposure to a loud 10-kHz tone for 2 or 4 h. In other words, they behave as though they hear a sound when no external sound is presented. The question is whether this result is due to tinnitus or can be explained by other factors.

The main alternative explanation is that the exposed animals responded differently because of hearing loss. Indeed, Heffner and Harrington (2002) found that simulating a unilateral hearing loss by training animals with the speakers placed around them and then testing them with all the speakers to one side did cause the animals to be more likely to continue drinking during silence. For this reason, the location of the loudspeakers was routinely varied during training to reduce the possibility that the hearing loss would affect the results. However, the most convincing evidence that hearing loss cannot explain the results of the tinnitus test is that the 1-h exposed animals had a hearing loss similar to that of the 2- and 4-h exposed animals, but tested negative for tinnitus. Thus, hearing loss alone cannot account for a positive score on the tinnitus test. This is also suggested by the demonstration that neural damage following noise trauma is not always related to audiometric thresholds (Liberman and Kujawa, 2006, 2009). The finding that the effect was strong at the frequency of the trauma tone argues against hyperacusis as the cause of the responses, as hyperacusis typically occurs in the frequency regions above and below the hearing loss region. The animal likely developed recruitment and hence the tinnitus would be quite loud in the hearing loss range (Chapter 3), which makes the test more sensitive.

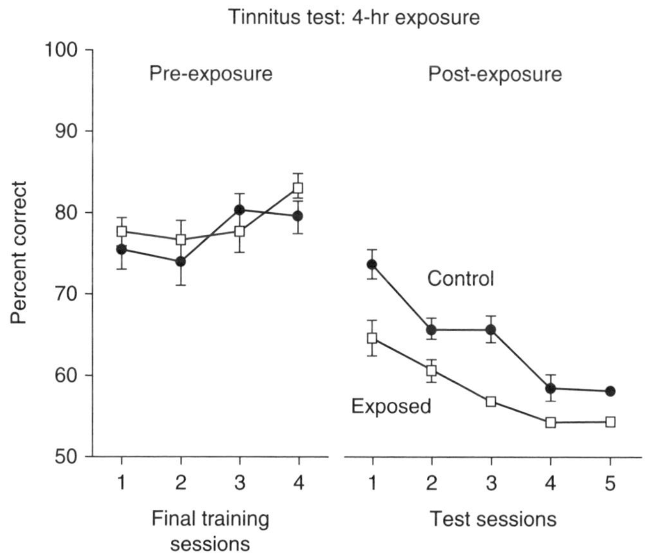

Figure 5.3 Effect of exposure of one ear to 10 kHz at 124 dB SPL for 4 h. The animals in the exposed group were more likely to continue drinking during silent trials than control animals and thus have a lower percent correct, an observation consistent with the hypothesis that they had developed tinnitus. Average scores are shown for the exposed group (open squares) and control group (filled circles) with bars indicating standard error. Note that the extinction rate is comparable in controls and exposed animals. Reprinted from Heffner, H.E. and Harrington, I.A. (2002). Tinnitus in hamsters following exposure to intense sound. *Hearing Research*, **170**, 83–95. Copyright (2002), with permission from Elsevier.

The procedure used by Heffner and Harrington (2002) differs from others in two ways. First, they demonstrated that it is possible to obtain a reliable assessment of tinnitus in a single 20-min session—other procedures require approximately five sessions to accumulate a sufficient number of trials (e.g. Jastreboff et al., 1988a). However, the ability to test in a single session comes at the cost of having to provide extensive training to the animals. Thus, animals are trained for approximately 30 sessions, whereas other procedures require as few as seven training sessions (e.g. Jastreboff et al., 1988a). Another difference is that individual animals can be ranked as to the likelihood that they have tinnitus, these rankings were previously restricted to group data. Thus, exposed animals scoring outside the range of the controls may be more likely to have tinnitus than those falling within the control range. The ability to assess animals individually is important in searching for the physiological basis of tinnitus because not all exposed animals may develop tinnitus.

The improved operant conditioning model has provided useful data but has limitations. The technique is still dominantly based on comparison of group data rather than data from individual animals. The behavior extinguishes quickly, therefore, tinnitus can only be assessed over a short time interval. With this model, it is not possible to determine if tinnitus is permanent or temporary, or to measure the time course of tinnitus onset and cessation.

5.3.3 The modified conditioning procedure by Bauer and colleagues

Bauer et al. (1999) trained food-deprived Norway-hooded rats to press a lever in the presence of broadband noise to obtain food, but to stop pressing the lever during silent intervals to avoid foot shock. Note that this is a minor variant of the Jastreboff method, which involved water deprivation, and used the same suppression ratio definition. The animals were tested by presenting in a 60-min period of 60 dB SPL broadband noise, four 1-min long intervals without noise but containing a tone, where no shock is given, and four 1-min long silent intervals followed by shock if the animal did not stop lever pressing. The tone was varied in frequency and intensity based on the assumption that animals with tinnitus will respond to the tones differently than control animals. Because the animals are always shocked for lever pressing during the silent intervals, their responses do not extinguish. In this study, four different frequencies (10, 15, 20, and 30 kHz) were presented at six different intensities (25–80 dB SPL). It was found that rats given salicylate (oral 8 mg/mL resulting in serum levels of 222 mg/L, corresponding to those reported to induce tinnitus in humans; Cazals, 2000) after training were more likely than control animals to continue lever pressing during tone intervals. Surprisingly, the salicylate animals differed significantly from the controls at only one sound level at each test frequency. The significant differences occurred at 70 dB during presentation of the 10-kHz tone, 50 dB for the 15-kHz tone, and 35 dB for the 20- and 30-kHz tones. The authors attributed this to the variation in the absolute sensitivity of rats, but the (hooded) rat audiogram hardly differs by more than 5 dB for those frequencies (Figure 5.2).

It was surprising that there was no salicylate-induced hearing loss based on the animal's click-evoked ABRs. Tone-evoked ABR studies for the salicylate-treated animals were also completely normal. Hearing loss is more or less linearly related to plasma salicylate levels with about 8 dB per 100mg/L plasma level (Cazals 2000). The reported plasma levels in the Bauer et al. (1999) study were between 145–382 mg/L and should thus have resulted in 10–35 dB hearing loss.

Hearing loss should make an animal less likely to respond to the tones, and therefore more likely to stop lever pressing during tone intervals, the opposite of the effect that was found. Bauer et al. (1999) explained this by assuming that an animal's tinnitus makes the tones sound "noisier" and thus more like the background noise (see Chapter 1 showing that humans tend to do the same). As a result, animals with tinnitus were more likely than control animals to continue lever pressing when tones were presented.

Bauer et al. (1999) also refuted a role of recruitment in the outcome of their study. They likely meant hyperacusis since recruitment is restricted to sensorineural hearing loss (Chapter 3) and these animals did not have a hearing loss. They considered the case where there was hyperacusis but no tinnitus. Their argument against hyperacusis was that it is supposed to affect all frequencies. This is likely the case after salicylate application, but not following noise trauma (Noreña et al., 2010). However, the largest change in the suppression of lever pressing was frequency specific at 15 kHz. In addition, hyperacusis would not only make the background noise louder, and thus falsely signaling tinnitus, but it would also result in better detection of tones compared to controls. The latter was indeed observed but only at one intensity level at each frequency, and this is difficult to attribute to hyperacusis.

Bauer and Brozoski (2001) exposed rats to a 105-dB noise band centered at 16 kHz in one ear for 1 or 2 h. The behavioral procedure in this study was different from that of their previous study in that the animals were exposed to the noise before training began and the number of test intervals was increased. The results showed that rats exposed to the loud noise before training were less likely to continue lever pressing during tone intervals than unexposed controls. They reported that ABR thresholds measured 90 days post-trauma were essentially the same as those measured immediately post-trauma, i.e. about 50 dB. This is unusual for two reasons; the hearing loss was the same from 4–32.5 kHz and the hearing loss did not change after trauma, suggesting that there was no TTS. Unlike the first study, the difference between the exposed and control groups were found at more than one tone intensity. Moreover, the difference in performance between the exposed and control groups appeared not only to be permanent, but to increase with time, in that one of the six exposure groups showed a larger effect at 17 months than at 2 months.

If the animals were mistaking external tones for their tinnitus, one would expect them to be more likely to stop lever pressing when presented with a tone of the same pitch as their tinnitus. Analysis showed that although the performance of the animals receiving the 2-h noise exposure did not vary with frequency, leading the authors to speculate that the rats' tinnitus was noise-like, the rats with 1-h exposure showed the largest difference at 20 kHz, suggesting that this was the pitch of their tinnitus.

The Bauer group excels in presenting well-documented studies of induced tinnitus with unexpected properties. Another example thereof is presented in Brozoski et al. (2002) where they combine behavioral testing with recording from the DCN. The exposure sound was a 4-kHz tone presented at 80 dB unilaterally for only 30–60 min. The maximum post-trauma hearing loss was as expected between 6–8 kHz and was about 30 dB. At 5 months post-trauma this TTS had completely disappeared (as also expected). One week after the trauma, psychophysical testing was done, immediately followed by electrophysiological recordings. Both behavioral (reduced suppression of lever pressing) and electrophysiological (increased SFR) evidence of tinnitus was found at a frequency of 1 kHz, and not at other frequencies. CF statistics were not provided in the paper. The psychophysical results were explained by suggesting that the auditory discrimination of subjects in the trauma group was affected by a trauma-induced chronic tinnitus with characteristics similar to that of a 1-kHz tone. The authors again considered the potential effect of hyperacusis and because the psychophysical effect was restricted in frequency did not fit their assumption that hyperacusis occurs at all frequencies. It may in fact have been induced by the temporary hearing loss, producing diminished lateral inhibition dominantly below 4 kHz similar to what has been described in Noreña et al. (2006) and Pienkowski and Eggermont (2009). However, hyperacusis at 1 kHz would have made the discrimination of the tones much more effective and the observed suppression ratio curve would then likely have been above that for the controls, which was not observed.

The fact that behavioral signs of tinnitus were not found at 2 kHz is also surprising in the light of tinnitus in humans sometimes having a pitch close to the edge frequency of the hearing loss (here a TTS), which would be 4–6 kHz (Chapter 3).

A potential solution to the paradoxical 1-kHz tinnitus after noise trauma was offered in a comprehensive follow-up study where they induced unilateral tinnitus in chinchillas in three different ways (Bauer et al., 2008). First, a 4-kHz tone at 85-dB SPL was presented into the ear canal for 1 h. Psychophysical testing resumed a minimum of 2 weeks after acoustic exposure. Secondly, carboplatin (4 mg/mL preservative-free normal saline, 2 μl, 15 min) or cisplatin (0.66 mg/mL preservative-free normal saline, 2 μl, 15 min) was applied to the round window membrane. The solutions were removed after 15 min using a saline rinse. Psychophysical testing resumed a minimum of 2 weeks after toxin exposure. Sound exposed cochleas showed no appreciable hair cell loss. Carboplatin exposed cochleas showed >50% IHC loss above 4 kHz and <5% OHC loss. In some of these animals, IHC loss above 4 kHz approached 100%. Cisplatin exposed cochleas had on average 40% OHC loss from 9–10 kHz and an average 10–20% IHC loss along the entire length of the organ of Corti. Both cisplatin and carboplatin produced more high-frequency than low-frequency damage, which perhaps is not surprising given their direct round-window application. In conclusion, cochleae from each treatment group displayed a distinctly different pattern and amount of cochlear pathology.

Most hearing loss at 7–9 months after exposure was found for the carboplatin group; it was larger for higher frequencies, i.e. >2 kHz. Noise trauma did produce potentially a very small high frequency hearing loss, <5 dB, and a 10-dB improvement of the hearing threshold at 1 kHz. Cisplatin resulted in a more or less flat hearing loss of about 20 dB (Figure 5.4). Despite these large differences in hearing loss, all experimental groups displayed significant and similar psychophysical evidence of tinnitus with features resembling a 1-kHz tone. The procedures used were similar to those described in Bauer and Brozozki (2001) and Brozoski et al. (2002). The psychoacoustic results basically confirm those in the previous study (Brozoski et al., 2002). Electrophysiological recordings from the IC found a subset of neurons that showed a significantly enhanced burst firing, enhanced neural synchrony and a prevailing spike firing frequency within bursts of about 1 kHz.

The authors conclude "the present experiment suggests that a subpopulation of IC neurons, characterized by the triad of increased bursting associated with low ISI variability and high within-burst spiking, may underpin tinnitus." If this is indeed the case it would validate the psychoacoustical technique resulting in unusual tinnitus frequencies that bear no apparent relationship to the hearing loss. It is still noteworthy that the lowest threshold in all conditions is at 1 kHz, i.e. at the tinnitus frequency. The reduced threshold at 1 kHz following noise exposure suggests hyperacusis in this frequency region and a potentially different explanation of the tinnitus at this condition (see Chapter 11).

Chen and Jastreboff (1995) had previously described similar bursting in the IC after salicylate application in Long–Evans rats (Chapter 6). The IC is the site in the auditory system where temporal coding of pitch is likely converted into a place code (Langner and Schreiner, 1988; Schreiner and Langner, 1988; Langner, 1992), so the occurrence of bursting with a within-burst firing frequency of about 1 kHz could potentially result in

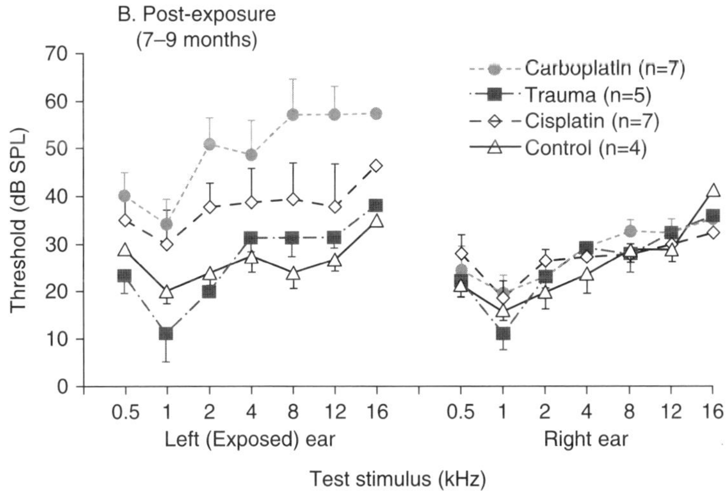

Figure 5.4 Acoustic brainstem evoked-response hearing thresholds obtained at the time of electrophysiological data collection, 7–9 months after unilateral (left) exposure (B). Error bars show 1 SEM. Note lowest threshold at all conditions at 1 kHz. Reproduced from Bauer, C.A., Turner, J.G., Caspary, D.M. Myers, K.S., and Brozoski, T.J. (2008). Tinnitus and inferior colliculus activity in chinchillas related to three distinct patterns of cochlear trauma. *Journal of Neuroscience Research*, **86**, 2564–78 © John Wiley and Sons, 2008 with permission.

activation of the pitch region in auditory cortex (Bendor and Wang, 2005) and give rise to a tinnitus pitch of that frequency. It remains surprising why such diverse mechanisms all result in a 1-kHz pitch and why, in normal situations, low frequency sound just results in firing intervals commensurate with the sound frequency.

Summarizing, Bauer's approach allowed for long-term assessment of tinnitus; however, a limitation of this technique was that differences in behavior attributed to tinnitus occurred only at elevated sound levels, and often at low frequencies, contrary to tinnitus reported in humans.

5.3.4 **Other modified conditioning models**

The caution needed in interpreting the conditioned response tests as evidence for tinnitus described in the previous three sections, obviously applies to the procedures discussed in this section as well.

Another behavioral technique also utilized "false positive" responses that occurred during a quiet interval to infer the presence or absence of tinnitus. The approach utilized a shock avoidance conditioning procedure in which rats learned to climb a pole during the presentation of a sound to avoid foot shock. Animals could remain on the cage floor during quiet intervals since the shock was turned off (Guitton et al., 2003). During salicylate treatment, rats climbed the pole (false positive) during quiet, which was interpreted as evidence of tinnitus.

A schedule-induced polydipsia (SIP) test (Lobarinas and Falk, 1998) was applied to the detection of tinnitus by Lobarinas et al. (2004). Polydipsia, refers to the excess intake of fluid, and is induced by delivering a small amount of food at regularly spaced time intervals (~1 min) to food-restricted animals. Although the animals are not water deprived, the scheduled delivery of food pellets spontaneously results in a high rate of water licking between pellets.

SIP training began by placing a food-restricted Sprague Dawley rat into a clear plexiglass chamber with a stainless steel bar grid floor through which electric shocks could be delivered. A loudspeaker was located in the center of the ceiling of the chamber. The chamber was placed within a sound-attenuating cubicle (56 × 64 × 50 cm) lined with sound-absorbing foam. During the SIP training phase a 30-s interval of a 100 Hz bandwidth noise centered at 4 kHz, and presented at 60 dB SPL, was followed by the delivery of a food pellet, which marked the beginning of a trial. The delivery of the food pellet was immediately followed either by 30 s of quiet or 30 s of the 4-kHz sound. This sound or quiet interval was then followed by 30 s of the 4-kHz sound. No shocks were delivered during this phase of training. Initially, the animal would only eat the food pellets but over time the animal started to lick for water following the delivery of the pellet resulting in SIP, which typically developed within 3–4 days.

Training continued as described; however, scrambled foot shocks (2–5 mA during initial training; ≤2.5 mA during data collection) were now delivered when licks occurred in the presence of sound (initially narrow-band noise centered at 4 kHz, 60 dB SPL). Training typically continued for 5–6 days until >90% of total session licks occurred in quiet (<10% licks-in-sound); during this phase of training the intensity of the 4 kHz sound was decreased to 40 dB SPL. Afterwards, additional sound stimuli were introduced (16 kHz tone or narrow-band noise 100 Hz wide centered at 4, 8, 12, 16, or 20 kHz, presented at 60 dB SPL or 40 dB SPL) so that the stimuli varied over each 30-s sound interval. During this phase of training, the animal learned to avoid licking to any sound stimulus and to only lick during quiet. Performance with >90% licks-in-quiet was typically achieved in 2–3 days.

Five rats were used by Lobarinas et al. (2004), and all achieved >90% licks-in-quiet within 2–3 weeks of training. Animals were given a series of saline injections so that the animals would acclimatize to the injections and to control for any non-drug effects associated with injections. The lick performance of a typical subject was evaluated for daily sessions that included baseline measures and treatments with saline, 50, 100, 150, and 350 mg/kg of sodium salicylate. During the baseline sessions, licks-in-sound were low (<10%) relative to licks-in-quiet. Licks-in-sound remained at a similar low level across all saline and salicylate sessions, which indicates that the conditioned avoidance behavior (licks-in-sound) remained under stimulus control throughout the entire experiment. When saline was administered, licks-in-quiet remained at a high level similar to that observed during baseline testing. However, when the rat was given a 350-mg/kg dose of salicylate, licks-in-quiet dropped to levels equivalent to those observed during 40 dB SPL sound trials. On the first day following the last dose of 350 mg/kg, licks-in-quiet were far below baseline levels and close to the licks evoked by real sounds. Licks-in-quiet increased on the second day post-treatment and by the third day of recovery the number of licks-in-quiet had returned to baseline levels. This suggests that 350 mg/kg salicylate only produces a transient tinnitus that ceases after the salicylate is metabolized. The 150-mg/kg dose of salicylate produced results similar to those obtained with the 350-mg/kg dose. In contrast, the 50- and 100-mg/kg dose of salicylate failed to reduce significant suppression of licking during quiet, suggesting that these doses did not induce tinnitus-like behavior. The other rats produced results similar to the example described here. The clear dose–response effects are strong indicators of a salicylate-induced effect.

Recent guidelines by animal protection agencies tend to aim at avoidance of fear, pain, and deprivation in animal studies. Thus altering current training procedures substantially towards a paradigm-based predominantly on rewards will be important in the future. Rüttiger et al. (2003) took this into account in their behavioral paradigm using Wistar rats for the acute effects of salicylate (Panford-Walsh et al., 2008). For salicylate, the acute effects on click-evoked auditory nerve CAPs and intensity-dependent summation potentials in the cochlea are maximal after 3 h (Stypulkowski, 1990). Thus, a time lag of 3 h between salicylate administration and recording was used.

Training and testing took place in a commercial conditioning chamber that was adapted for the purposes of this study. The layout was chosen to enhance high left–right locomotion activity. The conditioning chamber was housed in a small sound-attenuating chamber. Environmental noise within the conditioning chamber was between 3.5 dB and −0.2 dB SPL with a peak at 1 kHz and a minimum at 40 kHz. Electrical stimuli could be supplied via a stainless steel bar grid floor. Though the electrical impulses were meant to be unpleasant, they were far from noxious (0.1–0.5 mA) and since they were easily avoidable by the animal, they were not expected to induce stress or fear. Rats provoked with even many impulses in a short period of time (>1/min) were still motivated by the reward to perform the behavioral paradigm. Therefore, rats learned to avoid the punishment and to restrict their feeder accesses to the rewarded time periods. A resting platform with a mechanical sensor was mounted on one side of the cage, covering the bar grid floor and serving as a resting location for the animal. A wall separated the cage into two short parts. At both ends of the short parts, small amounts of fluid could be disseminated (sucrose in water, 3%), controlled by flow resistance- and vibration-muted magnetic shutter valves. A typical open time was 0.5 s, resulting in a reward drop of ~20 μl, supplied to the animal via a curved metal drinking spout. Reward drops not taken up by the animal were drained off. Photo sensors registered the animals' visits at the feeding sites. All sensors were monitored on a computer screen and a top-mounted camera gave pictures of the whole floor dimensions of the cage interior.

Animals were trained on auditory stimuli for 30–60 min/day for 5 days/week. Training session length was adapted to the animal's activity. Drinking water was withdrawn 15–18 h prior to behavioral testing. The conditioned rats were divided into two groups (one animal per cage for either group). Animals from the first group received a single intraperitoneal injection of sodium salicylate (350 mg/kg), while animals from the second group received an intraperitoneal injection of an equivalent volume of saline. Animals from either group were tested on the same day in a semi-random order 3 h after injection. On the next experimental day animals from the group previously treated with salicylate were injected with saline and vice versa and tested again.

Using this procedure, Rüttiger et al. (2003) demonstrated a significant behavioral change in animals treated with a dose of 350 mg/kg of sodium salicylate and a less pronounced but still significant behavioral change with a lower salicylate concentration (150 mg/kg). The change was based on the silence–activity ratio (SA ratio), defined as the ratio of the access rates during silence and during sound. This finding is comparable to the Lobarinas et al. (2004) results using a different test procedure and a different rat strain, and who showed in addition that lower doses of salicylate produced no effect.

Rüttiger et al. (2003) considered whether the behavioral changes were not due to an unspecific effect on locomotion and learning, or a masking effect, or a salicylate-induced hearing loss. They could rule out these possibilities. Masking of the external, physical sound by tinnitus, resulting in a lower access rate during sound, could have accounted for the behavioral change in the SA ratio. However, they estimated from the SA ratios during presentation of sound at various levels that salicylate produces an auditory phantom experience equivalent to a broadband noise of ~30 dB SPL regardless of the sound pressure level of the trained and tested sound, be it 70 or 45 dB. However, the access activity of a rat after salicylate treatment is still suppressed in periods of silence after the 45-dB SPL training and was suppressed to a larger extent after the 70-dB SPL training. Therefore, auditory masking cannot account for the behavioral changes after salicylate treatment.

Salicylate in a dose of 350 mg/kg induces a hearing loss of 15–20 dB on click or pure tone stimuli (Rüttiger et al., 2003). After salicylate treatment, rats could therefore experience less intense sound than after saline. However because of the sensorineural hearing loss (reduced activity of the OHC-based amplifier) there could be recruitment (especially at the higher sound levels) and hyperacusis (Chapter 6) that may counteract the putative reduction in sound level. The SA ratio was reduced when the sound was muted below 60 dB SPL even in untreated rats. However, if a hearing loss were responsible for the behavioral changes after salicylate treatment, one would have expected less activity during both periods of silence and sound, since the perceived sound intensity would be lower. This was not the case: the mean access activity during sound presentation was similar for saline and salicylate treatments. This indicates that a salicylate-induced hearing loss cannot account for the behavioral changes during silence after salicylate.

A limitation of Ruttiger's technique is the large amount of time needed to train the animals. However, this method can detect the presence, level, and time course of tinnitus.

5.3.5 Tests based on gap-detection deficits

The behavioral models previously discussed require training animals to respond distinctively to the absence of an acoustic stimulus. Tinnitus is then inferred by response deviations from control during "silent" trials, suggesting that the animals have heard something, i.e. tinnitus, when nothing was presented. These models also required complex behavioral manipulations such as food or water deprivation, finely-tuned shock parameters, variable reinforcement schedules, etc., and typically long behavioral training. In addition, these methods rely heavily on learning, memory, motivation, and absence of stress in the animals. It is also commonly assumed that tinnitus does not habituate in animals, since it does not in humans.

Turner et al. (2006) describe a novel method for tinnitus screening in Long–Evans rats by use of gap detection startle-reflex procedures. Startle is a fast response to sudden, intense stimuli and probably protects the organism from injury by a predator or by a blow. The acoustic startle response (ASR) is characterized by a coordinated contraction of the muscles of the eyelid, neck, and extremities (Figure 5.5) that can be recorded with piezo-electric sensors. In pre-pulse inhibition (PPI), a weak, non-startling stimulus (prepulse) of any type diminishes the response to a startle stimulus, which is presented 30–500 ms later. It is widely assumed that this prepulse activates the ascending auditory pathway in the IC and thereafter the superior colliculus, which in turn affects the startle pathway by activating an inhibitory cholinergic pathway from the pedunculopontine tegmental nucleus to the PnC (nucleus reticularis pontis caudalis) (Koch, 1999). There is thus no apparent reason for the need of changes in SFR in auditory cortex in order to affect the gap-startle response (Chapter 7). Activity from secondary auditory cortex and via the medial part of the medial geniculate body (not shown) may modulate the startle reflex via the amygdala and nucleus accumbens pathway to the PnC.

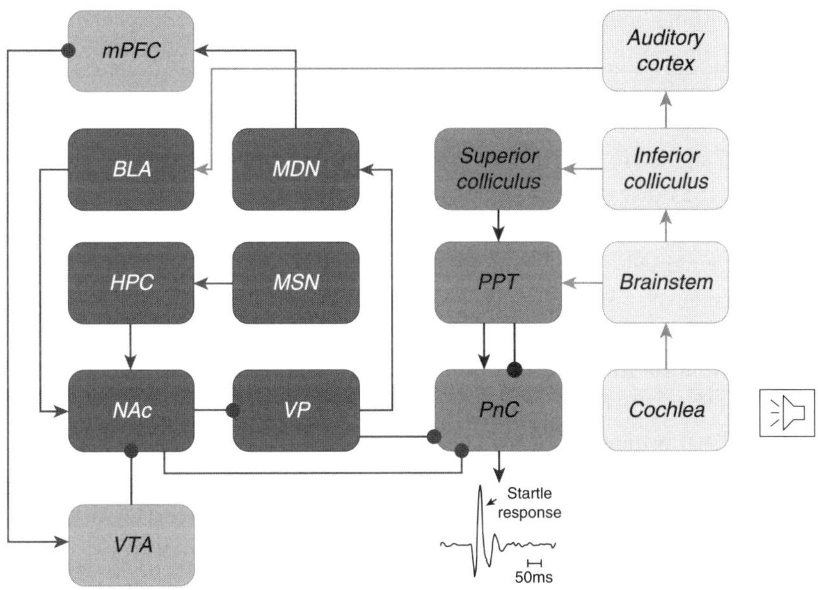

Figure 5.5 (Also see Color plate 2.) Startle response circuit with pre-pulse inhibition (PPI) circuit. The auditory pathway is indicated with olive colored boxes and connections. The startle circuit is indicated by pink boxes and red connections. The PPI modulating circuit is indicated in blue, it inhibits the PnC and is potentially affected by auditory cortex via the amygdala (BLA). Modulations from the mPFC via the VTA are indicated in orange-brown color. The arrowheads indicate excitatory connections, Round-dotted endings indicate inhibitory connections. BLA, basolateral amygdale; HPC, hippocampus; MDN, mediodorsal thalamic nucleus; MPFC, medial prefrontal cortex; MSN, medial septal nucleus; NAc, nucleus accumbens; PPT, pedunculopontine tegmental nucleus; PnC, nucleus reticularis pontis caudalis; VP, ventral pallidum; VTA, ventral tegmental area. After Swerdlow et al. (2001).

The Turner model is based on the ability to reduce the acoustic startle reflex by a preceding silent gap in an otherwise constant acoustic background. Benefits of the existing gap detection reflex technique include: (1) food or water deprivation is not necessary; (2) no training, learning, memory, or motivational demands are placed on the animal; (3) the startle neural circuit is well known, and its modulation using background sounds has been studied extensively (for reviews see: Swerdlow et al., 2001; Li et al., 2009); and (4) testing can be done quickly in a single 40-min session, allowing rapid assessment of acute manipulations. The authors hypothesized that if a background acoustic signal was qualitatively similar to the rat's tinnitus, poorer detection of a silent gap in the background would be expected because of the tinnitus filling in the silent gap to some extent.

Noise-trauma rats that displayed operant evidence (using the method of Bauer et al., 1999) of tinnitus ~10 kHz, were used in the gap detection testing procedure for tinnitus. Testing was conducted using Hamilton–Kinder startle reflex hardware and software. Gap detection testing was conducted with background noise presented through a speaker located in the door wall and startle stimuli presented through a speaker located in the ceiling of the testing chamber, 15 cm above the animal's head. The floor of the chamber, attached to a piezo-electric transducer, provided a measure of startle force applied to the floor. A clear animal holder, with holes cut for sound passage, was suspended above the floor, allowing the rat to freely turn around while minimizing excessive movement. An adjustable-height roof was set to a level that kept rats from rearing up, a behavior that adds variability to the startle response.

Background signals in the startle chamber consisted of broadband noise (BBN), or band-pass filtered noise centered at 10 kHz (1-kHz bandwidth) or 16 kHz (1-Hz bandwidth). The three test conditions were run sequentially, each lasting approximately 12 min. Rats remained in the chambers between tests. Test stimuli were calibrated at 60-dB SPL peak levels. Baseline noise levels in the test chamber (with background test noise turned off) were measured between 42–52 dB SPL in the 2–36-kHz range. The order of presentation for the three test conditions was counterbalanced across rats to control order effects. Each test consisted of 24 trials presented with a 20-s variable intertrial interval. Each session began with a 2-min acclimation period followed by two trials consisting of an abrupt startle-eliciting noise burst (115-dB SPL, 20-ms duration), which served to habituate the startle response to a more stable baseline. Data from the two initial trials were not used in the gap detection analysis. The remainder of the session consisted of 10 additional startle-only trials pseudo-randomly mixed with 12 gap trials. Gap trials were identical to startle-only trials, except for the inserted gap. Gaps always began 100 ms before the startle stimulus, were 50 ms in duration, and had a 0.1-ms rise/fall time. A pilot experiment using the present equipment and stimulus settings revealed no pre- to post-treatment testing ABR threshold shifts in Long–Evans rats, either immediately or 1 week after startle testing. There was a significant positive correlation between the 10-kHz results obtained from the two methods. In addition, neither method showed significant differences between tinnitus and control animals for either 16-kHz or BBN test conditions. It appears likely, therefore, that the independent operant-based and startle reflex-based methods are affected by the same phenomenon, although it is not a priori clear if this reflects tinnitus or hyperacusis.

The limited role of the auditory cortex in the acoustic startle is demonstrated by ablation experiments. Mice, 1 month after bilateral auditory cortex ablations were statistically indistinguishable from controls on all suprathreshold measures of the ASR. Averaged ABRs indicated no effects of these lesions on auditory sensitivity. If auditory cortex plays a modulatory role with regard to the ASR, it is apparently non-essential and/or readily compensated for after ablation (Hunter and Willott, 1993). The patency of the cortex is however important in gap-detection especially if the durations are short, <30 ms (Bowen et al., 2003). These aspects are important in interpreting gap-startle procedures to identify tinnitus.

5.4 **Summary**

Do animals have tinnitus? A positive answer to this question will allow in-detail investigation of various neural substrates at the single neuron and population level underlying tinnitus. Several behavioral models have been developed for guinea pigs, hamsters, rats, and mice that are based on the continuous presence of tinnitus. These are all reviewed in detail. The training typically occurs before the induction of tinnitus and may be focused on conditioning silence as a warning condition for the footshock, the presence of tinnitus is thus not punished or rewarded. Tinnitus may also be detected by its obliteration of a silent gap in noise used as a pre-pulse inhibitor for the startle response to a loud sound. Some surprising results have been obtained, i.e. detecting tinnitus in frequency regions outside those affected by the noise trauma, that lead to the suspicion that some of the tinnitus responses may actually reflect the presence of hyperacusis. As long as there is a one-to-one correspondence between tinnitus and hyperacusis this may not be a problem. However, this correspondence in humans is far from universal. Nevertheless, several of these behavioral tests have been cross-validated suggesting consistency in detecting some abnormality in animals subjected to agents that in humans frequently cause tinnitus. Therefore there is sufficient evidence that animals have tinnitus. How this manifests itself in changed neural substrates will described in Chapters 6–9.

Chapter 6

The salicylate model of tinnitus

Salicylate (aspirin) is likely the most used drug in the world. Salicylate, and other, non-steroidal anti-inflammatory drugs have several side effects: tinnitus, hearing loss, and changed sound perception are the three auditory effects described by human subjects after ingestion of large doses of salicylate. These symptoms develop over the initial days of use but may then level off, fluctuate, or decrease, and are reversible within a few days of stopping the drug use. Hearing loss is more or less linearly related to plasma salicylate levels with a hearing loss of about 8 dB per 100mg/L plasma level (Cazals 2000). It was already known at the end of the 19th century that tinnitus is the most constant and first appearing symptom of salicylate use that could occur even for plasma levels below 100 mg/L, which barely causes hearing loss (Sèe, 1877, cited in Cazals, 2000). However, although 1, 3, or 5 mM salicylate did not damage the sensory hair cells and caused no hearing loss in neonatal rats and guinea pigs, continuous application damaged the spiral ganglion neurons (SGNs) and their peripheral fibers in a dose-dependent manner (Chen et al., 2010; Wei et al., 2010a,b). If this loss of SGNs is diffuse and isotropic it would not be detectable in audiometric or ABR thresholds.

Numerous biochemical processes underlying the effects of salicylate have been identified: inhibition of prostaglandin synthesis through the inhibition of cyclooxygenase, inhibition of numerous metabolic enzymes, inhibition of free radicals, insertion into membranes and interference with ion transport, uncoupling of oxidative phosphorylation, and activation of heat shock transcription factor. Among them, the inhibition of cyclooxygenase activity is the best-known pharmacological effect of salicylate. Cyclooxygenase-1 (COX-1) and its inducible isoform COX-2 convert arachidonic acid to prostaglandin H2 (Vane, 1971; Vane et al., 1998). In normal conditions, arachidonic acid is competitively metabolized in the cochlea by COX to prostaglandins and tromboxanes, and by lipooxygenase to leukotrienes. Sodium salicylate application on the round window of the cochlea induces a reversible decrease of prostaglandins and tromboxane B2, with an increase of leukotrienes B4 and C4 (Jung et al., 1992).

Salicylate has been used as the preferred inducer of tinnitus in animal models because of its reliable effects in doing so. One of the drawbacks of using salicylate is that it acts on both the peripheral and the central auditory nervous system. In the periphery, salicylate affects first of all the motility of the OHCs. Most likely, salicylate primarily influences electromotility and the non-linear capacitance of the OHCs via a direct interaction with prestin (Greeson and Raphael, 2009). In addition, as described earlier, salicylate interferes with the arachidonic acid cycle in the IHCs and so upregulates the action of NMDA receptors in the cochlea, which increases the channel opening probability of the NMDA receptor and potentially leads to increased SFR for high doses (Guitton et al., 2003). A different mechanism for salicylate-induced hearing loss was recently suggested by Wu et al. (2010). Instead of an action on prestin, which would require a very high dose in humans compared to animals in order to cause a hearing loss, salicylate could also act via blocking the KCNQ4 outward current $I_{K,n}$. This would subsequently cause depolarization of OHCs, resulting in a reduction of the driving force (by a ceiling effect) for the transduction current and electromotility.

Besides these peripheral actions, salicylate has many effects on central auditory neurons ranging from downregulation of $GABA_A$ receptors, to reducing the effects of serotonin on the inhibitory activity of GABAergic neurons. We will discuss these aspects later.

Surprisingly, the most parochial claims for peripheral versus central tinnitus have been made on the basis of the peripheral versus central actions of salicylate. There is obviously an intricate interplay between the auditory periphery and the central nervous system in causing the type of reversible tinnitus induced by salicylate. Unraveling this will be the topic of this chapter.

6.1 Structural changes caused by salicylate

Structural changes in the cochlea following application of salicylate are modest at best and are generally reversible after application. Light microscopy showed no changes in monkey (Myers and Bernstein, 1965), in chinchilla (Boettcher and Salvi, 1991; Spongr et al., 1992), in guinea pigs (Cazals et al., 1988), and in rats (Zheng and Gao, 1996). Electron microscopy showed increased lysosomes and vacuolization in the subcuticular zone of OHCs (Deer and Hunter-Duvar, 1982), bent and flaccid stereocilia of OHCs of the outermost row with protrusion of phalangeal processes above the cuticular plate with distorted and reduced microvilli (Douek et al., 1983), and an increase in the luminal width of smooth and rough endoplasmic reticulum (Dieler et al., 1994). The minor structural changes that do occur are nicely illustrated by the study of Feng et al. (2010) in guinea pigs which compared the effects of sodium salicylate (400 mg/kg/d) or saline vehicle application for 10 consecutive days (Figure 6.1). They observed that this repeated (high-dose) salicylate administration activated caspase-3 and caused apoptosis in OHCs and SGNs ($P < 0.01$ vs. saline control for both measures and in both cell types). Cell counts showed a significant loss in OHCs ($P < 0.01$ vs. saline control), but not in IHCs. Transmission electron microscopy revealed chromatin condensation and nucleus margination in the salicylate-treated cochlea (Figure 6.1d). Scanning electron microscopy demonstrated breakdown and fusion of stereociliary bundles at the apical OHCs (Figure 6.1e), and villous matter was discovered to attach on the surface of SGNs (Figure 6.1b). These findings suggest that long-term administration of high-dose salicylate can activate the caspase-3 pathway to induce OHC and SGN apoptosis.

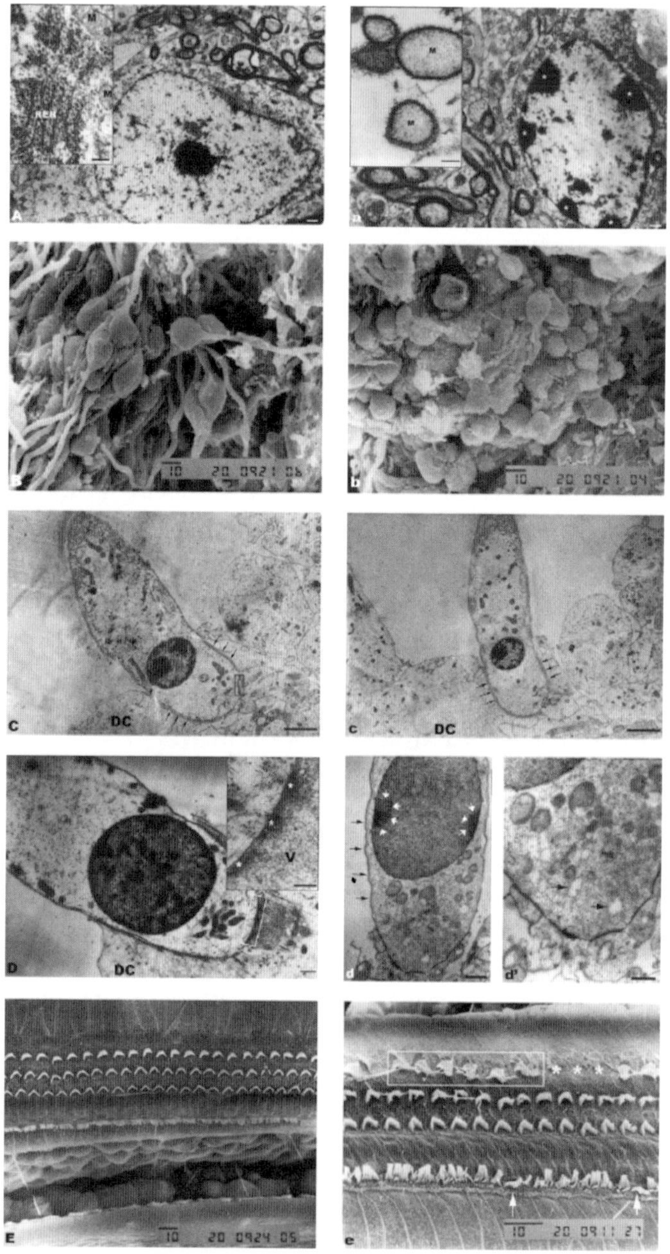

Figure 6.1 Electron microscopic observation for spiral ganglion neurons and hair cells. No morphological change occurred in the saline control (left column A–D). The salicylate-intoxicated neurons (right column) demonstrated apoptotic changes, characterized with chromatin condensation and nucleus margination. Villous matter was discovered to attach on the surface of SGNs (b). Salicylate-induced changes in OHCs lateral wall, featured with waviness (black arrows) (d and d′), while IHCs remained intact, with integrated tight junction (black arrow) and normal synapse at the basal pole (black box) (c). Montages of scanning microscope images from a cochlea with salicylate disposal revealed stereociliary bundles breakdown and fusion at the apical of OHCs (white box), accompanied by loss in the third row (white asterisks, e), while IHCs remained intact. Normal V-shaped stereociliary bundles showed staircase morphology in the saline control (E). Scale bars = 1 μm for A, a, d and d′; 500 nm for the box at the top left of A and a; 10 μm for B, b, E, and e; 2.5 μm for C and c; 0.5 μm for D. Reprinted from Feng, H., Yin, S.H., Tang, A.Z., Cai, H.W., Chen, P., Tan, S.H., and Xie, L.H. (2010). Caspase-3 activation in the guinea pig cochlea exposed to salicylate. *Neuroscience Letters*, **479**(1), 34–39. Copyright (2010), with permission from Elsevier.

6.2 Physiological and neural changes

6.2.1 Cochlea and auditory nerve

Basilar membrane movement The movement of the various cochlear parts in response to sounds at the threshold of hearing is smaller than the diameter of a hydrogen atom, i.e. $<10^{-10}$ m. These movements are nowadays measured using laser interferometry (Robles and Ruggero, 2001), but the initial measurements were done using a radioactive source placed on the basilar membrane (BM) and detecting BM-velocity-induced Doppler shifts in the frequency of the emitted gamma rays (Sellick et al., 1982). After some time, the radiation had a deleterious effect on the OHCs that amplify the mechanical response of the BM.

After perfusion of 10 mM salicylate for 5 min in the bathing medium a decrease of approximately 6 dB in the amplitude of BM movements was found by Mammano and Ashmore (1993). This effect occurred after about 5 min and was reversible within about 15 min. Murugasu and Russell (1995) measured BM tuning curves, the tip of which was elevated by 20–30 dB after 2.5 or 5 mM salicylate infusion and up to 40 dB after 10-mM salicylate infusion (equivalent to 1600 mg/L). Tuning-curve bandwidth at 10 dB above best threshold was broadened by a factor of about 3. The alterations in BM mechanics started within about 5 min, reached a maximum after about 30 min, and then recovered within the next 10 min.

The cochlear battery and receptor potentials A positive endocochlear potential (EP) of approximately 80 mV corresponds with the resting potential of the cochlear endolymph. The potassium channels in hair cells generate a negative membrane potential of about −60 mV. The resulting potential difference of about 140 mV is the battery that drives transduction currents through hair cells. In rats, guinea pigs, and chinchillas, the EP was not affected by either intraperitoneal (Tran Ba Huy et al., 1987) or intravenous injections (Stypulkowski, 1990) of 350–500 mg/kg sodium salicylate, or by administration into the scala vestibuli of up to 10 mM (again a very high dose; Fitzgerald et al., 1993).

The cochlear microphonic (CM) is considered to originate mostly from the compound oscillatory activity of OHC membrane potentials and shows surprisingly little sensitivity to salicylate. The summating potential (SP), likely originating from both the IHCs and OHCs, reflects the compound average membrane potential changes, i.e. without the oscillations, in these hair cells. For doses very similar to those used in studying the EP the same studies reported very little change in the CM or SP (Cazals et al., 1988; Puel et al., 1989, 1990; Stypulkowski, 1990; Kujawa et al., 1992; Didier et al., 1993). Murugasu and Russell (1995) recorded CM in response to tones of 15 and 1 kHz and perfused salicylate at various concentrations up to 5 mM in scala tympani. They showed either slight increases or clear decreases in CM for the 15 kHz tone, and no change or a slight decrease for the 1-kHz tone.

CM and SP as recorded extracellularly and typically at the round window are compound responses whose contributions from the hair cells appears to decrease by about 4–8 dB/mm along the cochlea depending on the species (Ponton et al., 1992) and in the case of the CM are dependent of the diverse phase relationships between the hundreds of hair cells that contribute to the round window response. This is affected by the coiling of the cochlea and whereas this is different across species it may explain some variability in the results. As we will see, salicylate affects the otoacoustic emissions generated by the OHCs and these are expected to behave like the CM (as was in fact demonstrated for the cubic distortion product component in the CM by Kujawa et al., 1992).

Otoacoustic emissions OAE are sounds emitted by the cochlea and recordable with a microphone that is tightly sealed into the external auditory canal. They were predicted by Gold (1948) based on a theoretical analysis of the frequency selectivity of the cochlea requiring a positive feedback mechanism, and first demonstrated by Kemp (1978). Gold, an astrophysicist, together with his astronomer colleagues Bondi and Hoyle proposed in that same year (1948) the now abandoned "steady state" hypothesis of the universe. Normal ears emit spontaneous OAEs as well as stimulus-induced ones; the latter are the result of non-linear interactions between low-level sounds of different frequencies that have overlapping activity patterns on the basilar membrane. OAEs are now routinely used in newborn hearing screening (Eggermont et al., 1996; Lonsbury-Martin and Martin, 2004; Chapter 4).

Spontaneous and stimulus evoked OAEs disappear shortly after the administration of salicylate in humans (Long and Tubis, 1988) and rhesus monkeys (Martin et al., 1988) but recover within a few days. Acute recordings in cat (Stypulkowski, 1990) showed that 1 h after salicylate administration, DPOAEs were notably reduced and after about 3 h one component was reduced by about 5 dB while another component had disappeared. Kujawa et al. (1992) found in guinea pigs that DPOAEs were reduced after intracochlear salicylate application, the reduction being greater at low sound-pressure levels. This low SPL effect was also found by Fitzgerald et al. (1993). Recently, Stolzberg et al. (2011) found that 2 h after intraperitoneal injection of 300 mg/kg salicylate the DPOAE was reduced by about 10 dB at 10 and 20 kHz, but only 5 dB at 16 kHz, the behavioral tinnitus frequency. At frequencies between 4–8 kHz the decrease in DPOAE level was between 2–5 dB.

There is an interesting time-dependent effect of salicylate first demonstrated by Huang et al. (2005b). A single injection of sodium salicylate (200 mg/kg) could reduce the amplitude of the DPOAE within 2 h. The reduction was significant at 20–50 dB SPL stimulus levels and recovered after 8 h. However, following daily injections of sodium salicylate (200 mg/kg), the DPOAE progressively increased. After injection for 14 days, the DPOAE increased about 2–3.5 dB SPL. The increase rate was about 0.2 dB SPL/day. The DPOAE input–output function remained nonlinear. The increase was greater at 40–70 dB SPL primary tone levels and reversible (Figure 6.2). This effect could be attributed to first suppression and then enhancing of prestin expression in the OHCs (Yu et al., 2008; Yang et al., 2009). However, while an enhanced or sustained DPOAE was seen, permanent reductions in the amplitude of the cochlear CAP and the ABR were often observed after chronic salicylate treatment (Chen et al., 2010). The reductions in CAP amplitude suggest that high-dose long-term treatment with salicylate may damage the neurites, soma, or axons of SGN. Previous *in vitro* studies with cochlear organotypic cultures indicate that high doses of salicylate damage the spiral ganglion neurites, but not hair cells. It is possible that the ribbon synapses in the IHC are affected, which could lead to loss of spiral ganglion cells.

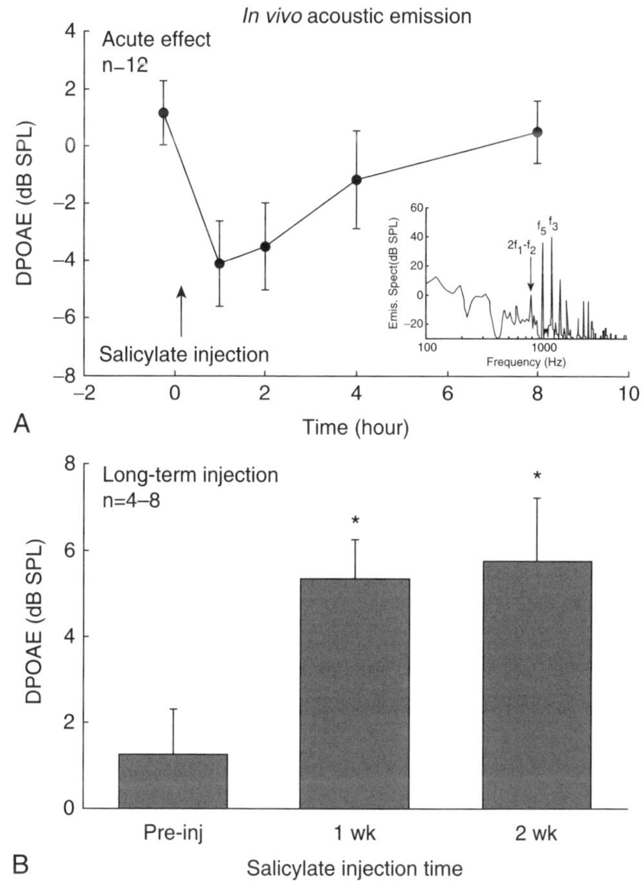

Figure 6.2 Dual effects of salicylate on distortion product otoacoustic emission (DPOAE) in measurement *in vivo*. (A) Acute inhibition of salicylate on DPOAE. The distortion product of $2f_1$–f_2 was measured at 3 kHz. A vertical arrow indicates a salicylate injection. The DPOAE recovered after 8 h. Inset: the DPOAE spectrum. The noise floor level was less than –15 dB SPL. (B) Increase in acoustic emission in long-term administration of salicylate. Asterisks indicate $p < 0.01$ (ANOVA). With kind permission from Springer Science + Business Media: Yu, N., Zhu, M.L., Johnson, B., Liu, Y.P., Jones, R.O. and Zhao, H.B. (2008). Prestin upregulation in chronic salicylate (aspirin) administration: an implication of functional dependence of prestin expression. *Cellular and Molecular Life Sciences*, **65**, 2407–18, Figure 1.

Thus, the most peripheral effect of salicylate is on the OHC motor that provides mechanical feedback to the basilar membrane, thereby amplifying its movement amplitude by up to 40 dB for low level sounds (Nobili et al., 1998). The loss of sensitivity caused by the acute salicylate action or the increase after chronic application subsequently is reflected in all downstream measures.

Auditory nerve fiber activity The effects of salicylate were initially observed for CAPs of the auditory nerve and they invariably indicated a loss of sensitivity in the order of 20–30 dB (Schreiner and Snyder, 1987; Cazals et al., 1988; Stypulkowski, 1990; Shehata-Dieler et al., 1994; Stolzberg et al., 2011). Evans et al. (1981) and Evans and Borerwe

(1982) recorded SFRs from ANFs in cats after injection of 400 mg/kg of sodium salicylate. They reported a significant increase of about 1–20 spikes/s on average for high SFR fibers, and a tendency for fibers with the highest CF to display the greater increase in SFR. This increase in SFR has quite frequently been interpreted as a correlate of tinnitus in the peripheral auditory system. However, one has to consider that, especially in cats, salicylic acid can cause non-specific toxic effects at relatively low dose (Chen and Jastreboff, 1995; Ochi and Eggermont, 1996). In cats the pharmacokinetics of salicylate is different from that in humans and small rodents. The metabolization rate of salicylate is small because cats lack glucuronyltransferase, which is necessary to metabolize salicylate. The half-life of salicylate in cats is 37 h, as opposed to 3–6 h in humans and small rodents (Knipper et al., 2010). Therefore, effects seen in cats for high doses of salicylate may reflect a drug action (e.g. fever with corresponding increase in SFR) that is non-specific to the cochlea.

Stypulkowski (1990) recorded from single ANFs in cats after a single intravenous injection of sodium salicylate at the dose of 200 mg/kg. Following salicylate application there was no significant increase in mean SFR. However, the author observed a few fibers that displayed increased SFR by about 20 spikes/s after salicylate infusion. Kumagai (1992) recorded from single ANFs of guinea pigs given an intravenous injection of 200 or 400 mg/kg of sodium salicylate. At 200 mg/kg, the 102 fibers recorded post-salicylate injection did not show any change in SFR, while after the 400-mg/kg dose the SFR of the 112-recorded fibers increased significantly. Shehata-Dieler et al. (1994) recorded spontaneous activity of eighth nerve fibers from pigeons after scala media perfusion with sodium salicylate at doses of up to about 18 mM (very high, equivalent to 2880 mg/L). Data obtained for 14 fibers indicated that after salicylate the mean SFR slightly increased or remained unchanged. Responses of single ANFs in gerbils were studied after systemic administration of 200 mg/kg salicylic acid (Müller et al., 2003). Two hours and more after application, single fiber thresholds were elevated by about 20 dB at all CFs. Sharpness of tuning was reduced. Mean SFR was significantly reduced at CFs below 5 kHz (mean: 44 vs. 28 sp/s). At CFs above 5 kHz the mean SFR remained unchanged. Perilymphatic perfusion of sodium salicylate (5 mM, equivalent to 800 mg/mL) caused the SFR of ANFs with CFs between 6–26 kHz to significantly increase from on average 17 to 40 spikes/s (Ruel et al., 2008). So it is not clear what really happens at the level of the auditory nerve. Species differences play a role, e.g. cats do not tolerate high-dose application. In rodents, acute high doses appear to increase the SFR of auditory nerve fibers. Perilymphatic perfusion at high concentration clearly results in an increase in SFR.

Compound SFR from the auditory nerve Attempts to estimate the compound SFRs of auditory nerve fibers, based on measuring putative neural noise on the round window or on the nerve surface, started in the 1970s. These methods may be potentially applicable for recording in humans. Kiang and colleagues (1976) were the first to show that the extracellular contribution of the spike waveform of auditory nerve fibers can be recorded from the round window of the cochlea in cats. For that purpose a spike triggered averaging technique was used, the trigger being spikes recorded from an individual ANF and the spike-triggered average (STA) was recorded from the round window. Later Prijs (1986) and Versnel et al. (1992) repeated these findings in guinea pig.

Schreiner and Snyder (1987) pioneered the analysis of the compound spontaneous activity of the auditory nerve using a gross electrode on the nerve trunk. Because there was no trigger from single nerve fibers available, they showed that spectral averaging of the spontaneous activity revealed the presence of underlying periodic activities, potentially reflecting spike waveforms, as peaks in the spectrum. After cats were given one intravenous injection of sodium salicylate of 200 mg/kg the average spectrum of the nerve spontaneous activity showed an increase in the spectral peak at 200 Hz. This was often accompanied by a decrease of the broader peak at 1 kHz observed before salicylate (see also Lenarz et al., 1993). One could interpret the 1-kHz peak as reflecting the spectrum of the action potential, whereas the 200-Hz peak could indicate salicylate-induced burst firing with relatively long 5-ms interspike intervals (but then the 1-kHz peak would not diminish). Martin (1994) used spectral averaging of spontaneous auditory nerve activity from 14 human subjects during cerebellopontine angle surgery. Seven of the patients had a pre- and post-operative history of continuous tinnitus. A spectral peak at about 200 Hz was found for these seven subjects. This spectral peak was also observed for three other subjects who had tinnitus before but not after surgery and for one subject who had tinnitus only after the surgery. One subject without complaints of tinnitus also showed a 200-Hz peak. Two subjects without tinnitus complaint did not show a 200-Hz peak.

Cazals et al. (1998) recorded the average spectrum of spontaneous eighth nerve activity from the round window of guinea pigs during and after intramuscular doses of 200 mg/kg sodium salicylate that were given twice a day for 2 or 3 weeks. In these guinea pigs, the average spectrum of spontaneous activity of the eighth nerve did not show a peak at 200 Hz but only the broad peak at about 1 kHz. This spectral peak was acutely decreased during a few hours following salicylate injections. In contrast, over days of treatment the 1-kHz peak increased progressively and returned to normal within 1 to several weeks after cessation of treatment. They interpreted the decrease and increase of the 1 kHz peak as a result of changes in synchronization of auditory nerve fiber activity; the more fibers fire synchronously, the larger the 1-kHz peak.

However, alternative interpretations are possible: the peak in the activity spectrum from the round window or auditory nerve surface at frequencies around 1 kHz may not reflect a change in neural synchronization, but is only a reflection of the waveform of the unit contribution at the round window (see Figure 9.5 and discussion in Chapter 9). In that case, the only explanation for an increased activity would be larger SFRs in the auditory nerve fibers as argued by McMahon and Patuzzi (2002). But as Patuzzi et al. (2004) remark, it is not clear which neurons are affected, since the round window noise spectrum is comprised of neural activity from the dendrites and cell bodies of cochlear neurons in the cochlear base, and from (all) axons in the basal modiolus. This also indicates that potentially recording from a promontory electrode in humans, which will give a fairly robust signal (Eggermont and Odenthal, 1974), will be subject to the same interpretational problems as they are in animals. Taking into account that the spectral amplitude only increases with the square root of the number stochastically firing neurons, but is proportional to that number if the neurons were indeed firing synchronously, the explanation favored by Cazals et al. (1998) is supported.

6.2.2 The central nervous system

The auditory brainstem The only measurements of a salicylate action on auditory brainstem neurons are from *in vitro* superfusion of slices from the cochlear nucleus (CN). Basta et al. (2008) used slices obtained from young mice after 20–45 days after the onset of hearing, and sodium salicylate at a concentration of 0.35, 0.7, and 1.4 mM (= 224 mg/L). The larger dose is known to induce tinnitus in rats (Bauer et al., 1999). Using extracellular single unit recordings, they found that the SFR increased by on average 40% (1.4 mM), 18% (0.7 mM), and 10% (0.35 mM). Wei et al. (2010b) repeated this experiment in slices of the DCN taken from Sprague Dawley rats (aged 13–20 days), by recording intracellularly. They used 1.4-mM sodium salicylate as well and were able to identify the cell types they were recording from. Fusiform cells were identified from their location within the fusiform layer of the dorsal cochlear nucleus (DCN), where they are the only large cell type, apart from cartwheel cells (Figure 8.2; Chapter 8). Fusiform cells typically have a large, spindle-shaped soma located in the deeper part of the fusiform cell layer; their apical and basal processes project towards the molecular layer and deep layer, respectively. In contrast, the cell bodies of cartwheel cells are round, generally smaller than fusiform cells, and lie close to the surface of the fusiform cell layer. Physiologically, cartwheel cells fire unique "complex spikes", i.e. spike bursts that reliably distinguish them from fusiform cells, which fire simple, uniform spikes. Five-minute superfusion with 1.4-mM salicylate suppressed spontaneous and evoked firing in fusiform cells; this decrease partially recovered after salicylate washout. In contrast, salicylate had no effect on the spontaneous or evoked firing of cartwheel cells. Salicylate's suppressive effects were not a consequence of increased synaptic inhibition from cartwheel cells onto fusiform cells because salicylate reduced spontaneous inhibitory postsynaptic current amplitude.

The inferior colliculus After intraperitoneally injection of 450 mg/kg of salicylate in guinea pigs, the mean SFR of the cell population in the central nucleus of the inferior colliculus (ICC) increased from 29 to 83 spikes/s and the median from 26 to 74 spikes/s. Control experiments with saline injections revealed no statistically significant differences in cell discharges compared to untreated animals (Jastreboff and Sasaki, 1986). A note of caution is in order here, as the firing rate was not measured by counting the number of spikes in a certain time window but as the "inverse of the modal interval (most frequently occurring interval) of a non-sequential interval histogram". This measure is extremely sensitive to burst firing and may not at all reflect increased and quite high SFRs. Subsequently Chen and Jastreboff (1995) evaluated the spontaneous activity of 471 units from the external nucleus of the IC (ICX), and showed that salicylate increases the SFR and caused the emergence of spike bursts with >4 spikes at short intervals. SFRs increased significantly from on average 6.2 spikes/s before salicylate to 9.2 sp/s after injection of 233 mg/kg of salicylate, Bursting activity with short (~1-ms) interspike intervals increased very saliently after salicylate application for units with CFs in the range of the behavioral estimate of tinnitus (Figure 6.3).

Manabe et al. (1997) recorded extracellularly from the IC (not clear if this included the ICX) of guinea pigs before and after an intravenous injection of 200 mg/kg sodium salicylate. SFR was on average around 5 spikes/s before salicylate. It reached a maximum of about 20 spikes/s at 2 h post-injection and then progressively decreased to pre-salicylate value within the next 8 h. Serum salicylate levels were found to decrease exponentially from about 450 mg/L a few minutes after injection to 100 mg/L at 10 h post-injection. In contrast, a significant decrease in SFR after acute salicylate treatment (200–300 mg/kg) was found in the ICC of CBA/J mice (Ma et al., 2006). They did not observe the increased burst firing observed by Chen and Jastreboff (1995) in ICX. The different results may indicate a species difference (rats, guinea pigs, vs. mice) in sensitivity to salicylate, potentially a difference in anesthesia (Nembutal or urethane vs. ketamine), and more likely recording from different parts of the IC (ICX vs. ICC).

Basta and Ernst (2004) investigated the direct effect of salicylate application on the SFR of mouse IC neurons in brain slices. Out of 92 neurons that were recorded extracellularly, 87% responded to the superfusate of 1.4 mM (= 224 mg/L) salicylate by significantly changing their SFR. Seventy percent increased and 17% decreased their SFR, respectively. Superfusing IC neurons in a slice preparation bypasses the effects of salicylate on the auditory periphery and indicates a purely central effect. It is not clear how this central effect interacts with peripheral input. The differences with the *in vivo* results might also result from different effects of salicylate on different cell types (see section 6.3).

The thalamocortical system Basta et al. (2008) used slices obtained from young mice (NMRI strain) after the onset of hearing (day 20–45) and superfused these with sodium salicylate at concentrations of 0.35, 0.7 and 1.4 mM (see brainstem and inferior colliculus sections). Using extracellular single-unit recordings, they found that the SFR in the MGB changed by about 78% for 1.4 mM and showed similar changes as in CN for lower

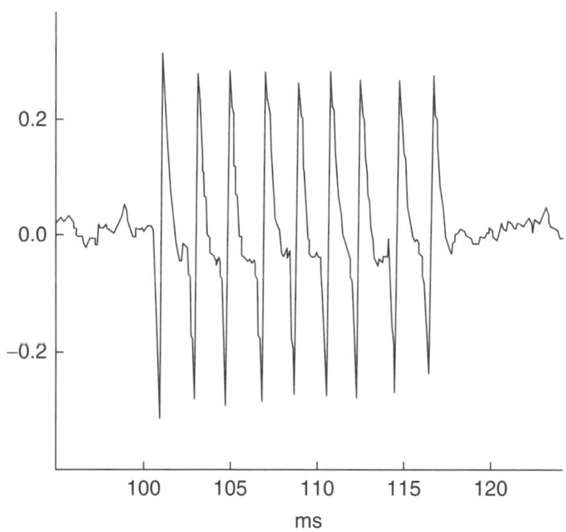

Figure 6.3 Burst-firing in the ICX of Long–Evans rats occurs after application of salicylate. The interspike interval here is ~2 ms. Reprinted from Chen, G.D. and Jastreboff, P.J. (1995). Salicylate-induced abnormal activity in the inferior colliculus of rats. *Hearing Research*, **82**, 158–78. Copyright (1995), with permission from Elsevier.

concentrations. This change at 1.4 mM is about twice the value found in the CN using the same techniques. In auditory cortex slice they found values that were somewhat in between: 60% for 1.4 mM and again the same values for lower concentrations as in CN and MGB. This suggests that changes observed for the lower concentrations were likely non-specific.

In the primary auditory cortex (AI) of cats we (Ochi and Eggermont, 1996) made recordings from the same single units prior to and continuously for, on average, 6 h after administration of 200 mg/kg salicylate. LFPs were used to track the threshold shifts. All animals showed 20–30 dB of threshold shift ~2 h after salicylate administration and showed no recovery during the following 6 h. This is in accordance with the very long half-life of salicylate in cats (mentioned previously). Significant changes were found in SFRs for two groups of units separately. Low-spontaneous rate units (pre-salicylate SFR <1 spike/s) showed a post-application increase in SFR and high-spontaneous rate units (pre-salicylate SFR >1 spike/s) showed a decrease in SFR. This likely is a real effect since the recordings were made from the same units before and after salicylate application, but overall there was a regression towards a mean value, which was the same before and after application. There were no significant changes in modal and mean values for interspike-interval (ISI) histograms, indicating no bursting. Peak cross-correlation coefficients for the firing patterns of simultaneously recorded cells showed no significant change but the correlogram's central peak was significantly narrower after salicylate application, suggesting changes in the synchrony pattern. Comparison of the effect of sodium salicylate at 200mg/kg in cats in different cortical areas showed increased SFRs for high CFs in secondary auditory cortex (AII), but not in primary (AI), and even significantly reduced SFRs in anterior auditory field (AAF) (Eggermont and Kenmochi, 1998; Eggermont, 2004). In awake cats Zhang et al. (2011) found that sound-evoked spike activities were significantly increased from 1 h after salicylate administration (200 mg/kg), and the increase of neural responses lasted >3 days with a peak at 12 h. The significant enhancement of tone-evoked neural responses was observed over the entire tested frequency range (0.1–16 kHz) with a relative peak in the band of 3.2–9.6 kHz. Salicylate-administration, however, decreased the mean SFR in A1 units, and the decrease of spontaneous rate was larger in the units with a high initial SFR, thereby confirming part of the findings of Ochi and Eggermont (1996). This study also shows that findings in ketamine anesthetized cats and awake cats do not differ qualitatively.

Recordings from auditory cortex (not specified if this was from the primary area or a mix of fields) in awake Sprague Dawley rats from relatively low-impedance (~100 kOhm) implanted microwire electrodes allowed Yang et al. (2007) to monitor the SFR in neuron clusters before and over a period from days to weeks following salicylate treatment. Spontaneous spike counts were sampled in a 100-ms window every 500 ms for 6 min (total sample time 72 s). The average SFR was 22 spikes/s before salicylate injection (mean of 2 days recording). Two hours after 150 mg/kg salicylate, the average SFR dropped to approximately 14 spikes/s, significantly below the pretreatment rate. The mean spontaneous spike rate measured at 1 day post-salicylate returned to pretreatment levels and remained constant for the next 2–3 days. However, 2 h after salicylate treatment, the behavioral tests (both scheduled polydipsia and gap inhibition of the startle reflex, see Chapter 5) suggested that both 150 and 250 mg/kg of salicylate induced the phantom sound of tinnitus. Clearly then, increased SFR in auditory cortex is not required for these positive behavioral tests that are supposed to indicate the presence of tinnitus. It should be noted that the LFP amplitude measured at 60 dB SPL was enhanced by at least 70% for frequencies of 16 and 20 kHz. Thus the changes in LFP gain (hyperacusis) correspond with the behavioral estimates of tinnitus. The reduction in SFR in these awake rats was also found in awake cat (Zhang et al., 2011).

In anesthetized Sprague Dawley rats with an implanted wire-electrode array in auditory cortex, Stolberg et al. (2011) tracked CFs following salicylate application. This showed that a population of neurons with original CFs, both above and below 16 kHz, shifted their CF towards the tinnitus frequency region of the tonotopic axis (~16 kHz). The authors suggested that salicylate-induced tinnitus results from an expanded cortical representation of the tinnitus pitch determined by an altered profile of input from the cochlea (i.e. the relative decreased reduction in the DPOAEs around 16 kHz compared to the neighboring half octaves, see section 6.2.1). No SFRs were recorded.

Systemic injection of 250 mg/kg of salicylate, a dose that reliably induces tinnitus in rats, significantly reduced the sound-evoked output of the rat cochlea as measured in the amplitude of the CAP (Sun et al., 2009; Noreña et al., 2010; Lu et al., 2011; Stolzberg et al., 2011). Paradoxically, salicylate significantly increased the amplitude of the sound-evoked LFP from the AC of awake rats (see earlier), but did not affect the LFP in the IC. Salicylate also caused reorganization of the tonotopic map in auditory cortex (Stolzberg et al., 2011). A behavioral correlate of the salicylate-induced LFP enhancement in AC was found in the increased amplitude of the acoustic startle response. Direct application of salicylate (25 mg/mL) to the round window of the cochlea, however, reduced the response amplitude of the cochlea, as well as IC and AC, suggesting that the AC amplitude enhancement induced by systemic injection of salicylate does not originate from the cochlea. It also suggests that high doses of salicylate increase the gain of the central auditory system. The enhanced startle response may be a behavioral correlate of hyperacusis that often accompanies tinnitus and hearing loss. There could be a link to reduced 5-HT levels, caused by salicylate (section 6.3.2; see also Chapter 3).

These findings also allow a different interpretation of the results obtained by Yang et al. (2007) presented earlier. The positive startle response and other behavioral tests may well indicate that hyperacusis is induced by gain changes in the auditory system, which are not affecting SFR and are not indicative of the presence of tinnitus. The alternative is that subcortical increases in SFR are sufficient for the tinnitus percept to occur and to affect the startle response (the reflex circuit is entirely subcortical) as well as the scheduled polydipsia test. This all under the working assumption that tinnitus results from increased SFR.

Homeostatic mechanisms may also stabilize neuronal activity in the face of large changes in synaptic drive. For instance, neurons can scale the strength of excitatory synaptic inputs up or down in response to changes in activity (Turrigiano et al., 1998). The LFPs, being the compound EPSP at the recording site, will reflect this as shown in the Sun et al. (2009) study. However, homeostatic changes in synaptic activity are also expected to affect SFRs (Schaette and Kempter, 2006), but in contrast to what is expected SFRs decreased in the same structures where the LFP is increased (Yang et al., 2007). This suggests that the salicylate effects are likely not due to homeostatic mechanisms.

Changes in auditory system metabolism Changes in metabolic activity in AC and subcortical structures were evaluated in gerbils using [^{14}C]2-deoxyglucose (2-DG) autoradiograpgy after a 4-day treatment with salicylate (200–350 mg/kg) (Wallhausser-Franke et al., 1996). Enhanced metabolic activity likely reflects the increased glucose uptake by neurons with increased SFRs. Gerbils received a 2-DG injection 2 h after the last dose of salicylate and were sacrificed 90 min later for autoradiography. Glucose uptake in these gerbils was increased in AC but decreased in IC and other subcortical structures including CN and lateral lemniscus compared to that seen in saline-treated animals. Based on these data, the authors suggested that salicylate-induced tinnitus was reflected in the AC without enhanced activation of the IC. Provided that the animals were in a soundproof room after the 2-DG injection this effect pertains to spontaneous use of glucose and does not necessarily relate to the increased gain of driven activity in the auditory system shown by Sun et al. (2009). It is interesting though that the LFP in IC was not affected by salicylate.

Paul et al. (2009) investigated the pattern of neural activation induced by salicylate in central auditory structures of rats using PET imaging. Awake Sprague Dawley rats were injected with the ^{18}F-deoxyglucose (FDG) once in a quiet state (baseline) and once after application of salicylate (250 mg/kg). Tinnitus was verified using schedule-induced polydipsia avoidance conditioning. Potentially this technique may also be sensitive to hyperacusis. Brain imaging was performed using a high-resolution microPET scanner. Rats underwent structural MRI and reconstructed MRI and microPET images were fused to identify brain structures. MicroPET imaging showed that FDG activity in the frontal pole was stable between baseline and tinnitus conditions, suggesting it was metabolically inert during tinnitus. Bilateral IC and auditory cortices showed significantly increased FDG activity during tinnitus relative to baseline; activity in the IC and auditory cortices increased by 17% $\pm$ 21% and 29% $\pm$ 20%, respectively. FDG activity changes in the thalamus were not statistical significant. These results show increased metabolic activity consistent with neuronal activation bilaterally in IC and AC of rats during salicylate-induced tinnitus. This is in partial contrast with the results of Wallhäusser-Franke et al. (1996) where the IC did not show changes in 2-DG uptake. It also contrasts with the decrease in SFR recorded from an implanted wire-electrode array in the AC of the same species after administering the same dose of salicylate by the same research group (Yang et al., 2007). It also suggests that metabolic activity imaging reflects mainly synaptic events (such as reflected in LFPs) rather than spike generation (see Chapter 4).

Manganese-enhanced imaging Manganese-enhanced MRI (MEMRI) with intraperitoneally administered $MnCl_2$ can be used to examine synaptic neuronal activity *in vivo*. The paramagnetic Mn^{2+} ion can enter active neurons through voltage-gated Ca^{2+} channels. Since Mn^{2+} remains in the cell for extended periods of time, activity-dependent accumulation of the Mn^{2+} ion in brain regions can be measured hours later at high spatial resolution (Duong et al., 2000) as a decrease in the tissue T1 relaxation time (see Chapter 4). In this manner, MEMRI has been used to measure sound-evoked activity in the midbrain from awake and free moving rodents. For instance, Xu et al. (2008) used MEMRI to show the tonotopic organization of the mouse IC.

Holt et al. (2010) examined rats with pre-pulse inhibition of the acoustic startle response and MEMRI with $MnCl_2$ administered 8 h prior to imaging. They were repeatedly administered salicylate (300 mg/daily) or exposed for 4 h to a 1/3 octave band noise centered around 10 kHz presented at 118 dB. Salicylate application resulted in widespread and larger Mn^{2+} uptake compared to noise exposure. Neither model demonstrated significant differences in the auditory cortex. The activation in control animals was lowest in auditory cortex as well. Yu et al. (2005) also could not show sound-evoked activation of auditory cortex in mice using MEMRI, whereas there was uptake in the brainstem, midbrain, and thalamus. It is possible that intraperitoneal administration of $MnCl_2$ may either need additional time to reach the cortex or that $MnCl_2$ needs to be administered intravenously accompanied by a blood–brain barrier breaking component (e.g. mannitol) to reach the cortex (Duong et al., 2000). So the interpretation of the cortex results remains uncertain.

In the DCN, Mn^{2+} uptake was similar in noise-exposed and control groups, but significantly elevated in the salicylate-exposed group. In contrast, in the dorsal cortex of the IC Mn^{2+} uptake was elevated in animals with behavioral signs of tinnitus; and there was no significant difference between salicylate- and noise-exposed groups. Besides these two areas, no group differences in Mn^{2+} uptake were found for any other subcortical structures (Holt et al., 2010). It is strange that even 5 days after the noise trauma there is no sign of hyperactivity in the DCN, whereas under these conditions the SFR is dramatically increased (Chapter 7).

6.3 **Molecular changes**

6.3.1 **Ion channels**

The two main potassium-channel currents in neurons, the transient outward current ($I_{K(A)}$) and the delayed rectifier current ($I_{K(DR)}$), have been identified in IC neurons by their activation and inactivation voltage ranges and kinetics and pharmacological sensitivities. Changes in the function of these potassium channels directly affect the resting potential, specifically the repolarization after spiking, and the shape and firing rates of the action potentials. Liu and Li (2004a) found that salicylate inhibits both $I_{K(A)}$ and $I_{K(DR)}$ in Wistar rat IC neurons. Depression of $I_{K(DR)}$ by salicylate might be causing the increased excitability of IC neurons and may play a role in salicylate-induced tinnitus. Liu and Li (2004b) reported that salicylate also causes a concentration-dependent blockade of voltage-gated sodium channels and shifts the inactivation curve to more hyperpolarized potentials, which could also be related to the increased SFR found in IC.

Both neonatal and adult rat IC neurons express high threshold voltage-activated calcium channels. One of those, the L-type calcium channel mediates long-lasting calcium currents in response to depolarization. Calcium influx through L-type calcium channels leads to activation of a cascade of intracellular signals and is also responsible for the after-hyperpolarization phase following action potentials. In addition, calcium currents through L-type calcium channels regulate neurotransmitter release (Murakami et al.,

2002). Salicylate alters auditory processing within the IC by depressing these L-type calcium channels, which results in a loss of GABA-mediated inhibition and thus might contribute to the development of tinnitus (Liu et al., 2005). Corroborating this, adding the L-type voltage-gated Ca^{2+} channel blocker nimodipine to the drinking water attenuates the manifestation of tinnitus produced by salicylate or quinine (Jastreboff and Brennan, 1988; Jastreboff et al., 1991), and abolishes the salicylate-induced increase of average SFR and of burst activity of single IC neurons (Chen and Jastreboff, 1995).

Perfusion of cortical brain slices with 1.4-mM salicylate (Wang et al., 2006; Su et al., 2009) caused no significant change in current-evoked firing rates in pyramidal neurons; however, it reversibly depressed those in fast-spiking interneurons by up to 50%. This suggests that salicylate may raise cortical excitability by suppressing the fast-spiking inhibitory interneurons, which could contribute to tinnitus.

Kizawa et al. (2010) used the rat behavioral model of tinnitus for salicylate (Guiton et al., 2003) to assess the expression of the transient receptor potential cation channel superfamily V-1 (TRPV1). Animals received a single injection of saline or salicylate (400 mg/kg intraperitoneally). TRPV1 expression was significantly upregulated in spiral ganglion cells 2 h after salicylate injection. These findings suggest that the resultant increase in arachidonic acid products has the potential to depolarize spiral ganglion cells by activation of TRPV1.

6.3.2 Receptor and transmitter systems

NMDA receptors Using cultured spiral ganglion cells, Peng et al. (2003) showed that salicylate produced a selective potentiation of NMDA-mediated responses, and had little effect on AMPA- and kainite-mediated responses. The potentiation was dose dependent with a half-maximal concentration of 2.2 mM and could be blocked by NMDA receptor antagonists and Mg^{2+} ions. This potentiation of NMDA receptor-mediated responses directly leads to a change in the opening kinetics of voltage-dependent L-type Ca^{2+} channels. Singer et al. (2008) suggested that the observed dose-dependent effects of salicylate on the calcium response transcription factor (CaRF1) result from an indirect dose-dependent effect of salicylate on L-type Ca^{2+} channels in cochlear neurons. Potentially this could also affect those in IC (see previous section).

After demonstrating that cochlear perfusion with 5-mM sodium salicylate increased the SFR in auditory nerve fibers, Ruel et al. (2008) showed that application of the NMDA antagonist MK-801 together with salicylate prevented this SFR increase. Furthermore only perfusing with MK-801 had no effect on the SFR, suggesting that NMDA receptors are not involved in determining the SFR. However, whereas the AMPA receptor antagonist GYKI 53784 abolished SFR, subsequent application of sodium salicylate did evoke spiking activity. Thus, the combined application of MK-801 and sodium salicylate suggests that the NMDA receptors are enabled by the application of sodium salicylate. This is potentially the result of an increase in the cochlear arachidonic acid content by salicylate through its inhibition of cyclooxygenase. The increased level of arachidonic acid subsequently enables the NMDA receptors to respond to glutamate released by the IHCs.

GABA activity is the major inhibitory neurotransmitter of the central nervous system. The GABAergic function is altered after sensory deprivation and following auditory-evoked and drug-induced seizures. Hypersensitivity of the central auditory pathway occurs after partial deafferentation. Preliminary findings suggest that the GABA system in the IC may be impaired under such conditions (Abbott et al., 1999; Bledsoe et al., 1995; Suneja et al., 1998a,b; Wang et al., 1996). As we have seen earlier, salicylate depresses the action of cortical fast-spiking interneurons that are the main suppliers of inhibition to cortical pyramidal cells. Glutamate decarboxylase or glutamic acid decarboxylase (GAD) is an enzyme that catalyzes the decarboxylation of glutamate to GABA and CO_2 and occurs in two isoforms, GAD65 and GAD67. GAD65 is considered to have a reservoir function for local synaptic requirements since it is stored in axon terminals in predominantly inactive form. Bauer et al. (2000) using Western blotting found a significant 63% elevation of GAD levels in the IC of chronic salicylate-treated Long–Evans rats (8 mg/mL salicylate-treated drinking water for 4 months). The Western blot is an analytical technique that uses gel electrophoresis to separate proteins that are detected using antibodies specific to the target protein. This GAD-level elevation occurred in rats demonstrating behavioral evidence of tinnitus, and was accompanied by a significant reduction in the $GABA_A$ receptor affinity and by a significant reduction in the number of muscimol (a $GABA_A$ agonist) binding sites in IC. Tinnitus perception may thus develop because of altered activity within the cochlea (OHC motor proteins, arachidonic acid cycle leading to increased NMDA activity), altered GABAergic activity within the IC (increased GAD, decreased $GABA_A$ receptor affinity), or a combination of the two.

Glycine activity is an inhibitory neurotransmitter mainly found in the spinal cord, brainstem, and retina. Lu et al. (2009) using a cell culture from the IC of newborn Wistar rats, found that salicylate effectively inhibited the current mediated by glycine receptors containing α1-subunits in a non-competitive manner. They speculated that salicylate serves as an allosteric modulator that putatively binds in the transmembrane region, although the pattern of the current–voltage curve may suggest direct interference of salicylate with ion flow as a kind of negative amphiphile. An allosteric modulator is a drug that changes the activity of a receptor indirectly via activation of an allosteric site (that is, a site other than the protein's active site) on the protein. Only when applied simultaneously with glycine did salicylate produce this antagonism. Salicylate is thus likely a non-competitive antagonist specifically on glycine receptors containing α1-subunits.

6.3.3 Hormones and neuromodulators

Serotonin activity (5-HT) transmission provides a major modulating factor within the auditory system. Caperton and Thompson (2011) showed that salicylate treatment (350 mg/kg) caused a significant increase in the number of *c-fos*-expressing serotonergic neurons in Mongolian gerbils. The increase was significant for three of the eight major serotonergic cell groups including B7, most of which are positioned along the midline or raphe of the brainstem. The DCN and the IC receive direct inputs from serotonergic neurons in the B7 cell group. In the forebrain, in general, ascending serotonergic innervation is greatest in thalamic nuclei associated with limbic functions including the amygdala and hippocampus. Serotonergic innervation is likely primarily active in the modulation of sound perception or determining the significance thereof. It has been shown in rats that 5-HT activation in sensory neurons increases with age (Cransac et al., 1996). These authors suggest that it may compensate for age-related dysfunction of sensory input and processing. Potentially, the perception of tinnitus could be linked to a dysfunction of 5-HT, inducing hyperacusis, at one or more levels in the CNS (Chapter 3). Liu et al. (2003) used microdialysis to measure extracellular concentrations of neurotransmitters in discrete brain regions. They found that 5-HT levels in IC and AC significantly increased 2–3 h after salicylate application, while saline did not produce any clear changes. This suggests that the increased 5-HT levels in IC and AC may be (indirectly) involved in salicylate-induced tinnitus generation. Extracellular glucose and lactate levels in IC and AC that reflect neural activity also increased after salicylate. They interpreted this as salicylate inducing an increase of spontaneous neural activities in IC and AC. However, this interpretation contradicts the findings by Yang et al. (2007) of reduced SFR in AC following salicylate administration.

Since serotonin-releasing fibers preferentially innervate inhibitory GABA neurons, it is possible that sodium salicylate causes an imbalance between inhibition and excitation through its effect on the serotonergic modulation of the GABAergic synaptic transmission. Wang et al. (2008) showed that perfusion of 40 μM 5-HT robustly enhanced both frequency and amplitude of GABAergic spontaneous inhibitory postsynaptic current (sIPSCs) and this 5-HT-induced enhancement of GABAergic sIPSCs could be suppressed by 1.4-mM sodium salicylate. Tetrodotoxin (0.5 μM) produced a similar effect as sodium salicylate, suggesting that sodium salicylate suppresses the 5-HT-induced enhancement of GABAergic sIPSCs through depressing spontaneous action potentials of GABA neurons. Thus, sodium salicylate may affect the normal level of GABAergic synaptic transmission maintained by the serotonergic system in sodium salicylate-induced tinnitus. Consequently, sodium salicylate could tip the balance between inhibition and excitation, and thereby causes hyperexcitability and facilitates the excitatory synaptic transmissions in the central auditory system with a decrease in the overall SFR of GABA neurons. This is accompanied by a smaller increase in the SFR of pyramidal neurons.

Serotonergic activity increases the perception of chronic pain and phantom limb pain and may play a similar role in the perception of tinnitus (Simpson and Davies, 2000; Holgers et al., 2003). The serotonin receptor agonist 1-(3-chlorophenyl) piperazine (mCPP) increases anxiety in humans and animals and exacerbates the behavioral perception of salicylate-induced tinnitus (Guitton et al., 2005). Serotonin has also been implicated in central hyperacusis (Marriage and Barnes, 1995). Importantly, serotonin activity is affected by the circadian rhythm (Thomas, 2006), suggesting that time of day could affect the tinnitus percept. The release of 5-HT increases during waking and decreases during sleeping. This may explain why tinnitus is commonly reported to be at its worst at waking up.

The relevance of the serotonergic system to tinnitus perception underlines the central aspects of tinnitus. Paradoxically, drugs (such as selective serotonin reuptake inhibitors,

SSRIs) that boost 5-HT concentrations in the brain can successfully relieve tinnitus in some patients (Folmer and Shi, 2004), suggesting that the downstream signaling pathway after activation of 5-HT receptors is somewhat impaired in tinnitus. The predominant projections from serotonin neurons to GABA neurons (Smiley and Goldman-Rakic, 1996) and the robust enhancing effects of 5-HT on GABAergic sIPSCs (Wang et al., 2008) raise the possibility that this downstream signaling pathway may be located in GABA neurons. Tinnitus may result if this signaling pathway is impaired by hearing loss by aging or by a tinnitus-inducing drug such as salicylate. In fact, impaired functioning of GABA neurons in tinnitus is suggested by the reduction in the number of GABA neurons in aged animals (Caspary et al., 1990) corresponding with the increased prevalence of tinnitus with aging. Thus, the serotonergic system may actually play a role in the perception of tinnitus through modulating the GABAergic system. This is supported by the report that 5-HT produces much more robust enhancement for inhibitory synaptic transmissions than for excitatory synaptic transmissions in the cerebral cortex (Zhou and Hablitz, 1999).

6.3.4 **Immediate early genes**

Immediate early genes (IEGs) are induced rapidly inside nerve cells by extracellular stimuli without the need of intermediate proteins. *c-fos* is expressed in neurons mainly after novel depolarizing stimuli, whereas glial cells do not express *c-fos* following depolarization. *c-fos* expression is, therefore, considered to be a marker of neuronal activity at the single-cell level. c-fos regulates the expression of the kainic acid receptor GluR6 and the brain-derived neurotrophic factor (BDNF). Fos is a transcriptional factor expressed by *c-fos*.

c-fos Mongolian gerbils that received salicylate (350 mg/kg), a few hours after salicylate administration, showed c-fos expression in auditory brainstem nuclei that was not significantly different from saline treatment. Pronounced differences between groups were found, however, in areas susceptible to stress, with many immunoreactive cells in the locus coeruleus, the midbrain periaqueductal gray, and the lateral parabrachial nucleus (Walhäusser-Franke, 1997). This suggests that salicylate may evoke tinnitus through a combined effect on auditory and non-auditory brain nuclei. More extensive data were presented in Walhäusser-Franke et al. (2003) for both salicylate and impulse noise (Figure 6.4) and also included structures above the brainstem.

In the auditory system, salicylate injections as well as noise trauma always initiated *c fos* expression in AC and sometimes in dorsal MGB (for salicylate), in IC (for noise and low dose of salicylate) or in DCN (noise). In IC slightly more immunoreactive neurons were observed in the group subjected to the low dose of salicylate whereas incidences in the group injected with the high dose were comparable to those of controls. More Fos-expressing cells in IC were also found shortly after noise exposure possibly due to acoustic activation of neurons. In dorsal MGB, labeled neurons were sometimes observed after injections of a high dose of salicylate, which coincides with the finding of increased activation evidenced in MGB with the 2-DG method (Wallhäusser-Franke et al., 1996).

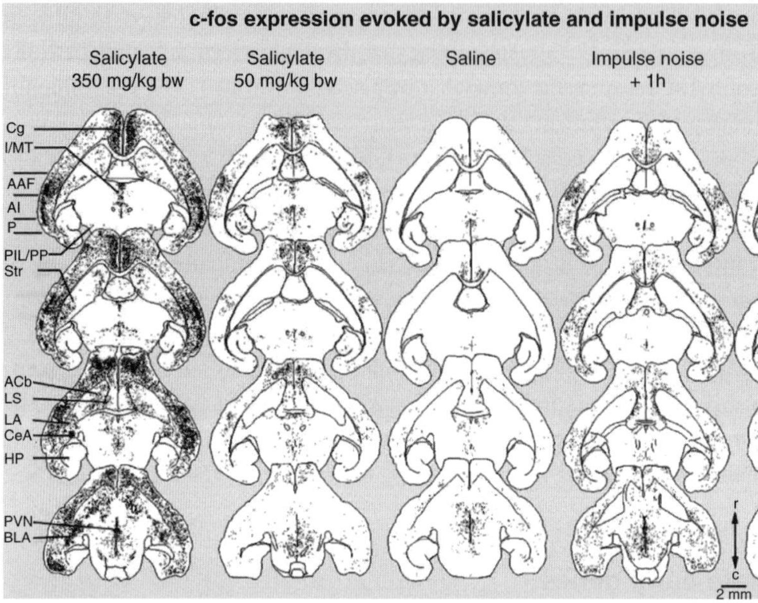

Figure 6.4 Computer-generated drawings of horizontal sections arranged in dorsoventral order from top to bottom. Each dot represents a *c-fos*-expressing cell. In the auditory system, numbers of labeled cells are consistently enhanced in auditory cortex (AI, AAF, P) after salicylate injections and noise exposure compared to saline controls. (CeA, LA, BLA central, lateral, basal nucleus of the amygdale; ACb, nucleus accumbens; LS, lateral septum; I/MT, thalamic intralaminar and midline nuclei; Cg, cingulate cortex; HP, hippocampus; PIL, posterior intralaminar nucleus; PP, peripeduncular nucleus; Str, striatum; c, caudal; r, rostral). With kind permission from Springer Science + Business Media: Wallhausser-Franke, E., Mahlke, C., Oliva, R., Braun, S., Wenz, G. and Langner, G. (2003). Expression of c-fos in auditory and non-auditory brain regions of the gerbil after manipulations that induce tinnitus. *Experimental Brain Research*, **153**, 649–54, Figure 1.

Staining intensity, however, did not differ much from background and stained cells were not consistently observed in all animals of this group. Also, Fos-immunoreactive cells were never observed in MGB after noise exposure. In contrast, the number of Fos expressing neurons always increased in the different fields of AC after salicylate injections as well as after noise exposure. Moreover, the number of stained neurons increased proportional to the injected dose of salicylate and Fos expression was still elevated 7 h after noise exposure. In the amygdale, Fos expression was negligible after injections of low doses of salicylate, which presumably are insufficient to induce tinnitus or cause a hearing deficit. Fos expression was elevated in limbic brain regions regulating emotion and attention (central, lateral, basal nucleus of the amygdala, nucleus accumbens, and lateral septum), in regions processing affective-motivational aspects of pain (thalamic intralaminar and midline nuclei, cingulate cortex); and in regions controlling autonomous and endocrine functions Walhäusser-Franke et al. (2003).

Wu et al. (2003) administered rats with five daily doses of sodium salicylate (intraperitoneal 250 mg/kg). On day 6 the rats were placed inside a soundproof room for 8 h before sacrifice. Immunohistochemistry showed a significant increase in the number of Fos-positive cells in IC, particularly its central division. Only a few Fos-stains were found at the DCN while no Fos-stain appeared at the VCN. Results are consistent with the hypothesis that salicylate-induced tinnitus is related to overactivity of cells at the IC. A discrepancy in these results is the absence of an increased Fos-expression at the DCN whereas hyperactive DCN cells were found in models of tinnitus following ototoxic drug treatment or loud sound exposures (Kaltenbach and Afman, 2000; Kaltenbach et al., 2000; Brozoski et al., 2002).

Arc/Arg3.1 The activity-regulated cytoskeleton-associated protein/activity-regulated gene *Arc/Arg3.1* is an IEG that is dynamically regulated by neuronal activity and is tightly coupled to behavioral encoding of information in neuronal circuits. *Arc/Arg3.1* expression has been directly correlated with BDNF-induced plasticity changes (Chapter 1). In particular, *Arc/Arg3.1* expression is upregulated, and transported to dendrites following synaptic stimulation. *Arc/Arg3.1* mRNA accumulates there at sites of synaptic activity and is required together with BDNF for long-term potentiation (LTP) and long-term memory (Plath et al., 2006). Arc/Arg3.1 particularly mediates activity dependent homeostatic synaptic scaling of AMPA receptors (Shepherd et al., 2006). Deletion of Arc/Arg3.1 has been shown to be essential for cortical homeostatic scaling and restoration of synaptic strength post sensory deprivation. In Arc/Arg3.1 knock-out mice, the basic mEPSCs are increased in pyramidal neurons of the cortex due to failure of scaling of AMPA receptors (Gao et al., 2010).

BDNF activity is a small protein that acts by binding to receptors such as tyrosine kinase TrkB. BDNF promotes survival through inactivation of components of the cell death machinery and also through activation of the transcription factor cAMP-response element binding protein (CREB), which drives expression of the prosurvival gene *Bcl-2*. Synaptic scaling, a homeostatic mechanism, is mediated in part by the activity-dependent release of the neurotrophin BDNF (Rutherford et al., 1998). BDNF is produced by cortical pyramidal neurons, and the high-affinity BDNF receptor TrkB is present on both pyramidal neurons and interneurons. Calcium plays an important role in the regulation of neuronal gene expression. The first step in this calcium regulation is the influx of calcium into the cytoplasm. The transcription of BDNF is preferentially driven by calcium influx through L-type voltage-sensitive calcium channels (West et al., 2001). NMDA receptors as well as L-type Ca^{2+} channels are expressed in cochlear neurons (Niedzielski and Wenthold, 1995). BDNF activates synaptic consolidation through transcription and rapid dendritic trafficking of messenger ribonucleic acid (mRNA) that is encoded by the IEG *Arc/Arg3.1* (Bramham and Messaoudi, 2005). BDNF is a key modulator of neuronal plasticity events. Up- or downregulation of BDNF is dependent on the degree of stimulation (Tabuchi et al., 2000). Spontaneous activity can also trigger BDNF trafficking to activated synapses and support its release from synapses. The *BDNF* gene consists of eight upstream exons (I–VII), each of which has a distinct promoter and can be independently spliced to the coding exon IX. *BDNF* exon IV expression is regulated via interaction between calcium-response elements CaRE1, CaRE2, and CaRE3/Cre (CaREs) that are bound by the transcription factor *CaRF1*, upstream stimulatory factors 1 and 2 (*USF1/2*), and *CREB*, respectively (Aid et al., 2007).

To determine whether the salicylate-induced changes in cochlear *BDNF* exon IV expression include a differential use of the CaRE binding proteins, Singer et al. (2008) studied the level of these binding proteins in the spiral ganglion neurons before and after systemic application of concentrated salicylate using *in situ* hybridization and RT-PCR. They found that *BDNF* exon IV and *CaRF1* expressions were upregulated after application of salicylate, whereas *USF1/2* and *CREB* mRNA expression remained unaffected. The changes in *BDNF* exon IV and *CaRF1* expression were also dose-dependent. The data show Ca^{2+} and CaRF1 as messengers of trauma- or salicylate-induced altered BDNF levels in the cochlea.

Salicylate application (Panford-Walsh et al., 2008) induced an IEG expression pattern similar to that found after noise trauma (Tan et al., 2007; Chapter 7): *BDNF* mRNA expression was increased in the spiral ganglion neurons of the cochlea and *Arc/Arg3.1* expression was significantly reduced in AC. Local application of the $GABA_A$ receptor modulator midazolam resulted in the reversal of salicylate-induced changes in cochlear *BDNF* expression and cortical *Arc/Arg3.1* expression, and reduced tinnitus perception in the animal model.

6.4 **Summary**

Salicylate is a tinnitus-inducing agent, either following a single high dose or following repeated administration of low dose. The result is predictable and salicylate has been much applied in animal experiments. Salicylate interacts with the auditory system in multiple ways in the cochlea and in the central auditory system (Table 6.1). In the cochlea it initially downregulates the action of prestin in the wall of the OHC and thereby causes a hearing loss. In addition salicylate interacts with the arachidonic acid cycle ultimately causing an increase in NMDA receptor activity and increased spontaneous firing rates in

Table 6.1 Changes after salicylate application

Structure	Cell density	SFR	2-DG	Glu	Gly/GABA	5-HT
OHC IHC	≈					
ANF	⇓ (chronic)	≈[4] ⇑[5]				
DCN		⇓ (FF)[1] ≈ (CW)[1]	⇓	⇑	⇓	
ICC		⇓	⇓(2-DG) ⇑ (FDG)			⇑
ICX		⇑	⇑			
AI		≈[2] ⇓[3,6]	⇑			⇑
AII		⇑	⇑			

[1]Superfusion in slice (Wei et al., 2010b); [2]cat (Ochi and Eggermont, 1996); [3]rat, (Yang et al., 2007); [4] (≤200 mg/kg, acute); [5] (≥400 mg/kg; chronic); [6]cat (Zhang et al., 2011) FF, fusiform cells; CW, cartwheel cells.

a subset of ANF. Long-duration application reverses its action on prestin and actually enhances its expression. Centrally, salicylate downregulates serotonin and GABA activity, and affects the conductivity of some K^+ channels. Cochlear perfusion of salicylate does not produce the central effects of systemically applied salicylate. This makes searching for neural substrates of tinnitus difficult at the least. Salicylate also increases the gain of the more central parts of the auditory system for sound, reflected in increased startle responses and potentially inducing hyperacusis. So it is not clear what enhanced gap-startle responses after salicylate application imply: tinnitus or hyperacusis? This also may depend on the presence or absence of auditory cortical activity modulation of the gap-startle reflex (Chapter 5). As far as spontaneous firing rates are concerned, high levels of salicylate result in increased rates in ANF but presents variable results in the IC and AC.

Chapter 7

The sensorineural hearing loss model of tinnitus

The sensorineural hearing loss (SNHL) model of tinnitus has a clearly defined peripheral substrate in hair cell damage, diffuse or localized ganglion cell degeneration, and frequency-specific reduced auditory nerve output to the central auditory system. This type of hearing loss can be due to noise exposure or the use of ototoxic drugs such as aminoglycosides (Kiang et al., 1976), but relatively few studies on ototoxicity-based tinnitus have been conducted. In addition, hearing loss can be induced experimentally by unilateral or bilateral cochlear ablation, and we will review several studies using this popular experimental method. Less severe forms of hearing loss can be introduced by removing middle ear ossicles, thereby creating a 40–60-dB conductive hearing loss without damaging the cochlea or auditory nerve. Cochlear ablation, which destroys the cochlea and the auditory nerve, is a model of a complete sensorineural hearing deficit. Noise-induced hearing loss (NIHL) and loss due to ototoxic drugs typically present a partial, mostly high-frequency, SNHL. Unilateral cochlear ablations create an imbalance between ipsi- and contralateral input to binaural nuclei in the brainstem and further along the auditory pathway. NIHL typically results in an imbalance of spontaneous as well as sound-driven cochlear output across frequency to the central auditory system. In profound conductive hearing loss, the occurring central gain change may affect the entire frequency range of the auditory system.

Laboratory studies have shown that tinnitus may develop in humans almost immediately after exposure to loud traumatic sounds. Animal studies can be used to discover the neural substrates related to such early onset, and often, transient tinnitus. After traumatic noise, prolonged exposure to occupational or recreational noise or following slowly acquired losses during aging, tinnitus may over time develop from an intermittent presence to a chronic status, and likely acquire a dominant central contribution.

7.1 Structural changes in the auditory system following noise trauma

7.1.1 Hair cells and auditory nerve fibers

Laboratory investigations into the effects of acoustic trauma started in the early 1900s (reviewed by Kemp, 1935) and showed, for instance, that cochlear pathologies were the primary cause of NIHL (Lurie et al., 1944). Noise-induced TTSs and the associated temporary loudness recruitment (Chapter 3) were also studied in humans (Davis et al., 1950). They already concluded that: “The audiogram alone is not an adequate measure of the impairment of auditory function.” One could make this finding specifically apply to tinnitus; the amount of hearing loss is likely not an adequate measure to predict either the presence or the loudness of tinnitus. We have seen in Chapters 3 and 4 that the distress aspect of tinnitus relates to factors outside the auditory system. The effect of loud noise on the cochlea also depends on the type of noise used. Impulse and impact noise, such as gunfire, are characterized by high intensity and short duration, and may produce immediate mechanical alterations to the cochlea. Continuous exposure at moderate-to-severe levels of noise typically produces more subtle changes. Guinea pigs exposed to a ~4-kHz pure tone with an intensity varying between 108–120 dB SPL for 22.5–360 min showed deformation of stereociliary bundles on OHCs and IHCs that were already visible 5 min after trauma, whereas after 4 h stereocilia were found missing. Hair cell loss was not obvious at that time but became clear at 5 days post-trauma (no intermediary measures were obtained; Fredelius et al., 1988).

Most of the 40-dB threshold shifts found in noise-exposed cat ears (~110 dB SPL for 2 h) could be accounted for by loss or damage to sensory cells that was clearly visible under the light microscope (Liberman and Beil, 1979). Of all the histological features that were evaluated for hair cells, the orderliness of the stereocilia, on both IHCs and OHCs, showed the closest correlation with single-unit thresholds. Cochleas with permanent NIHL of 40–60 dB were analyzed in detail, first at the light-microscopic level, and subsequently with transmission electron microscopy of serial sections. Transmission electron microscopy revealed no noise-trauma induced pathology in any parts of the organ of Corti except for the stereocilia. Thus, most structures that appear normal in a careful light-microscopic analysis, are also normal at the ultrastructural level (Liberman, 1987).

Scanning electron microscopy examination of guinea pig cochleas for structural damage immediately after a 1-h exposure to a pure tone ranging from 96–129 dB SPL showed little hair cell loss, but widespread damage to the stereocilia, especially those on the IHCs and the first-row of OHCs. The order of damage to receptor cells with increasing sound intensity was the first row of OHCs, followed by the IHCs, then the second and third rows of OHCs (Robertson and Johnstone, 1980). In general, the OHCs are more vulnerable to noise trauma than the IHCs, regardless of the type of noise. This susceptibility may be an inherent property of the OHC biochemistry (Saunders et al., 1985) since there is a relation between the pathology seen with noise exposures and the vulnerability of OHCs to ototoxic drugs (Slepecky, 1986). As stereocilia damage worsens from disarray, to partial fusion or loss, to total fusion or loss, the threshold shift increases. Further cellular impairment involves protein, lipid, and glucose synthesis needed for cell repair and survival, and such impairment would result in permanent cell injury or cell death, leading to permanent threshold shift (Lim, 1986). Neurotoxic aspects of noise trauma are caused by excessive release of glutamate, influx of large quantities of Ca^{2+} ions in the postsynaptic area, and disruption of the synapse. This can recover over a time period of about 1 week (Puel et al., 1997).

Retrograde degeneration of auditory neurons has been studied for cochlear damage caused by acoustic trauma, ototoxic drugs, hereditary-progressive deafness, and others. The nerve degeneration starts only when the peripheral dendrites to the IHCs are irreversibly damaged. This is a peculiar phenomenon in nerve degeneration that is apparently only found in the cochlear nerve (Spoendlin, 1976). The integrity of its peripheral terminal portion, and likely also the ribbon synapse (see later), seems to be essential for the neuron to survive. Without the peripheral dendrite synapsing on the IHCs there is no action potential activity in the ANF and that obviously is required for their viability. Retrograde degeneration starts almost immediately after rupture of the dendrites that synapse with the IHCs, and proceeds within a few days through the osseous spiral lamina to the spiral ganglion, where it is temporarily halted. After 3 weeks the great majority of nerve fibers in the osseous spiral lamina have entirely disappeared but there is only a slight reduction of ganglion cells in the spiral ganglion. It takes about 3 months before degeneration of the ganglion cells occurs on a larger scale and the number of surviving ganglion cells decreases in a relatively short time. When only the peripheral receptor is damaged, as in acoustic trauma, retrograde degeneration never affects all neurons (Spoendlin, 1976).

When acoustic trauma elevates the ANF thresholds, the shape of their frequency-tuning curves is generally abnormal (Liberman and Kiang, 1978; Salvi et al., 1983). A strong correlation was found between cochlear regions showing clumping of the IHC stereocilia and CF regions showing V-shaped tuning curves accompanied by a reduction in SFR. Selective damage to the OHCs, on the other hand, was typically not accompanied by an overall depression of SFRs, indicating that SFR only depends on spontaneous transmitter release in the IHCs. Bauer et al. (2007) examined the type and extent of cochlear damage that occurred after an acoustic trauma that was sufficient to induce chronic tinnitus in rats. The rats were exposed unilaterally for 1 h to an octave-band noise centered at 16 kHz with a peak level of 110 dB SPL. Cochlear damage was assessed 6 months after the exposure. There was minimal loss of IHC and OHCs in the exposed cochleas of rats that demonstrated behavioral evidence of tinnitus (Chapter 5). However, a significant loss of large-diameter nerve fibers, i.e. high SFR fibers (Geisler, 1998), in the osseous spiral lamina of traumatized rats was observed. This could explain the average reduction in SFR noted earlier.

Using confocal imaging of the inner ear in the mouse, Kujawa and Liberman (2009) again showed that acoustic overexposure for 2 h with an 8–16 kHz band of noise at 100 dB SPL, caused a moderate, but completely reversible, threshold elevation as measured by ABR. The absence of permanent changes in the otoacoustic emissions indicated that the exposure left OHCs, and therefore likely the IHCs as well, intact. They found that despite the normal appearance of the cochlea and hearing thresholds there was an acute loss of the ribbon synapses between IHCs and ANFs followed by a delayed progressive diffuse degeneration of the cochlear nerve (Figure 7.1).

7.1.2 The central nervous system

Noise trauma Adult chinchillas exposed to an octave-band noise centered at 4 kHz for 105 min at a level of 108 dB SPL showed, besides cochlear damage, partial deafferentation of the ipsilateral CN. Most of this loss presumably resulted from the above-mentioned degeneration of the ANFs. New growth of axons and axonal endings was observed in the CN following the deafferentation. The occurrence of neo-synaptogenesis in the adult auditory system is consistent with the observation that ablation of the cochlear nerve in adult guinea pigs may induce a reorganization of synaptophysin (a vesicle membrane protein)-stained endings in the ipsilateral anteroventral cochlear nucleus (AVCN; Bilak et al., 1997). The adult chinchillas that were allowed to survive after this exposure for 16 days or for 1, 2, 4, and 8 months showed axonal degeneration in the DCN that was only

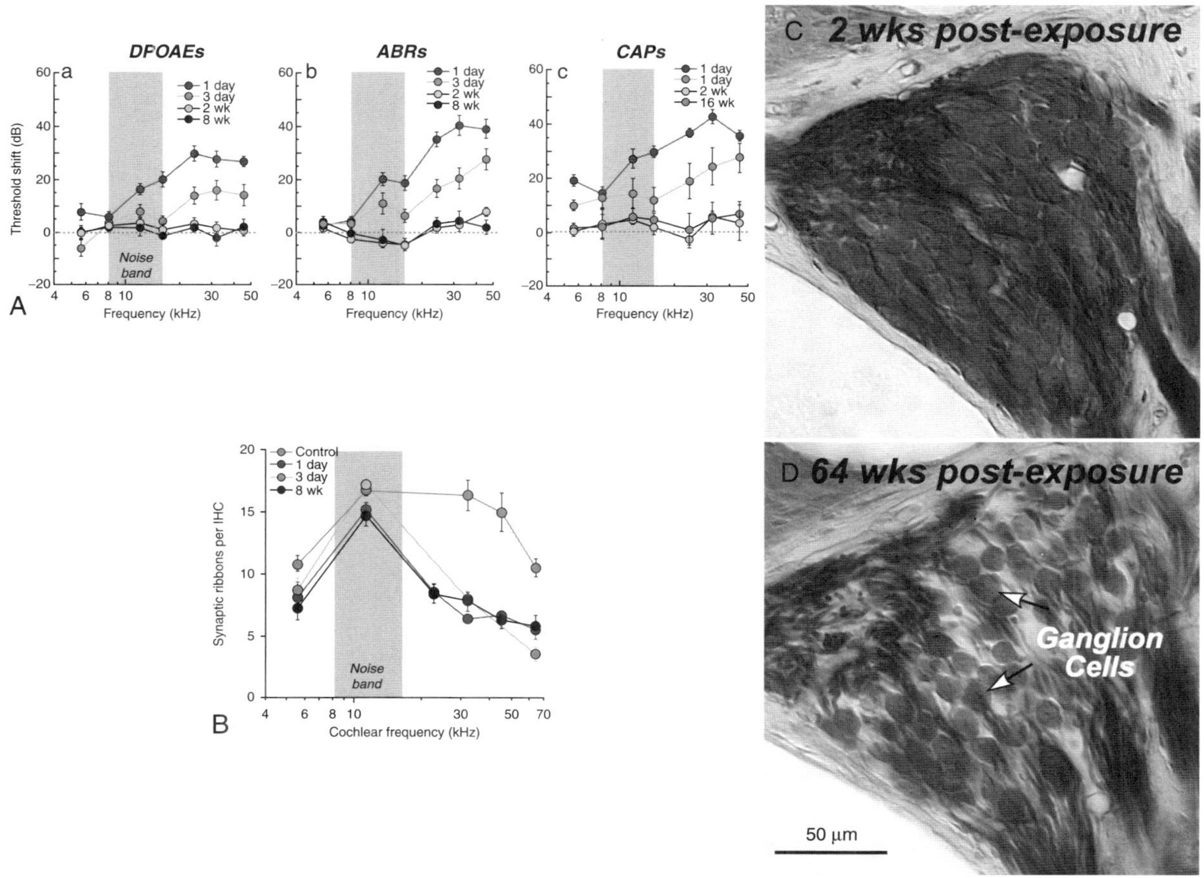

Figure 7.1 (Also see Color plate 3.) Noise trauma that only evokes temporary thresholds shifts and temporary changes in otoacoustic emissions (A) can still result in loss of synaptic ribbons (B) and delayed loss of ganglion cells (C and D). Compiled from Kujawa, S.G. and Liberman, M.C. (2009) Adding insult to injury: cochlear nerve degeneration after "temporary" noise-induced hearing loss. *Journal of Neuroscience*, **29**, 14077–85. © 2009, The Society for Neuroscience, with permission.

visible at 16 days post-exposure and not at longer survival times. Meanwhile, ANF degeneration continued to extend basally in the cochlea, and 2 weeks to 2 months later was followed by spread of axonal degeneration into the corresponding high-frequency region of the ventral cochlear nucleus (VCN; Morest et al., 1998). Following a 3-h exposure to the same sound, the cochlea and the cochlear nuclei exhibited degeneration of hair cells and axons that were investigated over periods of 7–150 days after the trauma. These findings could again best be explained by degeneration of synaptic endings followed by new growth of terminals (Muly et al., 2002). Freshly occurring synaptic degeneration appeared in the period from 1–16 weeks. After several months, however, all these changes reversed and eventually the endings recovered their normal appearance (Kim et al., 2004a). For periods of 6 and 8 months after a single exposure to a damaging noise level, a chronic, continuing process of neurodegeneration involving excitatory and inhibitory synaptic endings was observed. This neurodegeneration was again accompanied by newly formed synaptic endings, which repopulated some of the sites vacated previously by axosomatic endings on globular bushy cells in the AVCN. Noise-induced hearing loss thus may progress as a neurodegenerative disease with the capacity for synaptic reorganization within the cochlear nucleus (Kim et al., 2004b; Kujawa and Liberman, 2009). After noise exposure and recovery for up to 32 weeks, neuronal cell bodies lost both excitatory and inhibitory endings at first and later recovered a full complement of excitatory but not inhibitory terminals (Kim et al., 2004c). This pattern of change is consistent with and may provide a structural basis for the enhanced excitability of CN neurons, the relative deficits in inhibition, and the elevation of SFRs reported after noise-induced cochlear damage.

The role of glial cells in noise-induced hearing loss and recovery was elucidated by (Smith et al., 2002). After exposing adult mice with 4–16-kHz band-pass filtered noise at 115 dB SPL for 6 h. nearly all OHCs disappeared, while IHC and fiber loss was restricted to the frequency range of exposure. Fibroblast growth factor (FGF) staining in the CN showed hypertrophied astrocytes in the regions of nerve fiber degeneration only. The immunostaining peaked at 14 days, and was back to control levels by 60 days. FGF receptor staining of neurons occurred equally in all mice, exposed or not. This result is consistent with the hypothesis that the FGFs are upregulated in response to the synaptic degeneration following acoustic trauma and do play a role in the subsequent regrowth of neuronal processes in the CN.

Nerve degeneration continues central from the CN. Basta et al. (2005) exposed mice to noise (10-kHz center frequency at 115 dB SPL for 3 h) at 21 days of age under ketamine anesthesia (which, being a non-competitive NMDA antagonist, may potentially reduce the amount of neurotoxicity-based hearing loss). One week after the exposure the mice showed a significant threshold shift in the ABR over their entire frequency range. Cell density was significantly reduced in all subdivisions of the MGB and in layers IV–VI of primary auditory cortex (AI). No effects of the sound exposure were detectable in layer I of AI where projections of the magnocellular part of the MGB and the brainstem reticular activating system terminate. Effects on other layers were not reported.

Ossicular removal and cochlear ablation Unilateral ossicle removal also resulted in degeneration of the cochlear nerve and initiated fiber degeneration in central auditory nuclei (Potashner et al., 1997). Fiber degeneration in the CN was not evident at 7 or

70 days, suggesting that ossicle removal did little or no direct damage to the IHCs and the cochlear nerve. After 112 days, degenerating fine fibers and granulated axons were sparsely distributed in the CN bilaterally, but were more abundant on the ipsilateral side. Degeneration could not be shown in the superior olivary complex (SOC). Ablation of the left cochlea, in contrast, showed after 7 days an abundance of degeneration in the ipsilateral CN, consistent with the destruction of the left cochlear nerve. Dense degeneration of large, intermediate, and fine fibers was visible in the ipsilateral AVCN and PVCN. In the DCN, degeneration was heaviest adjacent to the dorsal acoustic stria, whereas most of the deeper part of the fusiform cell layer showed only moderate to sparse degeneration. The superficial part of the fusiform cell layer and the molecular layer showed no degeneration. Fiber degeneration was not apparent in the CN contralateral to the ablated cochlea.

This exemplifies that hearing loss in itself without cochlear damage (ossicle removal) still results in some modest degeneration in the CN, but only after several months. Ablation starts the degeneration process in the CN within a week.

7.2 Physiological and neural changes

7.2.1 Ossicular removal and cochlear ablation

Complete cochlear destruction results in an immediate disappearance of almost all activity in the ventral cochlear nucleus, while the activity in the dorsal cochlear nucleus is relatively unaffected even in chronic preparations (Koerber et al., 1966). Immediately after conductive hearing loss, there was a significant increase in the SFR of VCN neurons over the first 8 h (from 28.3 sp/s to 50 s/s) that declined with time but did not yet reach normal values by 14 days (Sumner et al., 2005). Bledsoe et al. (2009) found that, following partial cochlear ablation, after ~1.5 h in 61% of the recording sites in the VCN the SFR increased by a mean value of 49%. There is thus a delay in the increase in SFR following reduction of cochlear input to the VCN, even following a conductive hearing loss where the SFR of ANFs is likely unaffected. Thus, the increased spontaneous input likely results from a relatively fast central gain change.

7.2.2 Acute effects of noise trauma

In the distant past, when it was not yet realized that even short exposures to loud noise could have dramatic long-term effects such as ANF degeneration (Kujawa and Liberman, 2006, 2009; Bauer et al., 2007), humans were exposed for 5 min to a 110-dB SPL, 1/3 octave band of noise centered at 2, 3, 4, or 6 kHz respectively (Atherley et al., 1968). The subjects typically experienced tinnitus immediately after the exposure with a pitch at respectively 3, 3.9, 4.9, and 6.5 kHz. The tinnitus pitch was approximately one critical band lower than the frequency of maximum TTS found at 3.5, 4.5, 5.6, and 8.0 kHz respectively, and which in turn was approximately ½ octave higher than the trauma tone frequency (TTF). The critical band correspond roughly to the equivalent rectangular bandwidth $ERB = 24.7 \times (4.37 \times F + 1)$. Where the ERB is in Hz and F is the center frequency in kHz (Glasberg and Moore, 2000). The results were comparable to those obtained in another early human exposure study by Loeb and Smith (1967). There was no information about the potential long-term and remaining effects in these exposed people.

A comparable exposure of a chinchilla for 3–5 min to a 90–105-dB SPL tone located ½ octave above a VCN unit's CF that was recorded from, was used to investigate possible mechanisms of this immediate and often transient tinnitus (Boettcher and Salvi, 1993). The vast majority of cases of enhanced stimulus-driven firing rate occurred in units with sideband inhibition, and increased firing rates typically occurred when there was a concurrent loss of inhibition resulting in unmasking of excitatory activity. In a similar experiment but now comparing recordings from the same units before and after exposure in the ICC in chinchillas, cochlear trauma was induced by exposure to a 15–25-min, 95–115-dB SPL, pure tone at a frequency above the neuron's CF. Units with V-shaped tuning curves, representing ~90% of the sample, were generally unaffected by the traumatizing exposure. Overall, there was no significant trend in the SFR observed before and after the exposure (Wang et al., 1996). Van Heusden and Smoorenburg (1983) exposed cats for half an hour to pink noise at 105 dB SPL, producing an average threshold shift of 30 dB (maximum 50 dB) in the 2–6-kHz region. The post-exposure SFR in AVCN was not significantly different from the pre-exposure values.

Slightly longer exposure durations appeared to have effects on the SFR in cat auditory cortex. Kimura and Eggermont (1999) assessed the changes in SFR by simultaneous recording of the same units in AI, AAF, and AII of cats before and after a 30-min exposure to a 93–123-dB SPL (depending on the frequency) pure tone. Changes in SFR were most pronounced when the TTF and the CF were <1 octave apart. For this condition the mean SFR ratio (after/before) for AI was ~2 and significantly higher than 1. For AII the ratio was equal to 0.63 that was significantly lower than 1. In AAF, the mean ratio was 0.86 and not significantly different from 1. Thus the spontaneous activity in these three cortical areas was affected differently by the noise trauma. LFP amplitude in AI showed an enhancement after the noise trauma, those in AAF and AII did not as clearly.

Changes in the neural activity in cat AI occurring within a few hours after a 1-h exposure to a 120-dB SPL pure tone (5 or 6 kHz) were further assessed by recording, with two 8-microelectrode arrays, from the same multiple-single-unit (MSU) clusters before and after the trauma (Noreña et al., 2003; Noreña and Eggermont, 2003). Immediately after the exposure, the SFR was not significantly changed (Figure 7.2). The percentage of time that neurons were bursting, the mean burst duration, the number of spikes per burst and the mean inter-spike interval in a burst were enhanced. The cross-correlation coefficients, a measure of the degree of synchronous firing of two simultaneously recorded neurons (Chapter 9), were increased in neuron pairs with CFs above the TTF. A few hours post-trauma, the SFR was increased in units with CFs below and at least 1 octave above the TTF, whereas burst-firing properties returned to pre-exposure values. Typically, there was relatively little neural activity in the octave wide region immediately above the TTF. Moreover, the neural synchrony was further increased in nearly all neuron pairs comprising one or both neurons with a CF above the TTF. This increase in neural synchrony was significantly correlated with the increase in SFR. The left hand part of Figure 7.2 illustrates what happens to the response at one particular recording site where the pre-trauma CF of the neurons was ~10 kHz and had a threshold of 5 dB SPL. The frequency-tuning curve (FTC) was relatively narrow and indicated sharp tuning. Because the TTF

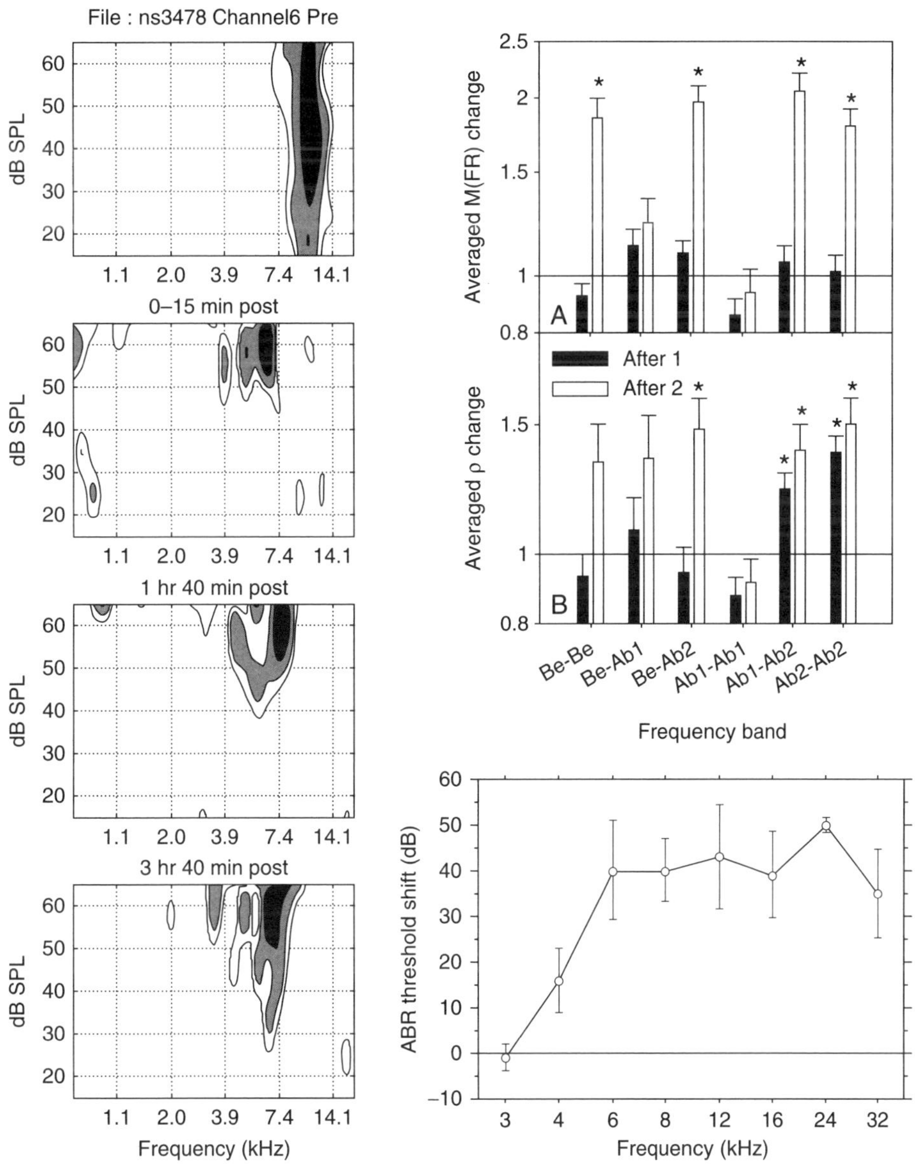

Figure 7.2 Effect of the acoustic trauma on frequency tuning, thresholds, neural synchrony, and SFR. After exposing to a 5-kHz tone for 1h at 120 dB SPL, neurons with CFs above the TTF change their tuning and threshold. Here a neuron with a CF of 10 kHz with threshold at 5 dB SPL was recorded before and after the trauma; the threshold was initially high (Left panel 2) at about 50 dB(A) but after 3 h 40 min was recovered to about 25 dB SPL. As a result, the neuron was now tuned at 6 kHz. The change in M(FR), i.e. SFR of the neuron pairs averaged (geometric mean) into six frequency bands is shown at the top right, and the change in neural synchrony averaged (geometric mean) into six frequency bands at the second panel right. Within 15 min (After1) after the acoustic trauma (black bars), one notes that the neural synchrony is significantly increased in the groups including neurons with CFs at least one octave above the TTF whereas M(FR) is not. At >2 hrs after the trauma (After2) SFR is significantly increased and the neural synchrony increased further as well. After 6 h post-trauma the hearing loss above the TTF was around 40 dB SPL (bottom right). Be indicates below the TTF, Ab1 stands for within 1 octave above the TTF, and Ab2 indicates 1–2 octaves above the TTF. Reprinted from Noreña, A.J. and Eggermont, J.J. (2003) Changes in spontaneous neural activity immediately after an acoustic trauma: implications for neural correlates of tinnitus. *Hearing Research*, **183**, 137–53. Copyright (2003), with permission from Elsevier.

was 5 kHz, the CF was 1 octave above the TTF and well in the frequency range where one expects a major effect of the exposure. Immediately after cessation of the trauma tone, the neural activity at that recording site was virtually absent in the frequency range covered by the pre-trauma FTC. Instead most activity was at frequencies below the original FTC range, notably below 7.4 kHz and with a threshold at CF of ~45 dB SPL, i.e. a threshold increase of 40 dB. About 100 min after the trauma, the threshold of the neurons at this electrode had improved to ~40 dB SPL and the major activity was found in the frequency range between 7.4–10 kHz. Finally 220 min after the exposure the CF had established itself at about 7 kHz with a threshold of 25 dB SPL, still an elevation of about 20 dB compared to the pre-trauma threshold. The average threshold elevation across 16 traumatized cats measure 6h after the trauma is shown in the bottom right of Figure 7.2 and amounts to about 40 dB for frequencies above 6 kHz.

7.2.3 Chronic effects of noise trauma

Auditory nerve fibers Cats were exposed for 1–4 h to narrow band or broadband noise with levels between 100–117 dB SPL, and ANF activity was recorded 15–305 days after the trauma (Liberman and Kiang, 1978). CF regions with normal CF thresholds typically retained a normal distribution of SFRs. Of the ANFs that retained some signs of sharp frequency tuning, two typical forms of abnormal tuning-curve shape were found: the V-shaped tuning curve for which both the low-frequency tail and tip (at CF) were elevated in level, and the W-shaped tuning curve for which the tip threshold was elevated while the low-frequency-tail threshold was at least as low as (and could even be lower than) that seen in normal ANFs. Units with reduced SFR appeared to have a predominantly V-shaped tuning-curve. The SFR distribution for the V-shaped tuning curve units had lost its normal bimodal appearance, characterized by a low firing region < 20 sp/s and a high firing region from 20–100 sp/s peaking at ~ 50 sp/s, and now showed an abnormally high proportion of units with SFR between 10–40 spikes/s. In one cat where the distribution of SFRs seemed shifted towards higher values, the tuning-curves were predominantly W-shaped. A large proportion of units that did not respond to sound had a very low or no SFR, and only a few had SFRs above 40 spikes/s. Non-responsive units were only common in ears with large CF-regions of non-responsive units. Most of the non-responsive units in traumatized animals had bursting SFR patterns or no spontaneous activity at all. An important finding is that SFRs were hardly ever increased after noise trauma.

Dallos and Harris (1978) studied the effects of systemically injected kanamycin in chinchillas. They found extensive OHC damage, but very little damage to IHCs. ANFs from the border region (where the behavioral audiogram has a steep slope between normal and hearing-loss regions probably corresponding to the segment where OHC loss progresses from <10% to >90%) had very complex response patterns. Their FTCs showed great variability. In general, the closer the fiber was to the fully developed lesion, the more abnormal its FTC became. However, the SFRs were not affected by the loss of OHCs.

Ventral cochlear nucleus Vogler et al. (2011) investigated SFRs in the VCN of guinea pigs exposed for 2 h to a 10-kHz tone presented at 124 dB SPL. After a 2-week recovery period, the mean SFR in noise-exposed ears (N = 189) was significantly elevated

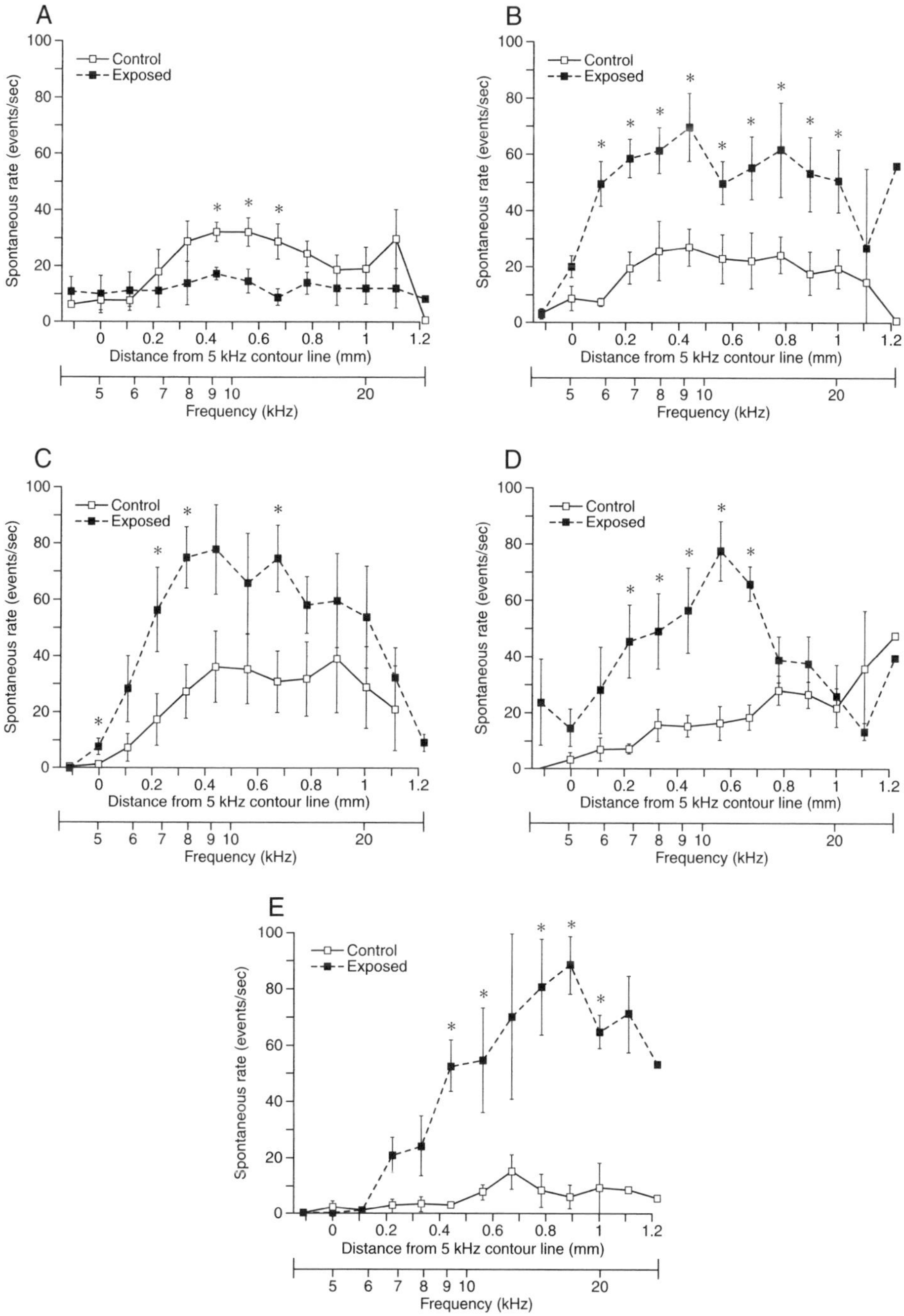

Figure 7.3 Comparisons of spontaneous rates in tone-exposed animals (filled symbols) with those in unexposed controls (open symbols) for each of the post-exposure survival times. A) 2 days post-exposure, B) 5 days post-exposure, C) 14 days post-exposure, D) 30 days post-exposure, and E) 180 days post-exposure. Asterisks placed above pairs of points, which showed statistically significant differences. Reprinted from Kaltenbach, J.A., Zhang, J. and Afman, C.E. (2000) Plasticity of spontaneous neural activity in the dorsal cochlear nucleus after intense sound exposure. *Hearing Research*, **147**, 282–292. Copyright (2000), with permission from Elsevier.

(by about a factor of two) compared to sham controls (N = 143). This was more evident in primary-like and onset categories of neurons. In addition, mechanical damage to the high frequency region of the cochlea (N = 258) showed similar significant results.

Dorsal cochlear nucleus A large series of recordings from superficial neurons, likely fusiform cells (Figure 8.2; Chapter 8) in the DCN was made by Jim Kaltenbach and colleagues from hamsters that were typically exposed to a 10-kHz tone at levels between 125–130 dB SPL for a period of 4 h. Kaltenbach et al. (2000) found increases in multiunit SFR, which were evident at 5 days after exposure, but were not observed in animals recorded from at 2 days after exposure. This time delay result contrasted with the effect of the intense tone exposure on neural response thresholds. That is, the shifts in response thresholds seen 2 days after exposure were similar to those observed in animals studied 30 days after exposure (Kaltenbach et al., 1998). Thus there is no strict correlation between SFR increase and hearing loss in this preparation. In hamsters, mean multiunit SFRs increased sharply from below normal levels at day 2 to higher than normal levels at day 5 (Figure 7.3). The mean magnitude of activity continued to increase more gradually over the next 6 months (Kaltenbach et al., 2000). Kaltenberg and Afman (2000) compared this trauma-induced increase in SFR with the tone-evoked activity in normal animals for tones with frequencies around the TTF. The results showed that the SFR distribution along the range of CFs or position along the DCN surface of exposed animals was strikingly similar to the distribution of the tone-evoked activity.

In two groups of exposed animals (10 kHz, 80 dB SPL, 4 h), which showed no sign of hair cell loss, SFR in the DCN was already increased above control levels at 1 h post-exposure and significantly increased at 2 days after exposure (Kaltenbach et al., 2005). It thus seems that extended overstimulation can result in the induction of hyperactivity in the DCN without loss of either hair cells or their stereocilia. It is interesting that when the exposure tone was only 80 dB SPL, hyperactivity had a more immediate onset, but with lower magnitude, than when the exposure tone was at 125–130 dB SPL.

Partial and complete cochlear ablations 30 days after the exposure had no significant effect on SFRs in the DCN, suggesting that the increased SFRs were not dependent on input from ANFs (Zacharek et al., 2002). However, the SFR was assessed only at 30 min after recovery from the ablation. This is reminiscent of the findings by Koerber et al. (1966) showing that SFR in DCN of normal hearing animals did not change after cochlear ablation.

By recording from single units of the DCN in hamsters, Finlayson and Kaltenbach (2009) showed average SFRs of 8.7 spikes/s in controls and 15.9 spikes/s after exposure to a 10-kHz tone at a level of 115 dB SPL for 4 h. The highest increases in SFR were found in the fusiform cell layer. Increases in SFR were significantly larger when the comparison was limited to a subset of units having type III frequency response patterns. Type III cells have V-shaped central excitatory areas that are usually flanked by inhibitory side bands (Young and Voigt, 1982). Approximately half of the increase in SFR in exposed animals was accounted for by an increase in bursting activity.

Noise trauma may only have a transient effect on the SFR of fusiform cells. SFRs were significantly higher than normal at 1 week following noise damage, whereas at 2 weeks post-noise damage SFRs were no longer significantly different from control (Zhou and Shore, 2006). This is not in agreement with the long lasting effects shown by Kaltenbach et al. (2000), which were also attributed to fusiform cells (Figure 7.3).

Ma and Young (2006) exposed cats to a 250-Hz band of noise centered at 10 kHz that was presented at 105–120 dB SPL for 4 h (to conform closely to Kaltenbach's procedure in hamsters). After a 1-month recovery period, neural activity was recorded in the DCN of a decerebrated preparation, which eliminates corticofugal activity towards the DCN among other effects. The threshold shift, determined from CAP audiograms, showed a sharp threshold elevation of about 60 dB for neurons with CFs above the 5–10-kHz lower-edge frequency of the hearing loss. In contrast to the earlier-described results in hamsters that were subjected to a similar exposure level and duration, SFRs in neurons with elevated thresholds were not increased over those in populations of fusiform cells in unexposed animals. This could suggest a species difference as the recovery period is in the range where increased SFRs were seen in hamsters. The different delays between the exposure and the recording may have had an effect as well; Zhou and Shore (2006) showed that there was only a transient elevation for 1–2 weeks after the trauma. Finally, the recovery of the cats could have been in a noisy acoustic environment, which may have prevented the increase in SFR (see Noreña and Eggermont, 2005, 2006). The various findings are not yet converging; identification of cell type may be needed as different neuron types may be differentially affected by noise trauma.

Spontaneous discharges were also recorded extracellularly in the DCN portion of brain slices from control rats and those exposed to a 10-kHz tone at a level of 115 dB SPL for 4 h at 7–11 days and 19–24 days post-exposure (Chang et al., 2002). Slices from exposed rats showed increased prevalence of spontaneous bursting activity accompanied by decreased tonic spontaneous activity. Since fusiform cells fire tonically, and cartwheel cells typically show bursts, intense tone exposure may lead to increased activity of DCN cartwheel cells and decreased activity of fusiform cells. However, this conclusion may turn out different if fusiform cells start bursting after exposure as found by Finlayson and Kaltenbach (2009). After exposure to a 4-kHz 80 dB SPL tone for only 30–60 min, Brozoski et al. (2002) found that the SFR of putative fusiform cells was significantly elevated bilaterally in the DCN of exposed chinchillas compared with controls. It is unlikely that this exposure caused hair cell damage, but it could have changed the central auditory system gain (Pienkowski and Eggermont, 2009) amplifying the SFR in the ANF and leading to the increased SFR in the DCN. This is supported by the findings of increased SFR in ICC following noise trauma followed by reduced SFR after subsequent cochlear ablation (Mulders and Robertson, 2009; see below). Surprisingly, in the Brozoski et al. study, both the behavioral evidence for tinnitus and single-unit electrophysiological SFR data suggested a maximum tinnitus-inducing effect of the 4-kHz exposure at 1 kHz (see also Chapter 5). This frequency was considerably lower than the TTF and the maximum temporary threshold shift that occurred at 6–8 kHz. In a follow-up paper it was suggested that the 1-kHz tinnitus could be the result of the increase of stereotyped bursting with ~1 ms interspike interval (Bauer et al., 2008, and next section).

Behavioral testing for tinnitus in noise-exposed hamsters (Chapter 5) showed that the increase in SFR was related to the strength of the behavioral evidence for tinnitus (Kaltenbach et al., 2004). After 4–6 weeks of recovery from the tone-induced trauma, surgical transections were made to isolate the DCN from its adjacent brainstem structures. The results showed that complete or nearly complete transection of descending inputs did not significantly affect the SFR in the DCN. However, hyperactivity in the DCN was enhanced by sectioning of the dorsal/intermediate acoustic striae, suggesting that this normally imposed an inhibitory action on the DCN (Zhang et al., 2006). It was not mentioned in the paper whether the section also included fibers from the trigeminal ganglion to the DCN. So either the DCN is an autonomous generator of increased SFR or this increased SFR is supplied by the trigeminal ganglion (Shore et al., 2008; Chapter 8). Under the autonomous generator assumption, the DCN fusiform cells would then be expected to feed the SFR in ICC neurons and cochlear ablation would then have no effect on ICC either, which is in sharp contrast to what has been observed (Mulders and Robertson, 2009).

The putative role of the DCN in the generation of tinnitus was also explored by behavioral tests for tinnitus in Long–Evans rats following unilateral exposure to a 60-min duration octave-band noise centered at 16 kHz, with a peak level of 110 dB SPL (Brozoski and Bauer, 2005). The animals showed the same behavioral responses that putatively indicated the presence of tinnitus at ~20 kHz before and after bilateral ablation of the DCN performed between 3–5 months after the acoustic trauma. This suggests that the increased SFR in the DCN is not the sole initiator of behavioral signs of tinnitus. Ipsilateral DCN ablation appeared to increase the behavioral evidence of tinnitus, likely through the reduced inhibition via the acoustic striae onto the contralateral DCN (see earlier).

The inferior colliculus The independence of SFR from cochlear input demonstrated in the DCN (see earlier) could not be replicated for recordings in ICC in noise-exposed (10 kHz tone at 124 dB SPL for 1h) guinea pigs; the increase in SFR ceased after cochlear ablation, cochlear cooling or perfusion with a pre-synaptic transmitter release inhibitor, or after destroying the postsynaptic receptors with kainic acid (Mulders and Robertson, 2009). As Figure 7.4 shows, acoustic trauma did not immediately evoke changes in spontaneous firing measured in the ICC. Recovery after acoustic trauma resulted in more neurons with high SFR compared to control animals, resulting in an increase in the average SFR. Subsequent cochlear ablation resulted in statistically significant decreases in the average SFR in ICC, from ~4.5 to ~1.4 spikes/s in the animals recorded 1 week post-exposure and from ~7.5 to ~2 spikes/s in the animals that were recorded from more than 4 weeks after the exposure. Thus, at all recovery times (up to 4 weeks) after the exposure, the increased SFR disappeared when cochlear input to the IC was destroyed. These data would suggest that the hyperactivity in the ICC after acoustic trauma is dependent on activity in the contralateral cochlea. How this could happen, with the persisting hyperactivity in the DCN after cochlear ablation at about the same post-recovery time, is somewhat of a mystery. Alternatively, the VCN may provide the dominant input to the ICC and determine the SFR this is likely as have seen that after chronic trauma SFRs are increased in VCN (Vogler et al., 2011).

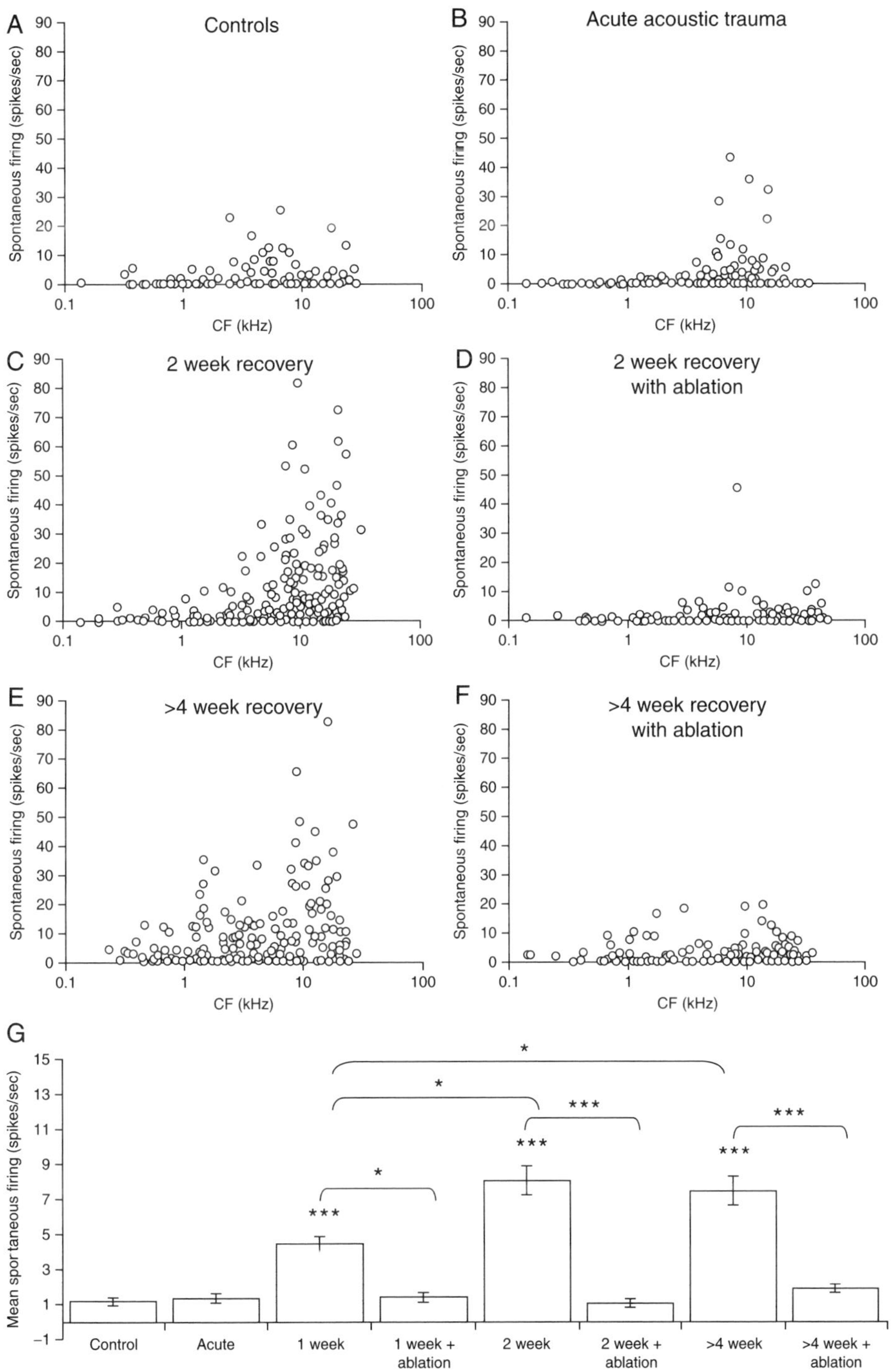

Figure 7.4 Spontaneous firing rates vs. CF of all single ICC neurons recorded in control animals (A, $n = 6$), after acute acoustic trauma (B, $n = 6$), after a recovery of 2 weeks (C, $n = 4$) and more than 4 weeks (E, $n = 3$) after acoustic trauma, as well as after cochlear ablation in 2 weeks (D, $n = 4$) and more than 4 weeks (F, $n = 3$) recovery animals. (G) Summary of mean ± SEM spontaneous firing rates for all experimental groups. * $P < 0.05$; *** $P < 0.0001$ (control and acute group, $n = 6$; 1 week 2 week and 2 week + ablation, $n = 4$; 1 week + ablation, >4 and >4 week + ablation, $n = 3$). Reprinted from Mulders, W.H. and Robertson, D. (2009) Hyperactivity in the auditory midbrain after acoustic trauma: dependence on cochlear activity. *Neuroscience*, **164**, 733–46. Copyright (2009), with permission from Elsevier.

Corroborating evidence came from electrically stimulating the olivocochlear bundle in noise-exposed animals, which is know to decrease ANF activity, and this also resulted in a decrease of the exposure-enhanced SFR in the ICC (Mulders et al., 2010). These findings suggest that partial deafferentation, at least initially, increases SFR in ICC by a mechanism different from that implied by cochlear ablation.

Robertson's group realized that even the type of peripheral lesion, mechanical (partial loss and permanent) versus ablation (complete loss and permanent), and noise exposure (TTS changing into a smaller PTS and thus varying) can have dramatic effects on electrophysiological outcomes. Therefore they compared partial mechanical lesions of the basal turn (as in Robertson and Irvine, 1989) with those following exposure to a continuous 10-kHz pure tone at 124 dB SPL for 1 h. Immediately following tone exposure there was no change in SFR in the IC, but there was after 2 weeks (Dong et al., 2010a), and also after 1 week following mechanical lesioning (Dong et al., 2009). Spontaneous hyperactivity was most marked in the frequency region of the peripheral hearing loss and occurred in contralateral as well as ipsilateral IC.

CBA/J mice were exposed bilaterally or unilaterally to a ½ octave band of noise with a center frequency of 16 kHz at 103 dB SPL for 1 h (Ma et al., 2006). Recordings were made 13–96 days after exposure, and bilaterally exposed mice (SFR = 6.0 ± 10.4 sp/s) overall showed no significant increases, relative to normal baselines (SFR = 4.1 ± 8.9 sp/s), in neurons with tuning near the TTF. However, the median SFR of units with CFs between 10–28 kHz in the bilateral exposure group significantly increased relative to that of units with CFs that were remote to the exposure frequency. In the contralateral ICC of unilaterally exposed animals, the number of responding units was greatly reduced and no firm conclusions about differences in SFR from control could be made.

Chinchillas exposed to a 4-kHz tone at 85-dB SPL for 1 h showed at 2 weeks post-exposure behavioral evidence of tinnitus with features (surprisingly) resembling a 1-kHz tone (Bauer et al., 2008). Recordings from the IC showed in exposed animals a statistically significant increase for SFR and spontaneous cross-fiber synchrony. Other animals received cisplatin or carboplatin, platinum-based chemotherapy drugs used to treat various types of cancers. Analysis identified a subpopulation of neurons that was more prevalent in animals with tinnitus. The firing activity of these units was characterized by a high rate of stereotyped bursting, and within-burst firing rates of approximately 1000 sp/s. The within-burst firing frequency found after noise exposure or cisplatin treatment was strikingly similar to the behaviorally obtained tinnitus frequency of 1 kHz. No studies from other groups have found tinnitus with frequencies well below the trauma tone frequency, however it is not unusual to find hyperacusis in that frequency range.

The auditory cortex We (Eggermont and Komiya, 2000) exposed juvenile cats in an anechoic room twice for 1 h to a 6-kHz tone of 126 dB SPL. During this exposure the animals were awake, confined in a small cage, and facing the loudspeaker. The first exposure was at 5 weeks after birth and it was repeated 1 week later. Recordings were made from AI at least 6 weeks after the exposure. The trauma caused a reorganization of the tonotopic map for frequencies above 6 kHz such that the original CFs were now replaced by CFs from the near normal low-frequency edge of the induced hearing loss. The mean SFR in the reorganized part of the primary auditory cortex (2.6 spikes/s) was significantly increased compared to the mean of the non-reorganized part of cortex (1.4 spikes/s). Recordings were done under ketamine anesthesia. For litter-matched control cats the mean SFR was 1.3 spikes/s and was not significantly different for units with CFs below and above 6 kHz. Seki and Eggermont (2003) presented findings in AI of cats exposed for 2 h to a 115-dB SPL, 6-kHz tone at 36 days, 56 days, or 118 days after birth. Recordings were made between 70–170 days after the trauma, when the animals were at least 4 months old. We found no effect of exposure age. Elevated SFRs were found in regions with reorganization of the tonotopic map compared to the neurons in the non-reorganized cortical regions in the same animals. A second finding was that in these regions the peak cross-correlation coefficients were also increased relative to those in the non-reorganized parts, and indicating increased neural synchrony. A third finding was that exposed animals showed higher SFR compared to controls for all CFs and not only in regions with cortical reorganization. Using the same exposure paradigm, Noreña and Eggermont (2005, 2006) showed again that NIHL and recovery in quiet induces reorganization of the tonotopic map in cat auditory cortex. Here the frequencies above 10–15 kHz were no longer represented. In addition the exposure increases the SFR (~6 sp/s and a factor 2 larger than in control cats) and neural synchrony (by a factor 1.2) in the reorganized part of AI (>6 kHz).

It has long been argued that a NIHL introduces central imbalances between excitation and inhibition, likely resulting from the decreased afferent input from high-frequency regions of the cochlea. We therefore aimed at equalizing the driven-firing rates of ANFs across frequency by presenting a high-frequency enhanced acoustic environment (EAE). This EAE was comprised of tone pips with frequencies between 4–20 kHz and presented randomly (Poisson-distributed) for each frequency with a rate of 3 pips/s (Figure 7.5). The pips were generated independently in each of the 38 frequencies spaced logarithmically at 16/octave, so that the overall presentation rate was 114 pips/s, and presented at 80 dB SPL peak equivalent. The idea was that this would be sufficiently above the expected 40 dB hearing loss in the frequency region >4 kHz and not too high to create additional hearing loss. Exposed cats (6 kHz for 4 h at 115 dB SPL) placed in this high-frequency (4–20 kHz) EAE immediately after the exposure for at least 3 weeks did not show significant tonotopic map changes (Figure 7.6), nor differences in SFR (~3 sp/s and same as in controls) or synchrony compared with normal hearing controls (Figure 7.7). This was interpreted as an absence of putative neural signs of tinnitus, and suggests potential benefits from targeted post-trauma sound exposure. In contrast, animals placed in a low frequency (0.6–5 kHz) EAE still showed increased SFR and even higher neural synchrony compared to controls (Noreña and Eggermont, 2006). Control cats exposed to a 4–20-kHz EAE that did not change ABR thresholds showed the same SFR in the 4–20-kHz range as non-EAE-controls but enhanced SFRs for units with CFs outside this range (Noreña et al., 2006), accompanied by increased LFP and driven neural activity in the non-EAE frequency range (Pienkowski and Eggermont, 2009).

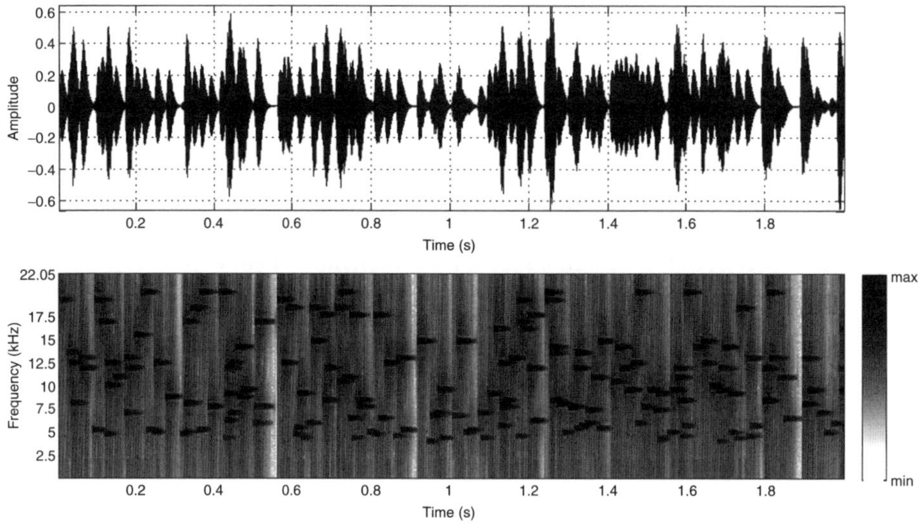

Figure 7.5 Waveform and spectrogram of the EAE for a 2-s-long sequence. The EAE was composed of tone pips randomly selected from 38 frequencies between 4 kHz and 20 kHz, separated by 1/16th octave and of equal SPL. The overall SPL of the EAE was around 80 dB. The peak SPL in the waveform (top) and spectrogram (bottom) was 82 dB. Tone pips at any given frequency were presented at an average rate of about 3 Hz, which gives an aggregate rate of 96 Hz when all stimuli are considered. Reproduced from Noreña, A.J., Gourévitch, B., Aizawa, N., and Eggermont, J.J. (2006) Spectrally enhanced acoustic environment disrupts frequency representation in cat auditory cortex, *Nature Neuroscience*, **9**, 932–39. © 2006, Nature Publishing Group, with permission.

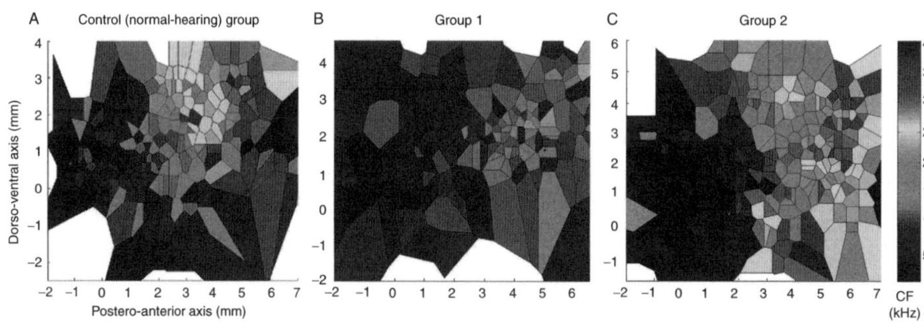

Figure 7.6 (Also see Color plate 4.) Compound CF maps in AI in control cats (A), cats recovered in quiet (B), and cats recovered in a high-frequency EAE (C). The center of each polygon, constructed using the tessellation method (MatLab), corresponds to the coordinates of a recording site in auditory cortex along the anteroposterior axis (abscises) and the ventrodorsal axis (ordinates). The tip of the posterior ecto-Sylvian sulcus was taken as the (0,0) coordinate. The CF is represented by color as indicated by the color bar. Reproduced from Kujawa, S.G. and Liberman, M.C. (2009), Adding insult to injury: cochlear nerve degeneration after "temporary" noise-induced hearing loss. *Journal of Neuroscience*, **29**, pp. 14077–85, © 2009, The Society for Neuroscience, with permission.

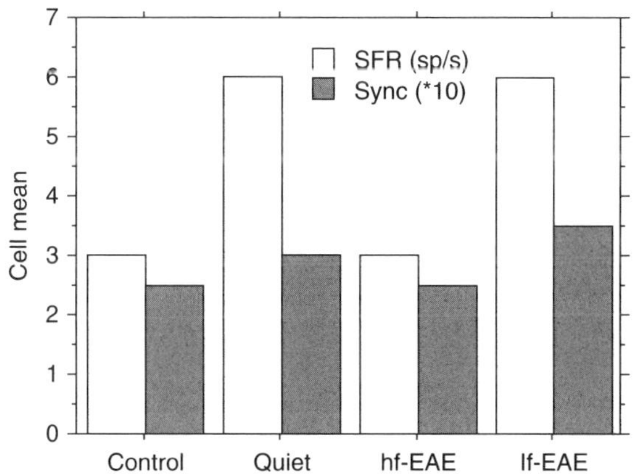

Figure 7.7 Averaged SFR in the four groups for CFs between 4–16 kHz (top panel). SFR was significantly larger for the no-EAE and low-frequency-EAE cats than the controls, whereas the high-frequency-EAE cats were not significantly different from the controls. The peak area of the cross-correlogram for CFs between 4 and 16 kHz (bottom panel) was significantly larger in the no-EAE and LF-EAE cats than in the control group, whereas the HF-EAE cats were not significantly different from the controls. Data from Noreña and Eggermont (2006).

Kotak et al. (2005) induced a complete SNHL by ablating the cochlea in gerbils bilaterally in order to assess changes in both synaptic and intrinsic properties of cortical layer II/III neurons. The neurons recorded from *in vitro* 3–13 days after the ablation displayed a depolarized resting membrane potential, an increased input resistance, and a higher incidence of sustained firing. NMDAergic currents were also larger and lasted longer in these neurons. They also exhibited significantly larger thalamocortically- and intracortically-evoked EPSCs. The decreased frequency and increased amplitude of miniature EPSCs in SNHL neurons suggest that a decline in presynaptic release properties is compensated by an increased excitatory response. The amplitudes of intracortically-evoked monosynaptic and polysynaptic GABAergic inhibitory synaptic responses were significantly smaller in SNHL neurons. These findings support the concept that excitability increases after deprivation of afferent activity. In addition, the decreased synaptic inhibition implicates that the strength of GABAergic synapses decreased after hearing loss. There are three major cellular changes that may account for increased excitability and support a homeostatic mechanism: passive membrane properties favor excitability, excitatory synapses become stronger, and inhibitory synapses become weaker after SNHL (Kotak et al., 2005). Thus hearing loss induces large changes in the strength of both inhibitory and excitatory synapses. This creates an imbalance of synaptic drive that can account for the observed triad of cortical tonotopic map reorganization, increased SFR and increased neural synchrony that accompanies NIHL (Chapter 9).

Similar to the gain changes found after salicylate application (Chapter 6), carboplatin also resulted in differential compound responses in the auditory nervous system (Qiu et al., 2000). Whereas the CAP was reduced, the LFP in the IC was unchanged or slightly reduced, and the LFP in the auditory cortex was unchanged or even enhanced.

This suggests that hearing loss regardless of its origin has the capacity to produce subtle gain changes in the central auditory nervous system, albeit smaller than those resulting from salicylate.

Engineer et al. (2011) induced noise trauma by exposing rats to 1 h of 115-dB SPL, octave-band noise centered at 16 kHz. This resulted in about 15–20-dB permanent hearing loss at 11 weeks post-trauma between 4–32 kHz. Eleven weeks after noise exposure, there were clear indications of tonotopic map reorganization. The average SFR was significantly increased by 23%. Note that this is far less than the 100% increase found in cat auditory cortex by Noreña and Eggermont (2003) and Seki and Eggermont (2003). The degree of synchronization during silence between multiunit activity recorded at nearby sites was significantly increased as well. Thus, the standard triad of cortical changes after noise trauma was present.

Eighteen out of 28 noise-exposed rats used in this study were significantly impaired in their ability to detect a gap, as evidenced by an increased gap-startle response (Chapter 5), in narrowband noise centered on 8 or 10 kHz, but they showed no impairment when the gap occurred in narrowband noise centered on 2 or 4 kHz or in broadband noise. This was considered as an indication for the presence of tinnitus with a pitch in the 8–10-kHz region in this subset of rats. Tonotopic map reorganization and tuning curve broadening, but not increased SFR or synchronization, were significantly correlated with the degree of gap-startle response strength in untreated noise-exposed rats. In addition, hearing loss and potential hyperacusis (as assessed from steeper rate level functions) were not correlated with gap-startle response strength.

Four weeks after noise exposure, vagus nerve stimulation (VNS) was repeatedly paired with multiple pure tones 300 times per day for 18 days in seven noise-exposed rats with impaired gap detection for mid-frequency sounds. The vagus nerve, arises from the medulla and carries both afferent and efferent fibers. The afferent vagal fibers connect to the nucleus of the solitary tract, which in turn projects connections to other locations in the central nervous system. Proposed mechanisms of VNS include alteration of norepinephrine release by projections of solitary tract to the locus coeruleus, elevated levels of inhibitory GABA related to vagal stimulation and inhibition of aberrant cortical activity by reticular system activation. VNS is currently used as a treatment for certain types of intractable epilepsy and treatment-resistant depression. Pairing VNS with tones is assumed to have the same effect as the well-known combinations of tone pairing with basal forebrain stimulation (Kilgard and Merzenich, 1998), and ventral tegmentum stimulation (Bao et al., 2001). After 10 days of pairing VNS with multiple tones the behavioral effect of noise exposure was reversed, which suggests that the rats' presumed tinnitus was no longer present. In addition, most of the A1 properties that were affected by noise exposure returned to pre-trauma levels. For example, the proportion of A1 neurons with characteristic frequencies between 12–23 kHz was indistinguishable from that in naive controls after VNS/multiple tone treatment (but this does not guarantee that the map returned to that of normal controls). These results support suggestions from human studies (Chapter 4) that reorganized cortical maps are causally related to tinnitus. VNS/multiple tone pairing also reversed the increase in cortical synchronization observed in noise-exposed rats to control levels, but surprisingly *did further increase the cortical SFR* observed in noise-exposed rats. Engineer et al. (2011) concluded that noise-induced increases in cortical SFR and local synchronization were not significantly correlated with behavioral correlates (gap-startle reflex) of tinnitus in individual rats.

This serves to illustrate a potential problem between the assumptions underlying the gap-startle response test as an indicator for tinnitus, and electrophysiological substrates of tinnitus in *auditory cortex* (cf. section 8.4.1). The model is based on increased SFR that occurs in the presumed gap-startle pathway. It is widely assumed that the gap activates the ascending auditory pathway in the IC and thereafter the superior colliculus, which in turn affects the startle by activating an inhibitory cholinergic pathway from the pedunculopontine tegmental nucleus to the caudal pontine reticular nucleus (Koch, 1999). Thus increased SFR in the pathway leading up to and including the IC may be sufficient to affect the startle response. Whether increased spontaneous activity limited to subthalamic nuclei is sufficient to lead to a tinnitus percept and can be decided based on a reduced gap-startle response remains an open question.

7.3 Molecular changes

Loss of auditory nerve fibers probably results from the death of cochlear IHCs or some of their ribbon synapses (Bauer et al., 2007; Kujawa and Liberman 2009). Additional cell degeneration in the CN could result from overstimulation of the system, which may induce glutamate excitotoxicity. During the first post-exposure week, before ANFs degenerated, glutamatergic release in the ipsilateral CN was elevated and glutamate uptake was depressed, consistent with hyperactivity of glutamatergic transmission and excitotoxicity. By 14 post-exposure days, when cochlear nerve fibers had degenerated, glutamatergic synaptic release and uptake in the CN were reduced. By 90 days, glutamate release increased again and so did AMPA-receptor binding suggesting an upregulation of glutamate transmission that resembled changes after mechanical cochlear damage (Muly et al., 2004). These complex structural-molecular interactions following noise trauma will be topic of the following sections.

7.3.1 Neurotransmitters and neuromodulators

AMPA Potashner et al. (1997) compared the changes in the release and uptake of D-[^{3}H] aspartate introduced by unilateral ossicle removal and unilateral cochlear ablation in albino guinea pigs. Both lesions produced severe unilateral hearing deficits and fiber degeneration as described in section 7.1.2. In the CN ipsilateral to the deficient ear, the changes were dependent on the type of lesion in the auditory periphery. Unilateral ossicle removal, after some lengthy delay, i.e. at 154 days, induced changes consistent with a weakening of excitatory glutamatergic transmission. Unilateral cochlear ablation showed 7 days later a deafferented ipsilateral CN and then induced changes consistent with a strengthening of the residual glutamatergic transmission. In the contralateral CN, the changes usually matched those in the CN ipsilateral to the deficient ear, irrespective of the type of lesion in the auditory periphery. This implied that the lack of activity of the auditory pathway that began in the lesioned ear may have initiated the regulation of synaptic strength in the contralateral CN. Finally, both lesions eventually induced changes consistent with abnormally strengthened glutamatergic transmission in the sampled nuclei of the SOC and the midbrain. Unilateral cochlear ablation in guinea pigs did induce a bilateral IC decrease of [^{3}H]AMPA binding 30 days post-lesion followed by an increase at 60 days (Suneja et al., 2000). This points again to the importance of the time delay after the ablation. An increase of AMPA receptor subunit expression (GluR2, GluR3, and GluR kainite) was detected by RT-PCR from 3 to 90 days following bilateral cochlear ablation (Holt et al., 2005). This suggests a delay in AMPA receptor changes occurring in the IC compared to the CN. It should be noticed that Argence et al. (2006) could not detect short-term (1–8 day) modulation of AMPA and NMDA receptors in the IC following unilateral cochlear ablation in Long–Evans rats.

Glycine In the AVCN and DCN, after unilateral ossicle removal, and in the DCN after unilateral cochlear ablation, [^{14}C]glycine release declined while uptake became elevated. These changes were bilateral and consistent with a down-regulation of the presynaptic component of glycinergic inhibitory transmission, accompanied by an accelerated removal of extracellular glycine. These effects imply a weakening of glycinergic inhibitory transmission (Suneja et al., 1998a,b). In young adult guinea pigs, the effects of unilateral cochlear ablation were determined on the specific binding of [^{3}H]strychnine, a glycine receptor blocker, measured in subdivisions of the CN, the superior olivary complex, and the auditory midbrain, after 2, 7, 31, 60, and 147 post-lesion days (Suneja et al., 1998a). Changes in binding relative to those in age-matched controls were interpreted as altered activity and/or expression of synaptic glycine receptors. Post-lesion binding declined ipsilaterally in most of the VCN and in the lateral superior olive (LSO). Binding was modestly deficient in the ipsilateral DCN and in the anterior part of the contralateral AVCN, but was elevated in the contralateral LSO. Binding was elevated transiently, between 2 and 31 days, contralaterally in parts of the AVCN, bilaterally in the medial superior olive (MSO), and bilaterally in most of the midbrain nuclei. Binding was deficient transiently, at 60 days, in most of the contralateral CN and bilaterally in the midbrain nuclei. At 147 days binding levels were again back to control levels.

Unilateral cochlear ablation decreased the expression of the α1 subunit of GlyR in the contralateral ICC; these downregulations were first apparent within one day of the lesion, maximized on day 8 and persisted until day 150 (Argence et al., 2006). These changes probably result from plastic neurochemical changes induced by asymmetrical auditory inputs converging on the ICC, since for GlyR α1 they disappeared when a second subsequent cochlear ablation was performed at 8 or 150 days on the opposite side for GlyR α1. Data suggest that synaptic transmission of inhibitory inputs to IC neurons through GlyR was severely and durably reduced by unilateral cochlear ablation.

These findings imply that unilateral cochlear ablation resulted in long-term deficiencies in glycinergic synaptic inhibition in most of the CN and the LSO on the ablated side, as well as in the AVCN and the DCN on the intact side. The mechanisms that contributed most prominently to these deficits include the downregulation of postsynaptic GlyR activity in the VCN and LSO, the downregulation of the synaptic release of glycine in the DCN and a faster removal of extracellular glycine. These mechanisms may contribute to sustaining the post-deafening hyperexcitability and elevated SFR over the long-term in the DCN and LSO, and may contribute to the symptoms of hyperacusis and tinnitus, and thus suggest they may generally co-occur.

Conductive hearing loss and unilateral deafness are among the etiologies that cause tinnitus in humans (Chapter 2). An unilateral conductive hearing loss of 20 dB produced by ear plugging, upregulated the GluR3 AMPA receptor subunit on auditory nerve synapses on bushy cells in VCN, while inhibitory synapses decreased the expression of the glycine receptor GlyR1 subunit on basal dendrites of fusiform cells of the DCN (Whiting et al., 2009). These changes were fully reversible once the earplug was removed, indicating that changes in activity affects the trafficking of receptors at synapses. Excitatory synapses on apical dendrites of fusiform cells (parallel fibers) with a different synaptic AMPA receptor subunit composition (compared to those on bushy cells) were not affected by sound attenuation. GlyR1 subunit expression at inhibitory synapses on apical dendrites of fusiform cells was also found unaffected. This suggests that the afferent auditory inputs (basal dendrites) are differentially affected by conductive hearing loss compared to the afferent multimodal and auditory feedback inputs to the apical dendrites of fusiform cells. Furthermore, fusiform and bushy cells of the contralateral side to the ear plugging showed upregulated GluR3 subunits at auditory nerve synapses.

GABA Noise exposure lowers GABA-mediated inhibition in the IC (Szczepaniak and Møller, 1995). Abbott et al. (1999) exposed rats to a 10-kHz tone at 100 dB SPL for 9 h showed an initial ABR threshold shift across all tested frequencies of 25–30 dB. By 30 days post-exposure, thresholds for clicks and for frequencies <10 kHz returned to near control levels but thresholds remained elevated at 10 and 20 kHz. IHC loss was only found at the apical and basal ends of the cochlea, and did not exceed 20% of the cells. Levels of the two isoforms of the GABA synthetic enzyme glutamate decarboxylase (GAD65 is more prevalent in nerve terminals, while GAD67 appears to be distributed throughout the neuron and especially the soma) in the IC were measured immediately post-exposure and at two and 30 days post-exposure using quantitative immunocytochemical and Western blotting techniques. Immediately post-exposure there was a significant increase in the level of GAD67 protein (+ 18%). By 30 days post-exposure, IC protein levels of both GAD isoforms were significantly below unexposed controls (−61% and −79% for GAD65 and GAD67 respectively). Since GAD67 levels are tightly coupled to the amount of GABA present in a neuron, this suggests increased GABA levels immediately following acoustic exposure followed by a decline to below control levels from 2–30 days post-exposure. Thus, the finding of reduced GAD levels at 30 days is consistent with reduced inhibitory neurotransmission in the auditory brainstem co-occurring with decreased output of the auditory nerve. Continuing this study, Milbrandt et al. (2000) exposed 3-month-old male F344 rats to a 12 kHz, 106 dB sound for 10 h, and GAD levels were measured immediately after exposure, at 42 h post-exposure, and at 30 days post-exposure, and compared to unexposed controls. Hair cell damage was primarily confined to the basal half of the cochlea. They found a significant decrease in GAD65 immunoreactivity in the IC membrane fraction compared to controls immediately (+ 241%) and 42 h (+ 228%) post-exposure, with complete recovery by 30 days post-exposure.

Unilateral cochlear ablation decreased the expression of GAD 67 in the contralateral ICC; these down-regulations were first apparent within 1 day of the lesion, maximized on day 8, and persisted until day 150 (Argence et al., 2006). These changes probably result from plastic neurochemical changes induced by asymmetrical auditory inputs converging on the ICC, since for GAD67, they disappeared when a second subsequent cochlear ablation was performed at 8 or 150 days on the opposite side. No statistically significant changes could be detected in the mRNA coding for different $\alpha 1$, $\beta 2$, $\gamma 2$ subunits of the inhibitory $GABA_A$ receptor, during the first week following unilateral cochlear ablation. These changes could be one of the cellular mechanisms involved in the tinnitus frequently observed in patients suffering from unilateral or asymmetric hearing loss.

After cochlear ablation (Suneja et al., 1998a,b), the changes in the contralateral ICC were consistent with an early weakening of GABAergic inhibition at 5 days post ablation that recovered by 59 days and was followed by a strengthening of inhibition by 145 days. This late strengthening may have developed in response to a similarly paced up-regulation of transmitter release from glutamatergic synaptic endings in the ICC. Although GABA release in the ICC was found to be deficient 30 days after bilateral deafening, the present changes after unilateral cochlear ablation indicated that [^{14}C]GABA release had recovered by 59 days and was elevated by 145 days. Thus, deficient GABA release in the ICC may have contributed to collicular hyperexcitability prior to 59 days. Between 59–145 days, collicular hyperexcitability may have been related to a strengthening of transmitter release from glutamatergic synaptic endings in the ICC, which was evident after unilateral ossicle removal and unilateral cochlear ablation (Potashner et al., 1997). The general effects of these treatments are summarized in Table 7.1.

Table 7.1 Effects of ablation, ossicle removal, and noise trauma on transmitters

Transmitter and trauma	1 day	1 week	1 month	2 months	5 months
AMPA unilateral ablation	⇓ (CN)		≈ (ICC)	⇓ (CN) ⇑ (ICC)	⇓ (CN) ⇑ (ICC)
AMPA unilateral ossicle	≈ (CN) ≈ (ICC)		≈ (ICC)	⇓ (CN) ≈ (ICC)	⇓ (CN) ⇑ (ICC)
Glycine unilateral ablation		⇓ (ICC)	⇓ (ICC)		≈⇓ (ICC)
GABA unilateral ablation		⇓ (ICC)	⇓ (ICC)	≈ (ICC)	⇑ (ICC)
GAD67 unilateral ablation	⇓ (ICC)	⇓ (ICC)			⇓ (ICC)
GAD67 noise trauma	⇑ (ICC)		⇓ (ICC)		
GAD65 noise trauma	⇑ (ICC)		≈ (ICC)		

Middleton et al. (2011) exposed mice unilaterally, for 45 min to 116-dB SPL, 1-kHz band noise centered at 16 kHz, which did not produce a permanent hearing loss based on ABR thresholds. They used the gap-startle reflex method to assess tinnitus before noise induction and 2–9 weeks post noise. Approximately 50% of noise-exposed mice displayed decreased gap detection (i.e. increased startle response) when the gap was embedded in 24-kHz bandpass (1-kHz bandwidth) background sound. Flavoprotein autofluorescence (FA) emanating from mitochondrial flavoproteins due to changes in oxidation state caused by neuronal activity, was used for imaging DCN brain slices. Imaging of slices in tinnitus mice showed an enhanced evoked FA response at the site of electrical stimulation. Blocking glycinergic and GABAergic inhibition did affect the center of the FA response similarly in control and tinnitus mice, however only GABA blockers showed a differential effect in the FA surround response between tinnitus and control mice. Blockers of excitation decreased the FA response to a similar extent in tinnitus and control mice. These findings indicate that auditory circuits in mice with behavioral evidence of tinnitus respond to stimuli in a more robust manner because of a decrease in GABAergic inhibition. Note however, that what is studied here is driven neural activity, which is a correlate of hyperacusis and not a direct neural correlate of the behavioral signs of tinnitus.

The type of peripheral lesion, mechanical (partial loss and permanent) versus ablation (complete loss and permanent) and noise exposure (TTS changing into a smaller PTS and thus varying) can affect molecular outcomes. Extensive assessment of gene expression following both mechanical lesioning (Dong et al., 2009) and noise exposure (Dong et al., 2010a) was done for CN and IC and is summarized in Table 7.2 and further explained in the following sections.

Changes in inhibitory activity Glutamate decarboxylase 1 (*GAD67* also known as *GAD1*) is a human gene. This gene encodes one of several forms of glutamic acid decarboxylase.

Table 7.2 Gene expression changes in contralateral IC compared to ipsilateral CN compared to control

Gene	Mechanical 1 week post		Noise trauma acute		Noise trauma 2 weeks		Noise trauma 4 weeks	
	CN	IC	CN	IC	CN	IC	CN	IC
GAD1	⇓	≈	≈	↓	≈	≈	≈	≈
GABRA1	⇓	≈	≈	⇓⇓	≈	⇓	⇑	⇓
GLRA1	⇓	≈	⇓	⇓⇓⇓	≈	⇓	⇑	≈
GRIA2	≈	≈	≈	↓	≈	≈	↑	≈
GRIN1	≈	≈	≈	⇓⇓⇓	≈	≈	⇑	≈
RAB3A	⇓	≈	≈	≈	≈	≈	≈	≈
RAB3GAP1	⇓	≈	≈	⇓⇓⇓	≈	≈	⇑	≈
KCNK15	⇓	≈	⇓	⇓⇓⇓	≈	≈	↑	⇓

↓,↑ clear trend for down, resp. up; ⇓,⇑ $P<0.05$; ⇓⇓ $P<0.01$; ⇓⇓⇓ $P<0.001$; ≈ no significant change.

The enzyme encoded is responsible for catalyzing the production of gamma aminobutyric acid from L-glutamic acid. Gamma-aminobutyric acid receptor subunit alpha-1 is a protein that in humans is encoded by the *GABRA1* gene. At least 16 distinct subunits of GABA-A receptors have been identified. Glycine receptor subunit alpha-1 is a protein that in humans is encoded by the *GLRA1* gene. It is a pentameric receptor composed of alpha and beta subunits. The decreases observed in the levels of the inhibitory neurotransmitter synthesis enzyme GAD1 and the receptor subunits GLRA1 and GABRA1 imply that both pre- and postsynaptic mechanisms contribute to deafness-related loss of inhibition.

Changes in excitatory activity Glutamate [NMDA] receptor subunit zeta-1 is a protein that in humans is encoded by the *GRIN1* gene. Glutamate receptor 2 is a protein that in humans is encoded by the *GRIA2* gene. This gene product belongs to a family of glutamate receptors that are sensitive to AMPA, and function as ligand-activated cation channels. The subunit encoded by *GRIA2* is subject to RNA editing within the second transmembrane domain, which is thought to block channel to Ca^{2+}. Although there was no statistically significant change in gene expression for GRIA2 subunit of the AMPA receptor, the mean values were much lower than those of the controls in both ipsilateral CN and contralateral IC. This conspicuous trend of reduction (Dong et al., 2010a) agrees with previous studies (Suneja et al., 2000). This reduction in the GRIA2 subunit, although counterintuitive, could be linked to neuronal hyperexcitability, because of its role in regulating calcium permeability.

Members of the RAB3 protein family are implicated in regulated exocytosis of neurotransmitters and hormones. RAB3GAP is involved in regulation of RAB3 activity. Dong et al.'s (2009, 2010a) observations in combination with the data from studies described earlier suggested to them that decreases in *RAB3A* and *RAB3GAP1* expression following partial deafness may result in downregulation of presynaptic transmitter exocytosis.

The *KCNK15* gene encodes the potassium channel subfamily-K member-15 protein, which is one of the members of the superfamily of potassium channel proteins containing two pore-forming P domains. In Dong et al.'s studies, the *KCNK15* mRNA level showed significant decreases ipsilaterally in the CN and bilaterally in the IC, suggesting that downregulation of the K2P subunit may contribute to the elevated neuronal spontaneous activity by dampening a mechanism that decreases membrane excitability.

Acetylcholine Acetylcholine receptors (AChR) come in two flavors: muscarinic (mAChR) and nicotinic (nAChR). The latter can be found on the cochlear hair cells and mediate the medial efferent olivocochlear bundle response. There are numerous efferent aspects to central auditory processing. Starting at the cortex, corticofugal activity cannot only be felt in the thalamus and the IC, but also in the olivary complex and the DCN (Winer and Lee, 2007). From the olivary complex these effects can be transmitted to the cochlea by the olivocochlear bundle. This modulates the electro-mechanical sensitivity of the outer hair cells and also regulates the sensitivity of the afferents at the synapse with the inner hair cells.

The nAChRs are ligand-mediated ion channels that regulate glutamate synapses (e.g. in auditory cortex to enhance learning and memory), and act as efferent system transmitters

(Lustig, 2006). AChRs play a role in homeostatic plasticity because they allow Ca^{2+} to enter through these channels (Morley and Happe, 2000). This could underlie the homeostatic changes proposed for the DCN following noise trauma that may underlie the increased SFR (Schaette and Kempter, 2006; 2008). Choline acetyltransferase (ChAT) activity has been mapped in the cochlear nucleus (CN) of control hamsters and hamsters that had been exposed unilaterally to a 10-kHz tone of 127 dB SPL presented for 4 h (Jin et al., 2006). ChAT catalyzes the acetylcholine synthesis from choline and acetyl coenzyme A inside cholinergic neurons. Eight days after this intense tone exposure, average ChAT activity increased on the tone-exposed side as compared to the opposite side, by 74% in the AVCN, by 55% in the granule cell region dorsolateral to it, and by 74% in the deep layer of the DCN. In addition, average ChAT activity in the exposed-side AVCN and fusiform cell layer of DCN was higher than in controls, by 152% and 67%, respectively. Two months after exposure, average ChAT activity was still 53% higher in the exposed-side deep layer of DCN as compared to the opposite side. This increased ChAT activity after intense tone exposure suggests plasticity of the descending cholinergic innervation to the CN, which might affect spontaneous activity in the DCN (Jin et al., 2006).

7.3.2 **Immediate early genes**

IEG are induced rapidly inside nerve cells by extracellular stimuli without the need of intermediate proteins. Activity-induced changes in synaptic efficacy and plasticity alter the expression pattern of IEGs such as brain-derived neurotrophic factor (*BDNF*), the activity-dependent cytoskeletal protein (*Arg3.1/Arc*), and *c-fos*. Spontaneous activity may also trigger BDNF trafficking to activated synapses and support its release from synapses, making it a suitable candidate for monitoring trauma-induced changes in SFR and neuronal excitability. *Arg3.1/Arc* expression is upregulated, and transported to dendrites following synaptic stimulation. It is required together with BDNF for long-term potentiation (LTP) and long-term memory in the nervous system. *c-fos* expression is linked to neuronal excitability.

In vitro studies have shown that the degeneration of spiral ganglion cells after loss of hair cells involves at least three mechanisms including: (1) the cAMP-dependent protein kinase and Ca^{2+} /calmodulin-dependent protein kinase II and IV systems, (2) pathways involving protein kinase C activation, Ca^{2+} signaling, and mitogen-activated protein kinases (MAPK), and (3) the c-Jun N-terminal kinase cell-death pathway (Lang et al., 2006). NMDA receptors play a regulating role in gene expression via a MAPK-dependent mechanism. This regulatory role results from mediating calcium influx and a transcriptional cascade that is gated by the CREB protein and immediate-early transcription factors such as activator protein-1 (AP-1). The two components of the AP-1 transcription factor, the IEGs *c-fos* and *c-Jun*, are the substrates of extracellular signal-regulated kinase (ERK) and the JNK/SAPK subfamilies of MAPKs. The c-Jun protein is a crucial bidirectional regulator for neuronal death, survival, and regeneration. c-Jun can be activated by c-jun N-terminal kinase (JNK)-dependent phosphorylation, which may occur in response to both physiologic or pathologic stimulation. Phosphorylation is the addition of a phosphate (PO_4) group to a protein or other organic molecule, resulting in its activation or deactivation.

Activity in the MAPK pathway ultimately phosphorylates ERK1 and ERK2, which, in turn, phosphorylate various effector or substrate molecules. After entering the nucleus, ERK1-P and ERK2-P may phosphorylate several transcription factors to exert control over gene expression. Through these actions, the ERK pathway is important in regulating neuronal growth, development, synaptic plasticity, survival, post-injury regeneration, and cell death. Injury of central or peripheral neurons or their axons increases transduction activity in the ERK pathway (Sweatt, 2001).

Canlon and colleagues looked at the different effects of TTS and PTS on the expression of MAPK and BDNF in the cochlea (Meltser et al., 2010) and inferior colliculus (Meltser and Canlon, 2010). To generate a TTS, a band-pass filtered noise (6–12 kHz) was presented at 100 dB SPL for 45 min. The more severe exposure to generate a PTS consisted of the same noise presented at 110 dB SPL for 2 h. TTS and PTS demonstrate different expression patterns and temporal aspects of MAPK, BDNF, and TrkB changes in the cochlea. In the acute phase of PTS an upregulation of phosphorylated p38, JNK1/2, and ERK1/2 was found while in the acute phase of TTS a downregulation of phospho-p38 occurred and no immediate change of pJNK1/2 and pERK1/2 was noted. After a 24 h recovery from TTS JNK1/2 and ERK1/2 was activated while the expression of phosphor-p38 was downregulated. In contrast, the PTS group showed complete recovery to control values for all three MAPKs by 24 h post. In the IC, the early effects of acoustic trauma that eventually result in a PTS, lead to a transient activation of BDNF and MAPK including ERK, JNK, and p38. The early effects of acoustic trauma that result in a TTS produced a reversible activation only of p38. The pattern of MAPK and BDNF activation in the IC was different from that described for the cochlea.

Calcium binding proteins (e.g. calmodulin, parvalbumin, calretinin, and calbindin) help regulate the intracellular calcium concentration. Following unilateral cochlear ablation in mature rats, Förster and Illing (2000) found: (1) a substantial increase in the number of calbindin-immunoreactive fibers and terminal boutons in the ventral subdivision of the ipsilateral CN; (2) a transient appearance of calbindin-positive astrocytes in the DCN and VCN; (3) a 10-fold maintained increase in the number of calbindin-positive neurons in the IC contralateral to the lesion; and (4) the emergence of calbindin-positive neurons in the LSO contralateral to the cochleotomy. These findings indicate that the cochlear ablation induces neuronal expression of calbindin in various cell types across several auditory brainstem nuclei.

Following acoustic trauma (2 h, 10 kHz, 115 dB SPL), Tan et al. (2007) observed in spiral ganglion cells an upregulation of the immediate early gene, *c-fos*, within hours and both *c-fos* and *BDNF* exon IV expression within days in spiral ganglion neurons. In auditory cortex, 6 days post-trauma the diminished levels of *Arg3.1/Arc* and *BDNF* exon IV following acoustic trauma suggested an overall reduced excitability within AI. Non-traumatic noise exposure increased Arg3.1/Arc levels in AI. A reciprocal responsiveness of activity-dependent genes became evident between the periphery and AI: *c-fos* and *BDNF* expression were increased in spiral ganglion cells, whereas *Arg3.1/Arc* and (6 days post-trauma) *BDNF* expression was reduced in AI. Since both *c-fos* and *BDNF* are activity-dependent genes, their augmented transcription in spiral ganglion cells could reflect enhanced glutamate release induced by traumatic noise and consequently NMDA/AMPA receptor activation (Puel et al., 1998; Duan et al., 2000). Reduced AMPA receptor trafficking is a direct correlate of deprivation phenomena, and thus could likely be correlated with reduced Arc/Arg3.1 levels, increased AMPA responses, and increased firing rate. The diminished levels of Arg3.1/Arc and BDNF following acoustic trauma in AI suggest reduced AMPA receptor trafficking and increased AMPA responses, and consequently increased firing rates.

Kraus et al. (2009) found that unilateral round window application of carboplatin (50 μl, 5 mg/ml) resulted in degeneration of IHC and OHC and a significant upregulation of growth-associated protein GAP-43 expression in fibers and presynaptic endings in the ipsilateral VCN in animals surviving for 15 or 31 days post-carboplatin. GAP-43 showed little or no change at the 3- or 7-day time points. At 15- and 31-day survival times, the increase in GAP-43 expression in VCN was positively associated with the degree of hair cell loss with the strongest extent in the high frequency region of the cochlea and VCN. No prominent changes of GAP-43 expression were found in the DCN.

7.4 **Summary**

The findings described in this chapter are summarized in Table 7.3, using the same format as in the summary of Chapter 6. The primary targets of noise trauma (and ototoxic drugs) are the cochlear hair cells. The most vulnerable are the OHCs in the first row followed by the IHCs. If the noise is not excessive, the minimal damage that correlates with hearing loss is related to changes in the hair cell stereocilia, which contain the transduction channels. Even if the result of noise exposure is just a temporary threshold shift, this may result in permanent loss of the Type I spiral ganglion cells that innervate the IHC. Central nerve degeneration may ensue. The trauma rarely caused increases in SFR of auditory nerve fibers and more generally a reduction. The result of reduced auditory nerve output is typically an imbalance between neural excitation and inhibition in the

Table 7.3 Changes after chronic NIHL

Structure	Cell density	SFR	2-DG	Glu	Gly/GABA	5-HT
OHC IHC	⇓					
ANF	⇓	⇓				
VCN		⇑				
DCN	⇓	⇓[1] ⇑[2]	⇑[3]	⇑	⇓	
ICC		⇑		⇑	⇓→⇑	⇑
ICX				⇑		
AI	≈	⇑		⇑→⇓		⇑
AII	≈					

[1]Fusiform cells (*in vivo*, Ma and Young, 2006).

[2]Fusiform cells (slice, Finlayson and Kaltenbach, 2009), Cartwheel cells (slice Chang et al., 2002).

[3]Middleton et al. (2011) using flavoprotein imaging.

central auditory system. This results in strong hyperactivity in the DCN, and can result in tonotopic map reorganization, likely only in cortical areas, accompanied by increased SFR and increased neural synchrony. This trio of changes is considered to comprise potential neural substrates of tinnitus. The balance between the excitatory and inhibitory transmitter efficacy in the CNS is only temporarily changed in the first few weeks to months after the trauma. It is believed that during that period restoration of the excitatory–inhibitory balance can prevent tonotopic map reorganization as well as increases in SFR and neural synchrony, and thus likely also tinnitus. Lesion studies suggest that the DCN may function as a source of increased SFR without cochlear input and descending input from the CNS. However, these studies also suggest that behavioral tinnitus persists in animals for which the DCN output is isolated from central auditory structures. In contrast the SFR in IC was dependent on output of the cochlea. Species dependence and recovery times may play a role in these discrepancies.

Chapter 8

The somatic tinnitus model

Tinnitus is most often initiated by cochlear disturbances affecting peripheral and central auditory pathways. However, there is growing evidence indicating that the anatomical location generating tinnitus can occur at sites different from the auditory system (Cacace, 2003). Support for this notion is found in individuals whose tinnitus can be triggered or modulated by inputs from other sensory modalities or sensorimotor systems. A clear example is found in the high percentage of tinnitus complaints following head and neck injury, e.g. caused by whiplash, and by dental problems (Chapter 2). Whiplash injuries can cause abnormal firing activity in the medullary somatosensory nuclei (MSN). The MSN normally provides inhibitory innervation to the DCN. During some whiplash accidents, MSN to DCN pathways are damaged. Subsequent disinhibition of the DCN and associated auditory structures could result in the generation and perception of tinnitus (Levine, 1999). Furthermore, jaw movements can modulate the loudness of existing tinnitus, either making it worse or sometimes completely alleviating it during the movement.

Additional evidence for incorporating auditory/somatosensory system interactions in the generation and/or modulation of some forms of tinnitus was provided by Møller et al. (1992). In adults whose tinnitus strength ranged from mild to severe, low-level electrical stimulation of the median nerve near the hand region could modify some of its perceptual characteristics, i.e. loudness or pitch. In adult control subjects without tinnitus, however, median nerve stimulation during sound presentation either had no effect on sound perception or produced only slight increases in loudness. Møller and Rollins (2002) provided additional insight into these multimodal interactions in normal subjects by showing that changes in loudness perception by concurrent low-level electrical stimulation was dependent on age. They showed that electrical stimulation of the peripheral part of the median nerve had an effect in children (7–8 years) and relatively little effect in adults (20–40 years). According to the authors "the change in function of the auditory system that we observed may be an example of specialization where auditory processing is shifted from the phylogenetically older non-classical system, towards the phylogenetically newer classical auditory system that performs finer analysis of sounds." In adults with tinnitus, the authors speculate that the non-classical auditory pathways become reactivated as an expression of neuroplasticity. It is known that around age 8 dramatic changes occur in the AEPs, particularly illustrated by the emergence of the N100 component and by increased immunostaining of the axon neurofilaments in the superficial layers of auditory cortex (Ponton et al., 1999). The unmasking of silent (ineffective) synapses, a mechanism to account for certain types of neuropathic pain, may be involved in this process (Møller, 1997). This may particularly involve connections between the dorsal thalamus and the basal lateral amygdala.

Somatic tinnitus includes the modulatory effects on the pitch and loudness of tinnitus resulting from neck and upper limb manipulation. Physiologically, this may correspond to modulation of neural activity in the DCN by activation of the dorsal column nuclei and cervical dorsal root ganglia (DRG). The majority of patients are able to modify their tinnitus by manipulating regions of the head and neck, such as jaw clenching (Levine, 1999; Levine et al., 2003; Abel & Levine, 2004). Other patients experience tinnitus after neck injuries or a tooth abscess (Lockwood et al., 1998; Pinchoff et al., 1998; Levine, 1999). These phantom percepts are likely a result of a disrupted or altered somatosensory input to the DCN.

Lockwood et al. (1998) studied a group of individuals that could alter (increase or decrease) tinnitus loudness by oral-facial maneuvers (OFMs; jaw clenching). Using PET and a between-group experimental design, two groups were scanned separately during OFMs, a tinnitus group and a normal control group without tinnitus or hearing loss, that also performed jaw clenching. Normal controls showed bilateral activation of somatosensory cortex and supplemental motor area in response to jaw clenching. In two patients where OFMs increased tinnitus loudness (i.e. where tinnitus was localized to the right ear in one patient and in the left ear in the other), increases in CBF were observed in somatosensory cortex, primary auditory cortex in the left superior temporal gyrus and in a region near the medial geniculate nuclei. To separate changes in cerebral blood flow due to increases in tinnitus loudness, group subtractions were performed between PET results obtained during jaw clenching in controls and OFM in tinnitus patients. The group subtractions showed residual activation in the left thalamic region (left medial geniculate nucleus) in the tinnitus group. This was interpreted as due to the increase in tinnitus loudness.

Multimodal integration of jaw protrusion and sound was studied in normal hearing and tinnitus subjects using fMRI (Lanting et al., 2010). The somatosensory system only responded to jaw protrusion, however, the auditory system responded to sound as well as jaw protrusion, in controls as well as tinnitus subjects. Compared to controls jaw protrusion responses were significantly enhanced in tinnitus patients both in the CN and the IC. There were no differences in thalamus or AC. Responses to sound alone were not significantly different between controls and tinnitus patients for all auditory areas studied. Again it is somewhat surprising that the behavioral effects of jaw protrusion on tinnitus loudness did not show a cortical correlate and may suggest important subcortical multimodal interaction effects that have no BOLD response correlate in auditory cortex.

A comprehensive series of animal studies, nearly exclusively from Susan Shore's laboratory, have elucidated the pathways involved in this remapping or unmasking of silent connections between the somatosensory and auditory systems.

8.1 The trigeminal ganglion and cochlear blood flow

The trigeminal ganglion innervates the cochlear blood vessels. Vass et al. (1997) showed this by using retrograde transport of wheat germ agglutinin conjugated to horseradish peroxidase (WGA–HRP) applied to the cochlea. A large number of labeled nerve cells were found in the anteromedial portion of the trigeminal ganglion at the origin of the ophthalmic division (Figure 8.1). When WGA–HRP was injected into the middle ear, cells were labeled in the middle posterolateral portion of the trigeminal ganglia and in the superior ganglia of the glossopharyngeal nerve.

Nearly all blood vessels in the body are surrounded by autonomic system nerve fibers containing among others substance P (SP). Stimulation of these sensory nerve endings is associated with the release of SP, and this may indicate a role of these nerve endings in controlling local blood flow. Treatment with capsaicin prior to WGA–HRP administration into the cochlea resulted in a significant reduction in the density of labeled cell bodies in the ipsilateral trigeminal ganglion. The action of capsaicin applied at high

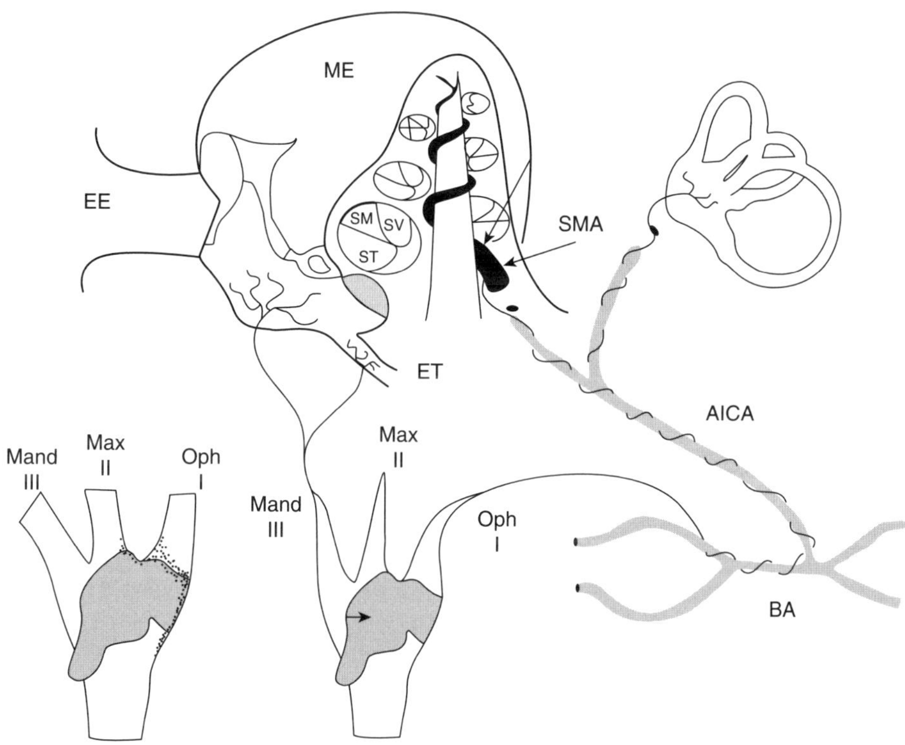

Figure 8.1 Schematic of the anatomical relationship of the trigeminal ganglia, cochlea, vestibular labyrinth, and basilar artery. The cochlea and vestibular labyrinth are innervated by the ophthalmic branch of the trigeminal ganglia through the basilar artery and anterior inferior cerebellar artery. In contrast, the middle ear and Eustachian tube receive their innervation from the mandibular branch of the trigeminal ganglia. Inset) Schematic representation of horizontal sections through the ipsilateral trigeminal ganglia after 2% WGA–HRP infusion. Each dot represents a single labeled neuron. EE, external ear canal; ME, middle ear; ET, Eustachian tube; Mand, mandibular division; Max, maxillary division; Oph, ophthalmic division; AICA, anterior inferior cerebellar artery; BA, basilar artery; SMA, spiral modiolar artery. Reprinted from Vass, Z., Shore, S.E., Nuttall, A.L., and Miller, J.M. (1998) Direct evidence of trigeminal innervation of the cochlear blood vessels. *Neuroscience*, **84**, 559–67. Copyright (1998), with permission from Elsevier.

concentration directly to the nerves causes irreversible impairment specifically of C-fibers. C-fibers are unmyelinated, have a slow (<2 m/s) conduction velocity, and are responsible for dull long-lasting pain. C-fibers can respond to thermal, mechanical, and chemical stimuli. Thus, Vass et al. (1997) suggested that the labeled innervation described in their study is sensory and are C-fibers.

Injection of biocytin into the trigeminal ganglion of guinea-pigs showed labeled fibers from the injection site as bundles around the ipsilateral spiral modiolar blood vessels, as individual labeled fibers in the interscala septae, and in the ipsilateral stria vascularis (Figure 8.1; Vass et al., 1998). Thus, trigeminal ganglion innervation may play a role in normal vascular tone and in some inner ear disturbances. For instance, it is not unlikely that sudden hearing loss may result from abnormal activity of trigeminal ganglion projections to the cochlear blood vessels. Vass et al. (1998) postulated that the anterograde labeled trigeminal neurons in the cochlea are the same as those retrograde labeled by cochlear infusion of WGA–HRP. The retrograde labeled sensory nerves originated in the medial part of the trigeminal ganglion, mainly at the origin of the ophthalmic branch (Figure 8.1, inset). The ophthalmic branch of the trigeminal nerve may participate in the pathogenesis of migraine headaches, neurogenic inflammation and vascular permeability due to the release of SP and calcitonin gene-related peptide at the nerve terminal. This pathway also innervates the extraocular muscles (Aigner et al., 1997), and may be involved in gaze-induced tinnitus (Chapter 2).

8.2 Somatosensory innervation of the auditory brainstem and midbrain

The potential contributions to tinnitus from the somatosensory systems result from the activity of two nerve tracts; the trigeminal nerve and the dorsal column. The trigeminal system innervates the TMJs, the tongue, the extraocular eye muscles (determining gaze), the oral mucosa, the teeth, and the facial skin, and is thought to mediate the suppression of body-generated sounds such as chewing and respiration (Shore, 2005; Shore and Zhou, 2006). The dorsal column system innervates the head and neck muscles, and the back of the pinna. These fibers likely provide proprioceptive information related to pinna and head position especially in cats and dogs, and other animals with movable pinna (Young et al., 1995). This body position information is combined with auditory information in the DCN, and so incorporate head orientation in the computation of the localization of a sound source (Wright and Ryugo, 1996; Kanold and Young, 2001). Bender and Trussell (2011) hypothesized that the dorsal column system may be more involved with temporal precision and neural plasticity and is limited in its innervation to the DCN, whereas the trigeminal system has a slower modulatory function that involves both the VCN and DCN.

In more detail (following the exposition by Shore et al., 2007), somatic sensations from the trunk, limbs, and neck are mediated by afferent fibers with their soma in the dorsal root ganglia. Sensations from the head, neck, and face are conveyed by the trigeminal nerve, with its cell bodies forming the trigeminal ganglion (TG). The central processes of the DRG form two pathways: the spinothalamic pathway of the anterolateral system and the dorsal column-medial lemniscal system. The anterolateral system mediates pain, temperature, and some deep touch. The dorsal column-medial lemniscal system primarily mediates proprioception and discriminative sensation, and consists of two nuclei: the cuneate nucleus, which serves the upper trunk and limbs, and the gracile nucleus, which serves the lower trunk and limbs.

The central processes of the TG terminate in the brainstem trigeminal sensory complex, consisting of three groups of nuclei that mediate different sensory modalities. These are the principal nucleus that receives information about discriminative sensation and light touch; the mesencephalic nucleus that receives proprioceptive information from the jaw; and the spinal trigeminal nucleus (Sp5) that receives information about pain and temperature as well as gentle pressure and vibrissa deflection. Sp5 also receives proprioceptive inputs from the vocal tract, the temporomandibular joint and tongue muscles. Both primary and secondary somatic sensory neurons project to the auditory system.

8.2.1 Primary somatosensory neuron projections to the cochlear nucleus

Shore et al. (2000) explored the neural connections between the trigeminal ganglion and the auditory brainstem by using retrograde and anterograde tract tracing methods. Iontophoretic injections of biocytin or biotinylated dextran-amine were made into the guinea pig trigeminal ganglion, and anterograde labeling was examined in the CN and SOC. Anterograde and retrograde labeling was also seen in the shell regions of the LSO and in periolivary regions. The labeling was seen in the neuropil, on cell bodies, and surrounding blood vessels. The tracing results demonstrated that neurons in the CN are innervated by the ophthalmic and mandibular divisions of the TG, which also project to the cochlea and middle ear, respectively as reviewed earlier (Figure 8.1). The ophthalmic division of the trigeminal ganglion also innervates the extraocular muscles. After vestibular schwannoma surgery a large number of patients develop gaze-induced tinnitus (Chapter 2), and this division of the trigeminal ganglion may be implicated in this phenomenon. A reciprocal connection exists between neurons in the shell regions of the LSO and the TG. So, potentially the TG could influence the olivo-cochlear bundle activity on hair cells and auditory nerve fibers. This may relate to the enhancement of otoacoustic emissions seen after head injury (Chapter 4). Trigeminal input to the CN and other auditory brainstem neurons could thus have a profound impact on both the metabolism and function of CN neurons by targeting both the neurons and their vascular supply.

The VCN, in addition to receiving afferent connections from the auditory nerve and efferent information from higher auditory centers, is also innervated by those somatosensory neurons that transmit tactile and kinesthetic sensations. The granule cell region in the VCN is innervated not only by the trigeminal ganglion but also by interpolar and caudal spinal trigeminal nuclei and the cuneate nucleus (Weinberg and Rustioni, 1987). The cuneate nucleus gives rise to mossy fiber terminals that synapse in both the DCN and the granule cell region of the VCN (Wright and Ryugo, 1996). These projections excite VCN neurons and inhibit DCN neurons (Davis and Young, 1997). Responses of VCN neurons to electrical stimulation of the TG consisted of one, two, or more phases of excitation, sometimes followed by a longer inhibitory phase (Shore et al., 2003).

Applying kainic acid or ablating the cochlea, which both eliminate ANF responses, abolishes SFR in the VCN (Koerber et al., 1966), decreases the firing rates of VCN units in response to trigeminal stimulation, and increased their first spike latencies (Shore et al., 2000). This suggests that trigeminal nerve activity not only provides independent inputs to the VCN but also modulates the effect of ANF activity. One possible route for the excitation of VCN neurons by the TG is its projection to the granule cell region of the cochlear nucleus. Activation of granule cells could, in turn, activate stellate cells in the AVCN and PVCN, which extend their dendrites into the granule cell region. Axosomatic trigeminal input in the magnocellular region of the VCN could cause the later inhibitory phase (Shore et al., 2000). The modulation of firing rate in second order auditory neurons by first order somatosensory neurons could influence central auditory targets and may thus be involved in the generation of "somatic tinnitus."

8.2.2 Secondary somatosensory projections to the cochlear nucleus

Secondary somatic sensory neurons that project to the CN are located in the Sp5 and dorsal column nuclei. The CN projection cells in the Sp5 have varied morphologies; their projection fibers and terminal endings are either small to medium *en passant* boutons, or the larger mossy fibers. The small Sp5 terminal endings are scattered across the entire CN, making synaptic contacts with granule cells or large principal cells (Zhou and Shore, 2004). The Sp5 mossy fibers are primarily located in the granule cell domain, making synaptic contacts with granule cells (Haenggeli et al., 2005). The CN projection neurons in the dorsal column nuclei primarily terminate in the granule cell domain. The postsynaptic targets of these mossy fibers include dendrites of granule cells (Wright and Ryugo, 1996). The secondary somatic sensory neurons indirectly affect DCN output via the mossy fiber–parallel fiber–fusiform cell pathway (as well as via inhibitory interneurons). Secondary neurons in another ascending pathway of the DRG, the anterolateral system that mediates pain and temperature, do not project to auditory structures (Shore et al., 2007).

The DCN is organized into a deep region that receives auditory inputs and a superficial region that receives input from a granule cell circuit (Figure 8.2). Often, the DCN is divided into three layers, molecular layer, fusiform cell layer, and deep layer (see also Chapter 7). Most numerous among the DCN interneurons are the cartwheel cells, which are glycinergic and perhaps also GABAergic, and inhibit fusiform cells and giant cells that both provide the output of the DCN to the inferior colliculus. Non-auditory projections to granule cells include afferents from vestibular receptors and the somatosensory dorsal column and Sp5. The fusiform cells in the DCN receive, in addition to inputs from auditory nerve fibers, various non-auditory inputs via the granule cell circuit (Oertel and Young, 2004). Activation of this circuit produces both weak excitation and strong inhibition of the fusiform cells, through granule cell axons (parallel fibers) that excite the principal cells as well as the inhibitory interneurons that in turn inhibit the fusiform cells. Note that the granule cells also receive direct input from auditory cortex.

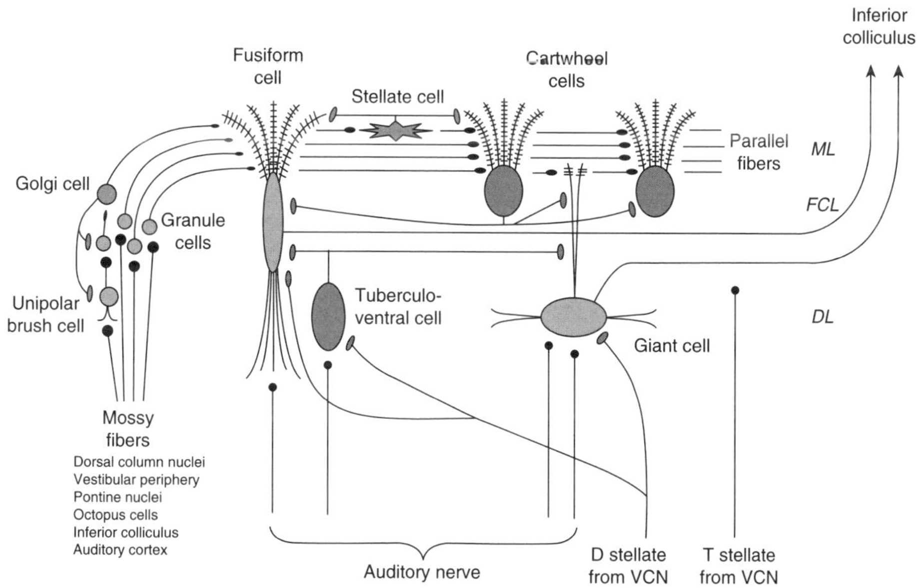

Figure 8.2 (Also see Color plate 5.) Schematic structure of the DCN. Granule cells lie in the fusiform cell layer (FCL) of the DCN, and are excited through mossy fibers originating from a large number of other regions of the brain (listed at lower left of the figure). The granule cell axons (parallel fibers) contact the spiny dendrites of cartwheel cells in the molecular layer (ML) and other inhibitory interneurons, i.e. superficial stellate cells and Golgi cells. Cartwheel cell axons contact fusiform cells, which are themselves contacted by parallel fibers on spiny apical dendrites. Auditory nerve fibers innervate the deep layer (DL), contacting the smooth basal dendrites of fusiform and giant cells, as well as tuberculoventral interneurons. Two groups of cells from the VCN, the D- and T-stellate cells, provide auditory inputs to the deep layer. Fusiform and giant cells, the principal cells of the DCN, project to the inferior colliculus. Glutamatergic neurons are shown in green and their terminals in black, glycinergic neurons and their terminals are shown in orange, and GABAergic neurons are shown in pink. Reprinted from Oertel, D. and Young, E.D. (2004) What's a cerebellar circuit doing in the auditory system? *Trends in Neurosciences*, **27**(2), 104–10. Copyright (2004), with permission from Elsevier.

The projections from Sp5 and the cuneate nucleus (dorsal column system) to the cochlear nucleus are glutamatergic. Four types of vesicular glutamate transporters are known, VGLUTs 1–3 and the glutamate/aspartate transporter sialin. These transporters move the neurotransmitter into synaptic vesicles so that they can be released into the synapse. The cuneate nucleus terminals as well as the Sp5 terminals colabel with the vesicular glutamate transporter VGLUT2. In contrast, the ANF terminals in the cochlear nucleus colabel with a different transporter VGLUT1. VGLUT1 and VGLUT2 are found in different regions of the VCN; VGLUT1 mostly in the magnocellular part and VGLUT2 in mostly in the granular cell division (Zhou et al., 2007).

8.2.3 **Somatosensory system projections to the inferior colliculus**

Somatic sensory neurons also project to other central auditory structures, including the IC. Injecting different tracers into both Sp5 and CN confirmed the overlapping areas of convergent projections from Sp5 and CN in IC (Zhou and Shore, 2006; Dehmel et al., 2008). This convergence of projection fibers from CN and Sp5 provides an anatomical substrate for multimodal integration in the IC. Single-unit recordings from the external cortex of IC (ICX) showed that trigeminal stimulation can suppress SFRs as well as produce bimodal integration of the type described earlier in the DCN (Shore, 2005; Jain and Shore, 2006). Units in both locations show mostly suppression, but also enhancement of acoustically driven responses by trigeminal stimulation. Some neurons in the Sp5 and the dorsal column nuclei project to both the CN and ICX by way of axon collaterals (Li and Mizuno, 1997). IC projection cells have not been found in the TG.

8.3 Physiological and neural changes

The dorsal cochlear nucleus Natural somatosensory stimulation, particularly involving pinna movements, as well as electrical stimulation in the MSN, can inhibit the spontaneous activity of DCN principal cells for tens of milliseconds (Young et al., 1995). Davis and Young (1997) demonstrated that cartwheel cells convey inhibitory polysensory (via their granule cell–parallel fiber input) information to DCN principal (fusiform and giant cells) cells. Kanold and Young (2001) found that only stimuli that activated pinna-muscle receptors, such as stretch or vibration of the muscles connected to the pinna, were effective in driving DCN units, whereas cutaneous stimuli such as light touch, brushing of hairs, and stretching of skin were ineffective. These results suggest that the largest somatosensory inputs to the DCN originate from muscle receptors associated with the pinna.

Shore (2005) stimulated the trigeminal ganglion in guinea pigs during the recording of spontaneous and sound-driven activity from DCN neurons. Thirty percent of DCN units showed excitatory, inhibitory, or excitatory–inhibitory responses to trigeminal ganglion stimulation. Trigeminal stimulation suppressed or facilitated the firing rate in response to paired broad-band noise pulses in 78% of units, reflecting multisensory integration. Activity in trigeminal pathways to the CN can alter the spontaneous (i.e. non-sound-driven) activity of neurons in both the VCN and the DCN. Change in the firing rate of CN neurons by incoming trigeminal neurons could be one means by which the loudness of tinnitus is modulated in individuals with somatic tinnitus. This modulation could potentially also change the synchrony between neurons (Shore et al., 2007).

Recently, Koehler et al. (2011) observed that preceding tonal stimulation with Sp5 stimulation significantly alters the first spike latency, the first interspike interval, and the average discharge regularity of firing evoked by the tone. They suggested that these spike-timing alterations may explain the ability of patients to modulate the pitch and loudness of their tinnitus by manipulations of their face and neck, regions that are innervated by the trigeminal nerve.

When postsynaptic depolarization is preceded by stimulation of parallel fibers in slice experiments, LTP or LTD is produced in fusiform and cartwheel cells (Fujino and Oertel, 2003; Tzounopoulos et al., 2004). Parallel fiber inputs to fusiform cells are strengthened (LTP) when co-occurring with subsequent postsynaptic firing, whereas in cartwheel cells they are weakened (LTD) (Tzounopoulos et al., 2004). The induction of LTP in fusiform cells and LTD in cartwheel cells has the net effect of increasing the firing rates of fusiform cells because cartwheel cells provide feedforward inhibition to fusiform cells. Reversing the order of pre- and postsynaptic activation produces LTD in fusiform cells, but does not change LTD to LTP in cartwheel cells, resulting in a net depression of firing rate in fusiform cells.

Another mechanism that could contribute to these effects is activation of $GABA_B$ receptors in the DCN that regulate dendritic excitability and excitatory inputs (Caspary et al., 1987). $GABA_B$ receptors in CN modulate glutamatergic neurotransmission between the granule cells and their targets, the fusiform, cartwheel, and stellate cells in the superficial layers. The sources of GABAergic inputs to these regions could be vertical cells or superficial stellate cells, which co-contain GABA as well as glycine (Mugnaini, 1985; Altschuler et al., 1991; Davis and Young, 2000). These actions would be reflected also in neurons in the ICX that receive direct inputs from DCN neurons.

Responses of DCN neurons to trigeminal and trigeminal plus acoustic stimulation were compared in normal and noise-damaged guinea pigs (Shore et al., 2008). Guinea pigs with NIHL had significantly lower thresholds, shorter latencies and response durations, and increased amplitudes of response to trigeminal stimulation than normal animals. Noise-damaged animals also showed a greater proportion of inhibitory and a smaller proportion of excitatory responses compared with normal-hearing animals. The increase in inhibitory responses to trigeminal stimulation after noise damage could be due to a redistribution of somatosensory fibers in the CN granule cell region, so that more inhibitory interneurons, such as the cartwheel or superficial stellate cells, are targeted. The increased inhibitory responses suggest that new inputs may target CN granule cells that project to inhibitory interneurons, such as stellate or cartwheel cells. Results suggesting that intense tone exposure leads to increased spontaneous activity of DCN cartwheel cells (Chang et al., 2002) are in line with this interpretation. One would thus expect a reduced SFR in fusiform cells after noise trauma as indeed found by Ma and Young (2006) in decerebrated cats recorded from at least 4 weeks after the trauma. However, Finlayson and Kaltenbach (2009; Chapter 7) found strong SFR increases from recordings in the fusiform layer using a similar exposure paradigm but recorded from at 5–6 days after the exposure.

Thus, noise trauma may have a transient effect on the SFR of fusiform cells. Zhou and Shore (2006) indeed found that fusiform cell SFRs were significantly higher than normal at 1 week following noise damage. At 2 weeks post-noise damage SFRs were still increased but were no longer significantly different from control. More importantly, the increased SFRs were detected only in units that responded with excitation (E and E/I) to trigeminal stimulation. Units that showed no response to, or were inhibited by trigeminal stimulation did not show increased SFR following noise damage. Noise damage leads to an increase in the number of VGLUTs in mossy fibers as well as axonal sprouting from the trigeminal system, resulting in stronger responses to trigeminal stimulation. The finding that only neurons activated by trigeminal stimulation showed increased spontaneous rates after cochlear damage suggests that somatosensory neurons may play a role in the potential pathogenesis of tinnitus by the DCN. The finding that the increase in SFR was transient suggests that compensatory mechanisms are in place and makes it unlikely that this enhanced activity of the trigeminal system could be the sole reason for chronic tinnitus.

After unilateral deafening guinea pig's cochleas with kanamycin, Zeng et al. (2009) found that VGLUT1 immunoreactivity in the magnocellular CN areas ipsilateral to the damaged ear was significantly decreased, whereas VGLUT2 immunoreactivity in regions that receive non-auditory input was significantly increased 2 weeks after deafening. This two-week period is consistent with structural changes such as sprouting of somatosensory nerve endings and formation of new synapses onto granule cells. This increased glutamatergic input observed after NIHL resulted in increased SFR in those CN neurons that received somatosensory input. In addition, the threshold for trigeminal stimulation in these cells was reduced after noise trauma (Shore et al., 2008). Spiral ganglion cell loss as a result of the deafening with kanamycin was found to correlate with the increases in VGLUT2 (Zeng et al., 2009). This is consistent with the results of Bauer et al. (2007), which suggested that loss of large-diameter auditory nerve fibers may be a reliable predictor of tinnitus after noise exposure. Zeng et al. (2011) subsequently demonstrated that the terminals of the cuneate nucleus in the DCN, also colabel with VGLUT2.

The inferior colliculus Aitkin et al. (1978) studied activity of neurons in the ICX of cats in response to auditory stimuli and electrical stimulation of the dorsal columns. 46% of units were strictly unimodal (26% auditory and 20% dorsal column stimulation); 18% was bimodally excitatory, the remainder was excited by auditory and inhibited by dorsal column stimulation. Thus, both auditory and somatosensory activity results in discharges of units in the external nucleus of the inferior colliculus. Aitkin et al. (1981) found units responding to tonal stimuli were more commonly observed in ICX than were units with identified tactile receptive fields. They also found that the entire accessible body surface was represented in ICX. However, only vague somatotopy or tonotopy were observed.

El-Kashan and Shore (2004) investigated the effect of electrically stimulating the trigeminal ganglion on various central auditory nuclei using [^{14}C]2-DG. Electrical bipolar pulse stimuli were applied to the left trigeminal ganglion. Negative control animals were not stimulated. A positive control animal was stimulated in the left ear using a 1-kHz tone burst with 200-ms duration and an amplitude of 80 dB SPL. 2-DG was administered by intramuscular injection. The cerebellum was also sampled as a gray matter indifferent intrabrain control region, but note that the cerebellum is often found to show increased activity in tinnitus patients (Chapter 4). Systematic and significant differences were found between 2-DG uptake in the CN and higher auditory centers (particularly the IC) between control and stimulated animals. Trigeminally-stimulated animals showed significantly higher uptake than non-stimulated animals in all auditory centers examined, especially ipsilateral to the stimulation site. In contrast to sound stimulation, which increased activity in the contralateral IC, trigeminal ganglion stimulation showed increased uptake in IC bilaterally. Thus, the projection from the predominantly somatosensory trigeminal ganglion can influence the activity of central auditory neurons in a manner distinct from acoustic stimulation, suggesting activation of non-lemniscal auditory pathways.

8.4 **Summary**

Inputs from the trigeminal and dorsal column systems innervate the granule cells in the CN. The output of the granule cells, the parallel fibers, activate both excitatory neurons such as the fusiform and giant cells, as well as the cartwheel cells which inhibit the fusiform cells and may also excite or inhibit other cartwheel cells. The resulting effect is a modulation of auditory inputs to the fusiform cells; this can be excitatory, inhibitory as well as a temporally offset combination thereof. This underlies the modulation of tinnitus loudness by, for example, jaw movements. Following hearing loss, the strength of the synapses of the somatosensory inputs increases and the activation thresholds decrease. This has been linked to increased SFR in the DCN and may explain the 2–5-day incubation time for these increased SFRs to develop after noise trauma. The suppression or enhancement of sound-evoked activity in DCN and ICX could be analogous to the suppression or enhancement of tinnitus by somatic stimulation experienced by some individuals with tinnitus. Manipulations that alter tinnitus, such as clenching of the jaw or lateral gaze, would be analogous to electrical stimulation of the TG or Sp5, as described earlier, that alters the neurons' spontaneous activity or responses to sound. The underlying mechanisms for these alterations could involve LTD and LTP. Such long-term plasticity of synapses occurs between parallel fibers and their targets, the fusiform, giant, and cartwheel cells.

Chapter 9

The neural synchrony model of tinnitus

Increased SFR in the auditory system may underlie tinnitus (Evans et al., 1981; Jastreboff and Sasaki, 1986). However, besides increases in SFR there are also changes in the synchrony of the spontaneous firings of neurons following application of tinnitus-inducing agents (Ochi and Eggermont, 1997; Noreña and Eggermont, 2003; 2006).

There is little doubt that tinnitus is a conscious percept (De Ridder et al., 2011) that can be affected by attention (Searchfield et al., 2007), and is not perceived during sleep. It has been proposed that consciousness requires synchronized neural activity (Melloni et al., 2007; Ward, 2011). The role of neural synchrony in perception is not an issue restricted to the study of tinnitus. There is a large body of literature, mainly originating from studies of the visual cortex that suggests (Singer and Gray 1995; Uhlhaas and Singer, 2006), but also disputes (Shadlen and Movshon, 1999), a role of neural synchrony in perceptual binding and conscious processing. Transient synchronization of neuronal discharges may be one possible mechanism to dynamically bind widely distributed sets of neurons. For example, the neural activity of those neurons that code for color and those that represent texture has to be bound into functionally coherent ensembles that represent the neural correlates of the different combinations of texture and color of the similar-sized red pool ball and the yellow tennis ball (Singer, 1999).

At least two different mechanisms can link a percept such as tinnitus to an increase in neural synchrony. First, an increase in firing synchrony for different neurons (parallel synchrony) within a limited frequency band corresponding to the tinnitus pitch could differentiate them from the other neurons. This local increase in synchrony within a neural population could result in "popping-out" activity that potentially underlies tinnitus. Furthermore, neurons that receive synchronized inputs generally show an enhanced probability of firing (Abeles, 1982) and thus make the transmission of neural activity downstream more reliable. Secondly, serial synchrony in the form of burst firing in individual neurons can also enhance the probability of firing in receiving neurons. If the tinnitus signal is generated in subcortical structures, increased synchrony in general may propagate this activity to the auditory cortex and frontal cognitive areas. The serial interplay of increased firing rates, increased synchrony, and therefore efficient propagation of neuronal firing to downstream neurons is potentially sufficient to maintain a phantom percept such as tinnitus.

This chapter aims to highlight the potential role of increased spontaneous neural synchrony, both serial (burst firing within a neuron) and parallel (coincident firing or bursting between neurons), in the perception of tinnitus.

9.1 What is neural synchrony?

I will from now on restrict the phrase neural synchrony to encompass only parallel synchrony. Burst firing will be discussed later on. Neural synchrony is a reflection of the nearly simultaneous firing of individual neurons (micro synchrony), of the synchronization of membrane potential changes in local neural groups as reflected in the LFPs (meso synchrony), or of the presence of oscillatory brain waves in the EEG (macro synchrony). A general review on large-scale integration in the brain and the role of neural synchrony therein is presented by Varela et al. (2001).

9.1.1 Micro synchrony

On statistical and geometrical grounds (Braitenberg and Schüz, 1991), the probability that two cortical neurons with the axonal arborization of one neuron overlapping the dendritic tree of the other neuron make no synaptic contacts is very high (~0.9), and the probability of making just one contact is thus relatively low (~0.1). Assuming independence of contacting (synapse formation), the probability for making two (~0.01) and more than two contacts are therefore negligibly small. Thus in the cortical system of pyramidal cell to pyramidal cell connections the influence of one neuron onto another (micro synchrony) is very weak.

9.1.2 Meso synchrony

The relationship of fast LFP oscillations with neuronal spike discharges consistently shows that these discharges are generally occurring at negative going deflections of the depth-recorded LFPs in auditory cortex (Eggermont and Smith, 1995a). The spontaneous firing of single neurons is also highly correlated with the LFP as reflected in spike-triggered averaging of the LFP (Eggermont and Smith, 1995a; Britvina and Eggermont, 2008). While it was originally thought that excitatory postsynaptic potentials are the most important source of LFPs (Mitzdorf, 1985), other sources such as inhibitory postsynaptic potentials, other subthreshold membrane potential oscillations, and action potential afterpotentials also contribute significantly to the LFP (Berens et al., 2008). Low-frequency (<10 Hz) oscillations in the LFP may reflect global neuromodulatory inputs (Steriade, 2006). Gamma-band (40–80 Hz) rhythms, however, originate more locally within cortical microcircuits consisting of pyramidal cells and interneurons (Berens et al., 2008). Thus, while pyramidal neurons may be the strongest contributors to the LFP due to their dipole size, geometry, and parallel-oriented dendritic trees, interneurons play an important role in generating the underlying gamma-oscillations. They act as oscillation generators via rhythmic inhibition of pyramidal neurons and their synchronized inhibitory synaptic potentials also contribute significantly to the membrane potential oscillations on pyramidal neurons (Bartos et al., 2007; Fries et al., 2007). GABAergic neurons play a pivotal role in the primary generation of gamma-band oscillations and their local synchronization, whereas glutamatergic connections appear to control the strength, duration, and long-range synchronization of these oscillations (Wang and Buzsaki, 1996). In order for the LFP to be recorded a large number of neurons should synchronously change their membrane potential. Thus the amplitude of the LFP represents meso synchrony at the population membrane-potential level.

9.1.3 Macro synchrony

The generation of detectable spontaneous oscillatory activity in the EEG requires synchronous activity of large numbers of neurons and in addition spatial alignment of the underlying LFP dipoles, which largely restricts the contributing elements to cortical pyramidal cells. Although the fast gamma- and beta-band (20–40 Hz) frequencies in the EEG spectrum are most clearly involved in the establishment of synchrony, they are occurring in the context of the slower alpha (8–14-Hz) and theta (4–7-Hz) bands (Varela et al., 2001; Young and Eggermont, 2009). The mutual influence between specific nuclei of the thalamus and the cortex is pervasive. These reciprocal relations can deploy across different frequencies (including the alpha range around 10 Hz; Steriade and Amzica, 1996; Llinas, 1988), which are involved in setting and resetting the cycles of excitatory postsynaptic potentials on pyramidal cells. A slower pacing occurs in the theta band in limbic structures during memory consolidation. These slower rhythms could provide the slower temporal framing (Buzsáki and Draguhn, 2004; Young and Eggermont, 2009) for successive cognitive moments of synchronous assemblies, a slower beat within which beta and gamma rhythms operate. In normal circumstances auditory or other sensory stimuli increase thalamocortical rhythms to gamma-band firing rates and reduce the amplitudes of alpha-range frequencies. In the deafferented state, however, the thalamocortical activity decreases in frequency to those of the theta band (Steriade, 2006).

9.2 Measuring neural micro synchrony

Correlated neural activity at the micro level results from above-chance level synchronous firings of groups of single units. Figure 9.1 shows in the left part, a 10-s section of spontaneous firings of four simultaneously recorded neurons in cat AI, three of which fire rather frequently, and the fourth rather sparsely. The right-hand side shows a 100-ms blow-up (around 17.5 s) elucidating the near synchronous firing of all four neurons in this timeframe.

The peak of the cross-correlograms obtained in auditory cortex is typically found around zero-lag time (Figure 9.2). This is usually taken as an indication of shared input by the two neurons involved. If the peak is displaced from the zero time point by one synaptic delay and is asymmetric, i.e. shaped as the time-derivative of the EPSP, then the cross-correlogram can be interpreted as due to monosynaptic excitation. This is very rarely observed for extracellular recordings in cortex, especially if they are done with separate microelectrodes. Correlated neural activity can be calculated for neurons recorded on single electrodes, but problems arise because of overlap of spike waveforms and of the dead time of the sorting, i.e. when one waveform is assigned to a particular class, the next item has to come one "waveform-duration" later. This results in a dead time, of the order of 1 ms, in the cross-correlogram (upper panel of Figure 9.2). LFP-spindles can produce a periodic correlogram (Figure 9.2 bottom panel). Secondary effects, such as

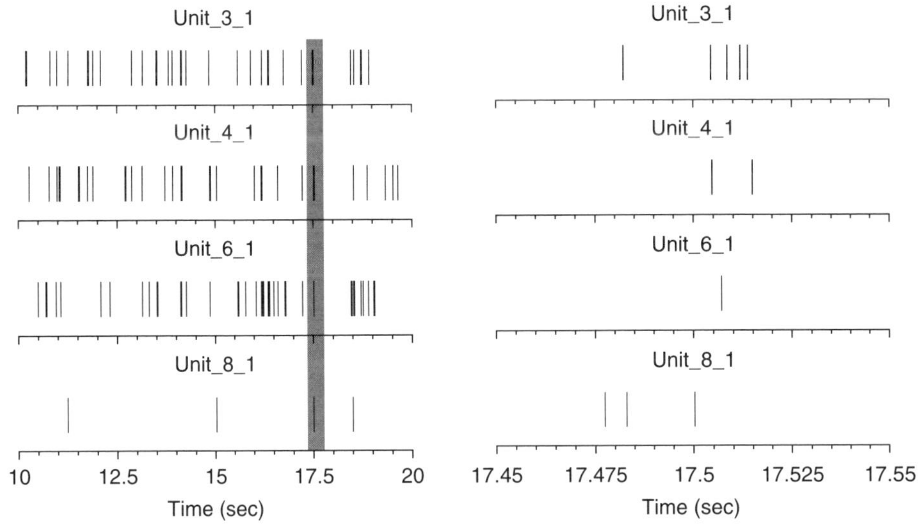

Figure 9.1 (Also see Color plate 6.) A 10-s section of spontaneous activity for four simultaneously recorded neurons from cat primary auditory cortex (left). On the right a 100-ms part around the time mark of 17.5 s (corresponding to the highlighted region on the left) shows synchronous firing within a 5–10-ms window.

those resulting from LFP spindles, can be effectively removed by deconvolving the cross-correlogram with the auto-correlogram of the trigger unit (Eggermont and Smith, 1996; Eggermont, 2006). The effect of burst firing could be estimated by comparing cross-correlograms obtained for "isolated" spikes (no spike preceding or following within 30 ms) with those for all spikes. Alternatively, one could only use the first spike of bursts together with isolated spikes.

A sleeping brain typically shows higher neural correlations that extend over larger distances than in an awake brain or a brain that is processing information (evidence reviewed in Eggermont, 1990). One could say that the neurons in a sleeping brain are part of several large assemblies. In contrast one may assume that in an information processing brain, regardless of whether it is under light anesthesia or not, neurons with different response properties will not very likely be doing the same thing at the same time and thus will show less synchronized firings (Cohen and Maunsell, 2009).

The strength of the correlation between the firings of cortical neurons, either as a result of thalamic common input or resulting from intracortical connections, is only modest. In Abeles's (1982) recordings from AI in the cat the average peak correlation coefficient was only 0.06. A subsequent literature survey revealed that even the strongest synapses did not exceed the peak strength of 0.25 postsynaptic spikes per presynaptic spike (Eggermont 1992a). From our own studies under ketamine anesthesia the mean correlation peak strength was 0.02 with only few values above 0.15. This suggests that only a very small fraction of the firings of two cells is synchronized because of either a common input or a direct unilateral excitation, and that the vast majority of the firings is due to asynchronous activation from a large number of other cells. Thus, despite the extensive horizontal,

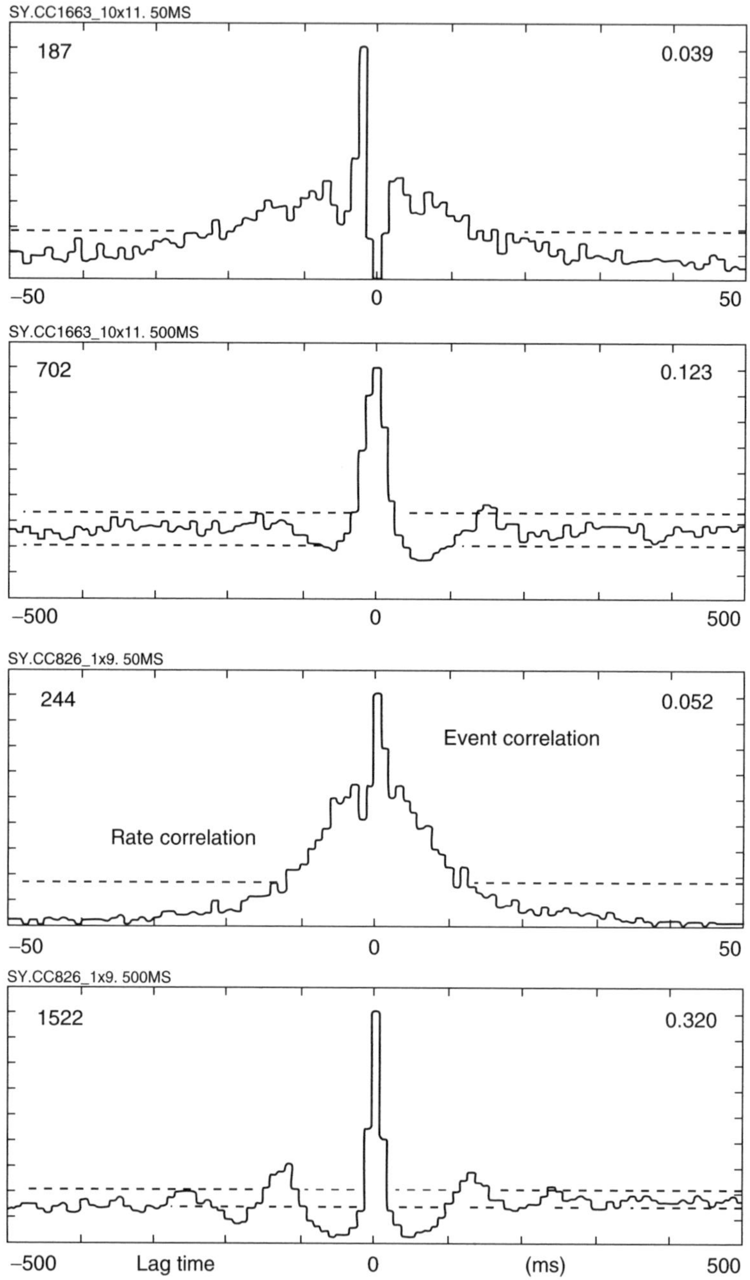

Figure 9.2 Examples of cross-correlograms for units recorded on single (top two panels) and separate electrodes (bottom two panels). The upper part of each block of two shows results on a time scale of –50 to 50 ms (1 ms bins) and the lower part shows a time scale of –500 to 500 ms using 10 ms bins. The dashed lines indicate ± 4 SD from the expected value under independence. Event-correlation relates to precise spike firing in the two neurons, whereas rate-correlation relates to covariance in SFR between the two neurons (see also Figure 9.4). The oscillation visible in the long time-scale correlogram in the bottom panel reflects periodicities of the spindle oscillation present in this anesthetized preparation. Reproduced from Eggermont, J.J. (1992) Neural interaction in cat primary auditory cortex. Dependence on recording depth, electrode separation and age, *Journal of Neurophysiology*, **68**, 1216–28. © 1992, The American Physiological Society, with permission.

i.e. within a layer, connectivity (see later), the active or awake cortex appears to be acting as a weakly coupled neuronal net that mostly produces uncorrelated firings and only occasionally seem to synchronize. In contrast, the "idling" or drowsy cortex is widely synchronized and produces detectable signs of this synchrony in the EEG waves and reflected as mostly oscillations in the lower frequency bands.

The shape of the cross-correlogram is affected by the auto-correlation structure of the spike trains, i.e. periodicities in the firing of the individual neurons will be reflected in the cross-correlogram as secondary features, see, for example, the spindle oscillations in the bottom part of Figure 9.2. Burst firing of neurons will also affect the width of the central peak in the cross-correlogram. In addition (Figure 9.3), the cross-correlogram may reflect the sum of the rate covariance between the two neurons (broad peaks or pedestals, often induced by a stimulus) and event-correlation (coincident firing). We found that secondary effects from burst firing, global synchrony related to LFPs, and stimulus-related effects largely obscure any underlying correlation produced by event correlation that would reflect direct connectivity (Eggermont and Smith 1996). The effect of rate covariance on the correlogram can typically be predicted from the overall firing rates (Figure 9.3). In this example, the oscillatory rate covariance produces the broad correlogram and subtracting that from the overall correlogram reveals the event correlation for this dual electrode pair.

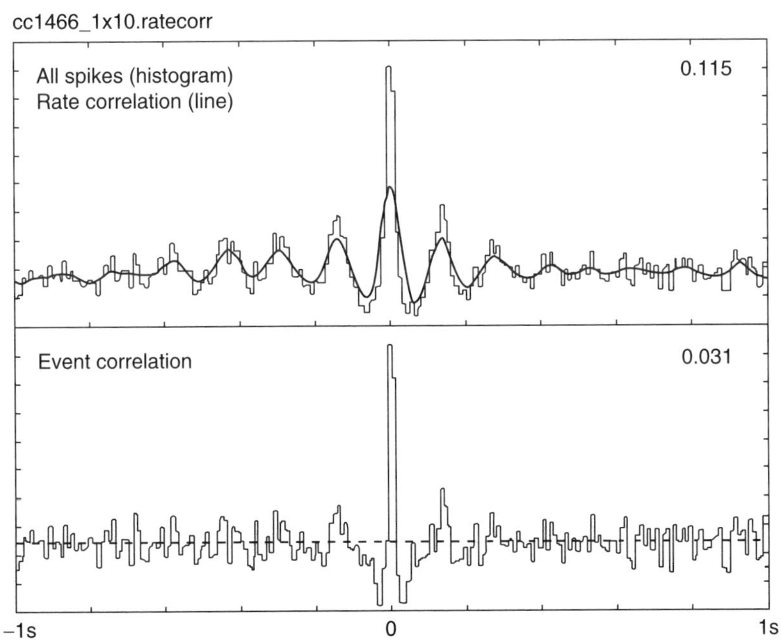

Figure 9.3 Oscillatory cross-correlogram (10-ms bin width) for which the secondary effects are largely due to rate covariation. The event-correlogram (lower panel) shows that only the first two side lobes of the oscillation pattern have some contribution from event correlation. Reproduced from Eggermont, J.J. and Smith, G.M. (1995) Rate covariance dominates spontaneous cortical unit-pair correlograms. *NeuroReport*, **6**, 2125–28. © 1995, Wolters Kluwer Health, with permission.

9.3 Where do the common inputs to cortical neurons originate?

Even in layer IV, the thalamocortical input layer, the thalamic input contributes no more than 20% of all excitatory synapses (Douglas and Martin, 1991). The terminal arborization of a single thalamic nerve fiber, however, covers an area several millimeters in diameter (Rausell et al., 1998; but see Miller et al., 2001 for a lower estimate in auditory cortex) and each thalamic afferent contacts many cortical neurons. Specifically, for the auditory system this comprises the tonotopically-organized projection from the ventral division of the medial geniculate body and the diffuse projection from the non-tonotopically organized medial division of the MGB. Potentially the input of the reticular activating system (RAS) to layer I of cortex may also contribute to neural synchrony. Because the RAS fibers are unmyelinated, and thus have slow conduction velocities, the variation in their spike firing will be large and the peak in the correlogram will be very broad. The horizontal spread of thalamic arborizations ensures that many neurons in a particular layer receive simultaneous activation from a single source. It is therefore not surprising that 60–90% of cell pairs in sensory and motor cortex show correlated firing behavior that is dominantly of the common input type (Eggermont, 1990a; Fetz et al., 1991). On the basis of the number of presynaptic, thalamic spikes that are correlated with a single postsynaptic, neocortical spike, Fetz et al. (1991) suggested that common input pairs in neocortex reflect a modest to good synchrony between the activity of 2–10 unobserved cells (presumably from the thalamus) that give rise to a spike in one of the observed cortical cells that is synchronized with a spike in the other cortical cell from which it was recorded.

Connections over large spatial divisions of auditory cortex are provided by the thalamic cell axonal divergence and convergence, often estimated to be between 2–5 mm at the cortical level (in cat and monkey, see earlier) and intracortically through horizontal fibers (Clarke et al., 1993; Wallace et al., 1991). These thalamocortical and cortico-cortical connections are for a sizable part heterotopic, i.e. do connect cell groups with CFs differing by more than one octave (Lee et al., 2004). Such neurons have generally non-overlapping receptive fields but still can have a sizeable correlation in firing (Eggermont, 2006).

Correlated neural activity in auditory cortex in the vast majority of cases is thus the result of common input either from divergent connections of thalamic cells, or from cortical interconnectivity via horizontal fibers. As a result, network properties such as oscillations synchronized over large (alpha and spindle rhythms) or more restricted (beta and gamma rhythms) cortical areas will affect the preferred timing of synchronous spiking and thus could be precursors for precise synchrony. Gamma-band oscillations in auditory cortex have been shown for unit recordings (Brosch et al., 2002) and for LFPs (MacDonald and Barth, 1995).

Experimental and clinical data suggest that convulsive epilepsy is often associated with an imbalance between excitatory and inhibitory neurotransmitter systems, causing enhanced excitability. On the one hand, interneuron activity reduces network excitability and thus the strength of the synchronization; on the other hand, inhibitory neurons contribute essentially to the oscillatory patterning and synchronization of neural activity. Recent evidence indicates that cholinergic modulation also plays a crucial role in the fast, state-dependent facilitation of gamma-band oscillations and the associated response synchronization (Steriade et al., 1991).

9.4 Neural synchrony measures depend on firing rates

Spontaneous neural synchrony in AI is correlated with the SFR with r^2 around 0.2–0.25 (Eggermont 1992a; 2000b). This can be interpreted in two ways; the first is that changes in neural synchrony are determined largely by changes in SFR (de la Rocha et al., 2007). This may be true for rate covariance but not necessarily for event correlation (Figure 9.3; Eggermont and Smith, 1995b). However, our previous studies into the immediate effects of an acoustic trauma (Chapter 7 and later in this chapter) indicate that initial changes in neural synchrony and SFR are uncorrelated, whereas, a few hours after the trauma, they became correlated. I interpret this to indicate that neural synchrony changes are priming subsequent changes in firing rate. Binding in auditory perception likely occurs through synchronous firing of different neurons to auditory stimulus contours (Eggermont, 2001). Auditory features can be categorized into contour and texture components of sound. Contour components are those temporal aspects of sound that covary across frequency and are likely exclusively coded in the temporal domain. Onsets, noise bursts, and common rates of slow (<20 Hz) AM and FM, i.e. the region where rhythm is dominant, are clearly contours that delineate, for instance, sound duration and separation between noise bursts and formant transitions. Texture comprises pitch, timbre, and roughness and slow changes therein. These texture aspects of sound relate to constant or slowly changing spectral representations in a cortical rate-place code. One expects texture and contour components thus to have largely independent neural representations in auditory cortex or potentially also in subcortical areas. After application of tinnitus-inducing agents, there is no evidence for increased neural synchrony between different cortical areas (Eggermont 2004), in contrast to the effect of sound stimulation, which did show preferential increases in synchronization between cortical areas (Eggermont 2000b). This could represent a way to distinguish between external sounds and internal ones such as tinnitus. So tinnitus is potentially the result of increased neural synchrony but it is likely not the cause of increased correlation in neural activity.

9.5 Consequences of neural synchrony

The effect of increased interneuron synchrony is generally a more efficient excitation of downstream neurons, especially at the cortical level, where synaptic strengths are weak. Abeles (1982) first drew attention to the power of synchronously arriving inputs to activate cortical neurons. The consequence of correlated neural input is that the amplitude of the postsynaptic potentials (PSPs) resulting from these inputs will be proportional to the number of inputs, whereas the PSP amplitude from asynchronously arriving inputs is only proportional to the square root of their number. The synaptic connection strengths, and firing rates for thalamic cells combined, are potentially a factor 3–4 higher than those of corticocortical cells (Gil et al., 1999). Then, even only 30 synchronously firing specific afferents (the putative number converging on a cortical cell, Miller et al., 2001), which produce compound PSPs that are 3–4 times larger than for corticocortical cells and effectively produce 90–120 amplitude units of PSP, would still be as effective as 9000 asynchronously firing non-specific corticocortical inputs that produce $\sqrt{(9000)}$ ~95 amplitude units of PSP. Given that a typical cortical cell receives ~10,000 inputs, the few synchronous ones are potentially as effective as all the other asynchronous ones together.

Simultaneous recording from cell pairs in the thalamus and a cortical target cell were used to investigate this hypothesis by studying the effect of synchronous inputs of thalamic neurons on the ability to fire a cortical cell (Alonso and Martinez, 1989; Roy and Alloway, 2001; Wang et al., 2010). These studies in the somatosensory system found that thalamocortical efficacy for synchronous thalamic activity was nearly twice as large as the efficacy obtained when pairs of thalamic neurons discharged asynchronously. Thus neural synchronization may play a critical role in the transmission of sensory information from the somatosensory thalamus to the cortex. It is likely that increased synchronization of auditory cortical neurons will also enhance the transmission of information to subsequent stages in auditory processing. Synchronous neural activity indeed ensures efficient propagation and preservation of the temporal precision of neural firing downstream along the auditory pathway (Kistler and Gerstner 2002; Kimpo et al., 2003; Reyes 2003).

Increased sound levels typically result in a higher firing rate of neurons and by virtue of that the covariance in firing rate between neurons will also increase. The emergence of such synchronized firing activity in neighboring ANFs as a result of increased stimulus levels may lead to stronger activation of certain CN cells, such as the bushy cells, whose activation partially depends on coincident input from different ANFs (Oertel et al., 1988). As a result, neural synchronization to the stimulus in CN neurons is enhanced (Joris et al., 1994). The strength of the resulting neural synchrony may even reflect the relative strength of the stimulus (Voigt and Young, 1980). Thus, the cooperative effort of several neurons enhances the accuracy of representing sound levels, and extends the dynamic range of the output neuron if the input neurons represent a certain range of threshold values.

A transformation of firing synchrony with the stimulus through "phase-locking" into firing rate, likely occurring at or below the level of the IC (Langner, 1997), may diminish the role of firing synchrony relative to the role of firing rate when advancing up the neuraxis. In this view, a sound sensation must be related to a "sufficiently strong" activation of the higher centers of the auditory nervous system. This "combination theory" requires a threshold mechanism that determines what "sufficiently strong" activation is in order to allow the decision that spontaneous activity does not represent sound.

Correlations between the spontaneous firings of neuron pairs in primary auditory cortex are found in at least 60% of the pairs recorded from within AI (Dickson and Gerstein, 1974; Eggermont, 1992a) and between different cortical fields (Eggermont, 2000b). These correlations may result from direct synaptic interactions between the neurons (although very few of these have actually been demonstrated in correlation studies from visual cortex;

Toyama et al., 1981). Thus, the phenomenon of correlated neural activity in AI in itself will not be sufficient to signal whether a sound is present or not. Changes in correlation strength must also be involved. This returns us to the problem of how much change is required to decide on the presence or absence of a stimulus.

Percepts such as tinnitus should be accompanied by measurable changes in auditory cortical activity. Thus, it is important to relate subcortical changes caused by tinnitus-inducing agents to those found in the auditory cortex under the same conditions to sort out the sequence of activity that leads to tinnitus. I have argued in the past (Eggermont and Sininger, 1995; Eggermont, 2000a; 2004) that the neural correlates of tinnitus should be similar to the neural correlates of audible sound. Thus, when a sound is made louder, the neurons in the cortex typically increase their firing rates, suggesting that increased SFR should indicate the presence of tinnitus. Raising the sound level from sub- to supra-threshold also increases the neural synchrony (Eggermont, 2000b), so one could interpret increased neural synchrony also as a correlate of tinnitus. Burst firing in cortex is also affected by sound (Bowman et al., 1995) but, typically, in AI the duration and number of spikes in a burst decrease with increasing sound level, likely as a result of post activation suppression. It is, however, not a priori clear if the changes need to happen in AI or that the neural correlate of tinnitus are dominated by the non-lemniscal pathway activity and thus by AII (Eggermont and Kenmochi, 1998).

9.6 **Burst firing: serial firing synchrony**

Bursts of action potentials, regardless of the nucleus of origin, have been implicated in various phenomena, including synaptic plasticity, selective communication between neurons, and pathological conditions such as epileptic seizures (Krahe and Gabbiani, 2004). Burst-firing within a certain time window (≤20 ms) is also efficient to ensure synaptic transmission by allowing temporal integration of the resulting postsynaptic potentials (Gourévitch and Eggermont, 2007). Although burst firing does not appear to increase the amount of information transmitted it appears to lower the detection threshold even when the overall firing rate has decreased (Bair et al., 1994; Sherman, 2001). Bursting may in fact enhance the overall firing rate of the recipient neurons, by up to a factor 20 (Dickenson, 1996). If spontaneous bursting manifests itself throughout the auditory system, it may be one of the substrates for tinnitus.

Typically, thalamocortical neurons possess two distinct firing modes. When depolarized, thalamic cells fire streams of single action potentials (tonic firing) and faithfully relay synaptic inputs to the cortex. When hyperpolarized, the activation of low-threshold T-type Ca^{2+} channels promotes burst firing, and the transfer is less accurate (Llinás and Steriade, 2006). However, Wolfart et al. (2005) found that this distinction no longer holds if synaptic background activity is taken into account. The combination of synaptic noise originating largely from cortex with the intrinsic properties of thalamocortical cells results in more linear mixed single-spike and spike-burst responses at all membrane potentials.

In ketamine-anesthetized AI, bursts come in two types—an irregular type (Figure 9.4a) and a stereotyped one (Figure 9.4b)—that can be recorded simultaneously and therefore

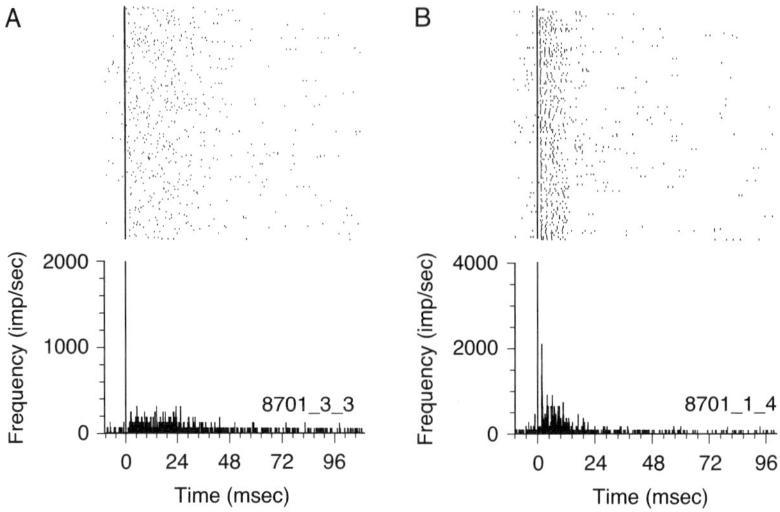

Figure 9.4 Peri-event rasters for (A) a non-stereotyped bursting unit recorded in AI, and (B) a stereotyped burster recorded in AAF simultaneously with the AI unit. The rasters and the PSTH were triggered on the first spike of the bursts. Ten ms of spontaneous activity are shown before the first burst spike. The burst in (A) has a duration of about 50 ms. The AAF unit in (B) exhibits a well-defined ISI of 2 ms (bin width 0.5 ms) and has an average duration of about 25 ms. Reprinted from Valentine, P.A. and Eggermont, J.J. (2001) Spontaneous burst firing in three auditory cortical fields and its relation to local field potentials. *Hearing Research*, **154**, 146–57. Copyright (2001), with permission from Elsevier.

are not reflections of a different state of the animal (Valentine and Eggermont, 2001). Stereotyped bursting shows preferred interspike intervals and was only observed in a few cases, and is shown here for a recording from AAF. As we have seen before (Chapters 6 and 7) stereotyped bursting in IC may be correlated with tinnitus.

Recalling that synchronous spiking activity from afferents causes much larger EPSPs than for asynchronous inputs, one might wonder what the effect of neural burst firing of input neurons might be. One observation that supports a selective communication role for bursts relates to the unreliability of synaptic transmission, especially in cortex. Facilitating synapses, which are typically found on interneurons in auditory cortex (Atzori et al., 2001), have a low probability of transmitter release when depolarized by a single spike. If one or more spikes follow within a brief time, accumulation of Ca^{2+} in the presynaptic terminal causes more transmitter to be released, and the postsynaptic response builds up over the course of the high-frequency input. This observation indicates that burst spikes can be transmitted across synapses more reliably than isolated spikes. This has been supported by multiunit recordings from cat striate cortex (Krahe and Gabbiani, 2004), in which bursts of spikes showed a significantly increased effectiveness of eliciting a response in a postsynaptic neuron. Furthermore, effectiveness was positively correlated with burst length. However, the role of bursts in synaptic transmission from thalamic relay cells to pyramidal neurons in the auditory cortex differs from this model. Auditory thalamocortical synapses typically have a higher probability of transmitter release for single presynaptic spikes and show depression if spikes follow each other rapidly (Atzori et al., 2001); still burst inputs could be more efficient compared to single spike inputs.

Increased burst firing is also common in patients with amputation pain. And because pain and tinnitus potentially share common mechanisms (Møller, 1997) it is relevant to give a brief account of some findings. Lenz et al. (1998) observed burst firing in the human ventral caudal nucleus of the thalamus in awake patients during the physiologic exploration that precedes surgical procedures for treatment of stump pain and movement disorders. All patients with amputations showed increased thalamic representations of the stump in both receptive and projected field maps. Three different firing patterns were observed: bursting activity, Poisson-like (i.e. random firing), and a group with firing patterns other than the two previous ones. Burst-firing neurons were significantly more common in thalamic regions outside the stump area (41%) and stump areas (33%) than in control areas obtained from movement disorder patients (15%). The burst-firing patterns were consistent with the spike-burst pattern of low-threshold Ca^{2+} spikes. This spike-burst pattern was most common among cells with receptive fields in the stump area. However, in these cells the firing rate between bursts was significantly higher than for other cells in the region of the ventral caudal nucleus, suggesting that in the stump area the spike-bursts were not due to hyperpolarization, as in low-threshold Ca^{2+} spikes. Therefore the increased spike rate in amputation patients is consistent with persistent membrane depolarization. In this situation spike-bursts might result from dendritic calcium spikes, which can result from depolarizing pulses applied to putative dendrites, i.e. high threshold spikes.

9.7 **Increased neural synchrony in animal brains**

9.7.1 **Serial synchrony in putative tinnitus**

In most studies involving damage to the cochlea, either after NIHL (Liberman and Kiang, 1978), application of ototoxic drugs (Kiang et al., 1970), or after the introduction of an endolymphatic hydrops (Harrison and Prijs, 1984), an increase in the number of spike pairs and short bursts over and above that expected from a Poisson process is observed in those fibers with reduced SFR. In tone-exposed animals, Finlayson and Kaltenbach (2009) observed an increase in the SFR of simple spiking activity as well as in the incidence of spontaneous bursting activity that accounted for about half of the increase in spontaneous activity. Increased burst firing in the ICX also occurs after salicylate application (Chen and Jastreboff, 1995). After noise trauma burst firing increased in the DCN (Finlayson and Kaltenbach, 2009), whereas in primary cortical neurons that tended to burst in synchrony with EEG-spindles (Eggermont, 1992a; Eggermont and Smith, 1995a,b; Britvina and Eggermont, 2008), the amount of bursting observed after salicylate or quinine application did not change (Ochi and Eggermont, 1996; 1997). The majority of non-lemniscal MGB cells exhibits bursting responses, whereas lemniscal neurons discharged mainly single or spike doublets (He and Hu, 2002). This favors the extralemniscal pathway in the propagation of tinnitus related activity.

Increased burst firing has been observed to co-occur with increased SFR in IC after both salicylate application (Chen and Jastreboff, 1995) and noise trauma (Bauer et al., 2008).

In primary auditory cortex, there was some increase in burst firing but only in the first minutes after the exposure (Noreña and Eggermont, 2003) and not for longer survival times, regardless the nearly twofold increase in SFR. The reason for this may be that all our recordings were from putative pyramidal cells and these have depressing synapses. We cannot exclude that inhibitory interneurons in auditory cortex show increased burst firing following noise trauma or salicylate.

9.7.2 Parallel synchrony in putative tinnitus

Møller (1984) proposed pathological neural synchrony, thought to result from disruptions in the myelin sheath of auditory nerve fibers, to explain tinnitus in patients with vestibular schwannoma. At the same time, Eggermont (1984) introduced interneural synchronization as a more general phenomenon underlying any sound sensation, stimulus induced or pathological, as follows:

> This [reorganization of firings] actually means that there are instantaneous rate changes in the individual nerve fiber firing patterns, but above all it means that activity patterns of small groups of nerve fibers become synchronized. This is a cooperative effect and it is now postulated that such cooperative effects in the central nervous system or auditory periphery may give rise to spontaneous sensations of sound, to tinnitus.

This idea was specifically elaborated to include both the myelin disruption mechanism and a mechanism for the induction of synchrony among the 20 or so auditory nerve fibers innervating the same IHC (Eggermont, 1990b). The proposal was based on the observed calcium spikes in saccular hair cells for increased extracellular calcium concentrations. Modeling showed that increased calcium concentration in the IHCs could produce temporal changes in individual neuron spike trains as well as synchrony between the firings of neighboring nerve fibers.

Experimental findings relevant for the peripheral neural synchrony model of tinnitus were reported first by Schreiner and Snyder (1987) and Martin et al. (1993). They recorded ensemble activity from the surface of the exposed cat cochlear nerve, and the exposed auditory nerve in human subjects during intraoperative monitoring (Martin, 1995). After salicylate application, the normally present 1-kHz peak in spectrum of the spontaneous activity disappears and a clear peak emerges around 200 Hz (see also Chapter 6). This was interpreted as a sign of changed synchronized rhythmic ensemble activity (Lenarz et al., 1995). Cazals and Huang (1996), after salicylate application, also found a strong 1-kHz peak in the average activity spectrum recorded from the round window or from the auditory nerve surface of guinea pigs, which was decreased by contralateral masking. They suggest that "some kind of synchrony on a short time basis, of about 1 ms, should exist among a group of fibers innervating a limited cochlear area centered around 16 kHz." In this author's opinion, the emerging 200-Hz peak could also represent the increased appearance of oscillatory neural components with a typical duration of 5 ms such as calcium-spikes accompanied by stereotypical fast-spike bursts of that duration. Because spontaneous periodic 200-Hz or 1000-Hz activity, has never been observed in individual auditory nerve fibers it is difficult to account for synchronized activity with that signature (except under stimulation with a 1-kHz tone). Lenarz (1996) subsequently stated that this 200-Hz component does not originate in the auditory nerve.

Cazals and Huang (1996) suggested that the activity spectrum recorded from the nerve surface would reflect synchronized ANF firings and they provided evidence for an influence of olivo-cochlear feedback that was dependent on drowsiness versus awakeness of the guinea pig. The spike-like waveforms recorded by McMahon and Patuzzi (2002) from the round window in guinea pigs in response to sound were very similar and independent of the sound level in the range of 20–60 dB SPL (Figure 9.5). They assumed therefore that these waveforms were an estimate of the shape of the contribution of the firing of a single neuron to the gross round window signal (the unitary potential). They could model the production of these signals and the compound action potential in response to 0.25-ms and 25-ms tone-bursts at 20 kHz by including only a damped 900-Hz resonance in the unitary potential without refractory effects. This suggested to them that the spectra of these waveforms did not reflect preferred intervals or synchronization in the timing of neural spike generation. Searchfield et al. (2004) showed that this round window spectrum is dependent on the sensitivity of the auditory nerve and requires intact auditory

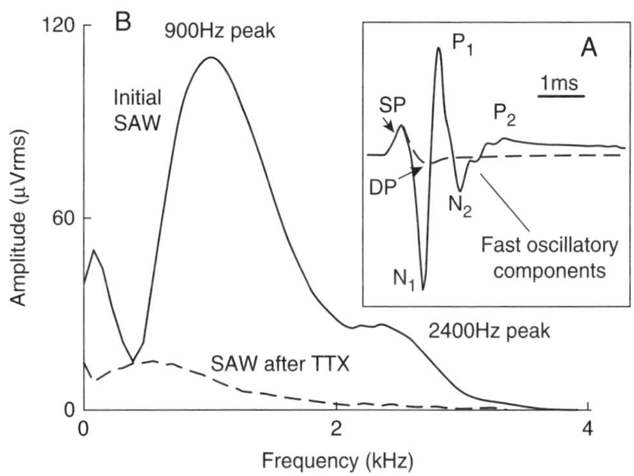

Figure 9.5 A) Average waveforms and B) their spectra evoked at the RW by a 0.25 ms tone-burst at 20 kHz presented at 40 dB SPL. The waveforms obtained before blockage of neural spiking by TTX (solid curve) included the SP from the inner hair cells, and the N1, N2, P1, and P2 peaks, which are known to be of neural origin. Notice also the fast oscillatory components between the N2 and P2 peaks. After application of TTX (dashed curve) the SP remained, as did the DP, thought to be due to summed Na^+ currents entering the primary afferent nerve terminals. The spectrum of the average waveforms (SAW) before TTX (solid curve in B) has a major peak at 900 Hz and a minor peak at 2400 Hz. After TTX (dashed curve) the 900 Hz and 2400 Hz spectral components were abolished, and the spectrum was dominated by a broad 500 Hz peak presumably due to the combination of the remaining SP and DP. Reprinted from McMahon, C.M. and Patuzzi, R.B. (2002) The origin of the 900 Hz spectral peak in spontaneous and sound-evoked round-window electrical activity. *Hearing Research*, **173**(1–2), 134–52. Copyright (2002), with permission from Elsevier.

neurotransmission. Sendowski et al. (2006) pointed out that experimental conditions such as anesthesia, body temperature, and ambient noise modified the spontaneous neural outflow of the cochlea and must be taken into account when studying this compound activity.

Noise-induced hearing loss NIHL causes a reorganization of the tonotopic map in cat primary auditory cortex and increases both SFR and neural synchrony. Noreña and Eggermont (2003) showed that the peak cross-correlation coefficient (R) for spike-pair firings was increased immediately after the traumatizing exposure. This increase in R was local, affecting only the multiunit recordings for which the frequency-tuning properties were changed (emergence of new inputs and shift in CF). Interestingly, for long-standing NIHL, Eggermont and Komiya (2000) showed that the "reorganized part" of AI (units with a CF above the TTF) presented an increased proportion of significant correlation peaks. Seki and Eggermont (2003) and Noreña and Eggermont (2006) found in exposed cats, recorded from between 2 weeks and 4 months after the exposure, a significant increase in R (about a factor 1.4) in the reorganized parts of AI. These frequency-dependent changes in R in long-standing tone-induced hearing loss are then consistent with the acute ones seen in Noreña and Eggermont (2003).

The hearing loss induced by the acoustic trauma mostly affected the frequency band above the TTF. We observed that the frequency tuning of neurons shifted dominantly toward the TTF, and that the response areas became somewhat broader (Noreña et al., 2003). Consequently, the amount of response area overlap between neurons is increased after the trauma. Assuming that these neurons receive their excitatory inputs from the same thalamic sources (or at least a part of them), an increase in overlap of their response areas should then be paralleled by an increase in their common "synchronized" inputs. A few hours after the trauma, R was further increased, and nearly independent of the pre-trauma CF difference. The changes in R a few hours after the trauma were significantly correlated with those in the SFR. This overall increase in R could then be accounted for at least partially by the large increase in SFR. In this context, it is important to note that the increase in R immediately after the trauma did not show any dependence on changes in SFR.

These changes in tonotopic maps, SFR, and increased neural synchrony (a commonly occurring triad) could be prevented by rearing the noise-trauma cats in an EAE that compensates for the reduced firing activity in the high-frequency neurons as a result of the NIHL (Noreña and Eggermont 2005; 2006; Chapter 7). This suggests that the changes in tonotopic maps, SFR, and neural synchrony are likely all due to an across-frequency imbalance of neural output activity from the cochlea.

NIHL is not the only condition that may cause increases in neural synchrony and increased SFR. We also found these after exposing adult cats to a 4–20-kHz non-traumatizing sound for >5 months at a level of 80 dB SPL peak equivalent. No changes in ABR threshold were found, but neurons in AI ceased to respond to frequencies between 4–20 kHz. Neurons whose CFs would normally be in the 4–20-kHz range, based on their location in AI, now all responded to either frequencies >20 kHz, <4 kHz, or to both. For the neurons outside the reorganized range the SFRs were increased and their pair correlations showed greatly increased cross-correlation coefficients (Noreña et al., 2006).

Salicylate and quinine induced effects Quinine and salicylate application at safe doses produced the same average hearing loss in cats. However, the peak unit-pair cross-correlation in AI was significantly increased for quinine (Ochi and Eggermont, 1997) and not changed after salicylate, albeit that the cross-correlogram peak became narrower (Ochi and Eggermont, 1996). The amount of salicylate that cats can tolerate is limited to about 200 mg/kg, less than half of what rats tolerate (Knipper et al., 2010). SFR increases were only found in AII for both salicylate and quinine (Eggermont and Kenmochi, 1998).

Neural synchrony was also calculated between electrodes in different cortical areas (Eggermont, 2004). Overall, there was no significant effect of salicylate or quinine on the peak cross-correlation coefficient; none of the pair-wise correlations between neural activity recorded in the AI, AAF, and AII were significantly changed. This lack of increased neural correlation between cortical areas after salicylate or quinine application at a dose that potentially results in tinnitus, is in contrast to the effect of noise trauma, which produces both increased firing rates and increased neural correlation (Noreña and Eggermont, 2003).

9.7.3 Electrical kindling model of epilepsy

Anticonvulsant drugs have often been tried to alleviate tinnitus (Chapter 12) based on the assumption that hypersynchrony underlies both epilepsy and tinnitus. Kindling refers to a highly persistent modification of brain functioning in response to repeated application of initially subconvulsant electrical stimulation, typically in the limbic system, which results in the development of epileptiform activity (Goddard et al., 1969). The effect of electrical kindling could also be demonstrated by stimulation of primary auditory cortex (Valentine et al., 2004). Kindling by stimulating twice daily for 1 s with 60 biphasic pulses of 200–400 μA peak value for 2–3 weeks, resulted in approximately two-thirds of the animals reaching a fully generalized convulsive state in 40 stimulation sessions. After the animals reached this state, multiunit recordings were obtained under ketamine anesthesia from primary auditory cortex contralateral to the kindled site. A 40% enhancement of spontaneous neural synchrony, as measured by the peak spike-pair cross-correlation coefficient, was found. Hearing thresholds as measured by auditory brainstem response, were not affected by the kindling sessions. A profound alteration of the tonotopic map in AI was observed where a large area became tuned to similar high characteristic frequencies related to the approximate CF of the electrical stimulation site.

Other evidence from correlation studies in auditory cortex (Eggermont, 1994; 2000b; Brosch and Schreiner, 1999) indicates that cells with common response properties are more likely to exhibit significantly correlated activity even when anatomically distant. The very similar CFs for large parts of AI in kindled animals may similarly provide the basis for the enhanced correlated activity. Thus, electrical kindling resulted in substantial alterations in unit firing characteristics and reorganization of cat auditory cortex.

The role of GABA and glutamate receptors in kindling and NIHL is also comparable. Kindling is associated with a loss in GABAergic inhibition (Lopes da Silva et al., 1995). It has also been hypothesized that kindling stimulation in cortex produces an accumulation of glutamate. This may trigger altered NMDA channel functions and long-lasting changes in synaptic efficacy of long-range horizontal connections (Racine et al., 1995; Chapman, 1998).

In cat AI horizontal connections generally follow a course parallel to the isofrequency contours (Read et al., 2001) but others project in directions orthogonal to the isofrequency contours and may even extend as far as anterior and posterior auditory fields (Wallace et al., 1991). Strengthening of the horizontal connections after kindling may explain the larger percentage of double-tuned frequency-tuning curves that we observed (Valentine et al., 2004). A similar emergence of multituned neurons was found after long-term exposure to a spectrally (5– 20-kHz) enhanced acoustic environment that did not cause a hearing loss but still resulted in a dramatic cortical reorganization and increased neural synchrony (Noreña et al., 2006). If these horizontal connections become stronger than those of the specific thalamic inputs, they could elicit an outward migration of glutamate hyperactivity and subsequent synaptic changes. This process would likely stop when all the neurons in the stimulated cortical area (and its projection region) were combined into one large neural assembly.

One could thus entertain the notion that correlated neural activity may result in cortical map reorganization. Corticocortical connections for a large part connect cell groups with characteristic frequencies differing by more than one octave (Lee and Winer, 2004). Such neurons have generally non-overlapping receptive fields but still can have sizeable peak cross-correlations (Tomita and Eggermont, 2005). Correlated neural activity and heterotopic neural interconnections are potentially the substrates for cortical reorganization; increased neural synchrony and tonotopic map reorganization go hand in hand. This links cortical reorganization with hypersynchrony that can be considered as an important driving force underlying tinnitus (Eggermont, 2007b).

9.8 **Macroscopic synchrony in tinnitus patients**

The frequency of network oscillations in cortex as reflected in the EEG covers more than three orders of magnitude, from slow oscillations in the delta (0.5–3-Hz) and theta (3–8-Hz) ranges to fast oscillations in the gamma (30–90-Hz) and ultrafast (90–200-Hz) ranges. Within this spectrum, gamma oscillations have among others been proposed to represent reference signals for sensory binding of features into a coherent percept.

A key requirement for the generation of network oscillations is regular and synchronized neuronal activity. If several neurons fire action potentials both regularly and synchronously, the rhythmic synaptic activation pattern results in a fluctuating field potential signal, which can easily be measured using extracellular recording electrodes (Bartos et al., 2007). If large numbers of neurons are activated simultaneously, the corresponding currents then add up in phase and become so large that they can be detected at the scalp (Eggermont, 2007a). Although the amplitude of EEG or MEG signals correlates with the degree of synchrony of neuronal responses, there are numerous confounding variables that make it difficult to draw firm conclusions on synchrony by considering only amplitude measures. In particular, the degree of precision with which the neuronal discharges are synchronized is important. This reflects itself mostly in the higher frequency bands of EEG activity.

Spontaneous EEG or MEG activity relies on neural synchrony; without synchrony the EEG would just be noise. Neural synchrony appears to be confined to several frequency bands. In general, there is a correlation between the distance over which synchronization is observed and the frequency of the synchronized oscillations. Short distance synchronization tends to occur at higher frequencies (gamma-band) than long-distance synchronization, which often manifests itself in the beta-band but also in the theta-band and alpha-band. The latter, when occurring over the *temporal* lobe, is typically referred to as tau (τ) rhythm (Hari and Salmelin, 1997).

A chronically slowed occipital alpha rhythm during ongoing wakefulness is a definitive indicator of underlying pathology and several neurological and psychiatric disorders (Niedermeyer 1997). These include depression (Nystrom et al., 1986), and neurogenic pain (Llinas et al., 1999). Indeed, it has recently been shown that a shift in alpha rhythm frequency to the theta band is a common finding across a population of patients exhibiting a diverse selection of neurological and psychiatric disorders and is highly correlated with the presence of positive symptoms. This phenomenon has been given the general term "thalamic dysrhythmia" to reflect the belief that the alpha rhythm disturbances arise as a result of abnormal disfacilitation at the thalamic level (Llinas et al., 1999).

The spontaneous MEG in a group of individuals with tinnitus was characterized by a marked reduction in tau-band power together with an enhancement in delta-band power as compared to normal-hearing controls. This pattern was especially pronounced over the temporal cortex. Moreover, correlations with tinnitus-related distress revealed strong associations with this abnormal spontaneous activity pattern, particularly in right temporal and left frontal areas (Weisz et al., 2005a). This could be related to the often comorbid depression in tinnitus patients. Reduction of tau-rhythm power is a normal cortical reaction to sound presentation (Lehtelä et al., 1997) just as visual stimulation (eyes open) reduces the occipital alpha-rhythm. Thus, the results of Weisz et al. (2005a) could suggest that a decrease of tau-rhythm power in tinnitus patients may be related to the sound perception itself. Kahlbrock and Weisz (2008) found that changes of tinnitus intensity induced by residual inhibition (RI) were remediated by alterations in the pathological patterns of spontaneous brain activity, specifically a reduction of delta activity. Delta activity is a characteristic oscillatory activity generated by deafferented/deprived neuronal networks (Llinás et al., 2005). This implies that RI effects might reflect the transient re-establishment of balance between excitatory and inhibitory neuronal assemblies, via reafferentation, that have been perturbed (in most tinnitus individuals) by hearing damage.

High-frequency (gamma-band) oscillations have been linked to conscious sensory perception and positive symptoms in a variety of neurological disorders (Llinás et al., 2005). With that in mind, Weisz et al. (2007) examined gamma-band activity during brief periods of marked enhancement of delta-band activity (see Weisz et al., 2005a). Results revealed that both control and tinnitus groups showed significant increases in gamma-band activity after the onset of slow waves. However, gamma-band activity was more prominent in tinnitus subjects than in controls. The hemispheric lateralization of the gamma-band activity did correlate with the lateralization of the perceived tinnitus. In contrast, Ashton et al. (2007) identified discrete localized unilateral foci of high-frequency activity in the gamma-band range (>40–80 Hz) over the auditory cortex in eight patients experiencing tinnitus during recording. These hotspots were not present in subjects without tinnitus.

In unilateral tinnitus patients, source analysis of resting state EEG gamma-band oscillations shows a strong positive correlation of the source strength in the contralateral auditory cortex with VAS loudness scores (van der Loo et al., 2009). Auditory phantom percepts thus show similar sound level dependent activation of the contralateral auditory cortex as observed in normal audition.

Recently, the reciprocal relationship between reduced tau-activity and increased gamma activity in tinnitus patients was further elucidated (Lorenz et al., 2009). We have seen that depending on the frequency band of the EEG oscillations, local or more global neural synchrony is implicated. In particular the long-range synchrony coupling appears to be affected in tinnitus patients. In particular long-range coupling in the 9–12-Hz frequency band was reduced and that in the long-range 45–54-Hz frequency band was increased (Schlee et al., 2009a). In patients with a tinnitus history of <4 years, the left temporal cortex was predominant in the gamma network whereas in patients with tinnitus duration of >4 years, the gamma network was more widely distributed including more frontal and parietal regions.

9.9 Role of neural synchrony in tinnitus perception

Although, in the auditory cortex, neural cross-correlation studies are not performed routinely, one can summarize the findings as showing a significant increase in neural synchrony between neural spiking in separate cortical areas during stimulation compared with spontaneous firing. The amount of cortical neural synchrony in the cat auditory cortex (Eggermont, 2000a) or, for that matter, in the frog auditory midbrain (Eggermont, 1989) did not depend on the stimulus level. Thus, the abrupt change in neural synchrony between sub- and suprathreshold conditions may help, together with the increase in firing rate, in signaling the presence of sound. Neural synchrony has also been implicated in signaling the presence of a continuous tone in the absence of any difference in firing rate from the spontaneous condition (deCharms and Merzenich, 1996; Eggermont, 1997). Thus, neural synchrony may signal both a change from silence to sound and the presence of an ongoing stimulus. The latter may be important in the perception of tinnitus. What does this mean in relation to macro synchrony in the EEG?

Synchronized gamma-band activity in general is proposed to bind sensory events into one coherent conscious percept (Singer and Gray, 1995). Tinnitus, as a constant auditory phantom percept is potentially correlated to persistent gamma-band activity in the auditory cortex (van der Loo et al., 2009). In normal hearing there is a sound level dependent activation of the primary auditory cortex in humans as investigated with EEG and fMRI (Mulert et al., 2005). Gross et al. (2007) provided direct evidence for a close association between induced gamma oscillations and the conscious and subjective perception of pain. Single-cell recordings for close-to-threshold somatosensory stimuli suggested that stimulus intensity is coded in the primary somatosensory cortex whereas the conscious percept per se is coded in the prefrontal cortex (de Lafuente and Romo, 2005). Following the earlier-mentioned ideas van der Loo et al. (2009) hypothesized that if tinnitus is a symptom of TCD, and if there is a sound intensity dependent activation of the primary auditory cortex, spontaneous high frequency oscillations at the level of the primary contralateral auditory cortex should correlate with subjective reports of tinnitus loudness in patients with unilateral tinnitus. Van der Loo et al. (2009) did not find correlations in prefrontal, parietal, or cingulate cortices between resting brain activity and tinnitus loudness perception. Thus, the gamma-band oscillations, which are present in auditory cortex in tinnitus, are not related to the conscious perception of tinnitus, but only code the intensity of the perceived phantom sound.

Thus, for a sensory stimulus to be consciously perceived, activation of the early sensory areas is a prerequisite but is not sufficient (Dehaene et al., 2006). Even in the absence of sensory inputs, cortical and thalamic neurons can show structured patterns of ongoing spontaneous activity (Luczak et al., 2009). These observations are also reflected in an implementation of the neural workspace model (Dehaene and Changeux, 2005) in which ascending brain stem nuclei (e.g. cholinergic among others) send globally depolarizing neuromodulatory signals to a thalamic and cortical hierarchy. Simulations help characterize two main states of activity. First, spontaneous gamma-band oscillations emerge at a precise threshold controlled by ascending neuromodulator systems. Second, within a spontaneously active network, we observe the sudden "'ignition"' of one out of many possible coherent states of high-level activity amidst cortical neurons with long-distance projections. This is not unlike the ignition of a neural assembly defined as that point where the level of activation by external sources reaches such a level where the assembly as a whole becomes active (Braitenberg, 1977). When an assembly is active as a whole, the activity within it is envisaged to serve the function of representing "meanings" in the cerebral cortex. Since the component cells of an assembly are likely scattered diffusely over the cortex, the representation of meaning will also be distributed. Abeles (1982; Abeles et al., 1993) has derived the concept of "synfire chains," which are seen as organized sequences of firing, distributed amongst many cells, the sequence as a whole having a very precise temporal structure. Abeles has assumed that the temporal dependencies in such chains arise from the concatenation of very many synaptic relays, with multiple synaptic delays as the major contributor to the observed delays between correlated firing of different neurons. He also assumes that neuronal firing requires the coincidence of activation by many synapses. Villa and Abeles (1990) have reported that simultaneously recorded single units in the auditory thalamus of the anaesthetized cat do demonstrate favored spike patterns.

9.10 **Summary**

In humans with tinnitus, changes in the macro synchrony are found with large-scale cortical networks reflecting higher synchrony in the gamma band and reduced synchrony in the tau band. At the level of local circuits in cat auditory cortex increased spike firing synchrony is observed as well. Such local synchrony could facilitate synchronous activity along fiber tracts connecting widely spaced cortical areas and hence result in more efficient information transfer. Tinnitus could thus be the consequence of the formation of a large-scale neural assembly in cortex that is especially pronounced during silence, but can be disrupted by external sound and may be modifiable by attention.

Chapter 10

Tinnitus and aging

Epidemiological studies (Chapter 2) show that tinnitus is about twice as frequent in the elderly as in young adults. The increase in tinnitus prevalence in older patients does not necessarily mean that aging in itself is causative to tinnitus (Hoffman and Reed, 2004) but more likely that aging facilitates its perception. The important factor in this is may be the interaction between age-related hearing loss (ARHL) and noise-induced hearing loss (NIHL). Age-related tinnitus may potentially present itself as a distinct pathology being the result of degeneration at all levels of the auditory system with age (McFadden, 1982). Common underlying mechanisms for tinnitus and aging are likely a downregulation of central inhibition as the result of partial peripheral deafferentation either from NIHL or ARHL. In this chapter the role of the ARHL/NIHL interaction model and the age-only model will be contrasted with respect to the prevalence of tinnitus, at least as far as they are separable. Therefore, a short overview of potential mechanisms of aging and ARHL will be presented first, followed by an in-depth survey of structural, electrophysiological, and molecular findings in animal aging models that potentially relate to tinnitus. The chapter will be concluded with a comparison of "pure ARHL" findings with those of NIHL that were presented in Chapter 7.

10.1 Causes of aging

10.1.1 General mechanisms

Three of the proposed mechanisms underlying aging are featured in the following models: the telomerase theory, the dysdifferentiation hypothesis, and the membrane hypothesis. These were reviewed in detail by Seidman et al. (2002) and are partially summarized here.

The telomerase theory attributes aging to changes in the chromosome. Specifically the theory suggests that there is a reduction in telomere length over the lifespan. Telomerase is an enzyme that adds DNA-sequence repeats to the end of DNA strands in the telomere regions. The end of a chromosome comprises the telosome, the tip of which consists of DNA-repeat sequences and associated proteins and is called the telomere. DNA transcription and replication are mediated by the telomere. Reduction in the length of the telomere and alterations in its chromatin assembly may explain the instability in the chromosome that occurs during senescence. Chromatin is the combination of DNA and proteins that makes up a chromosome. The telomerase theory specifically proposes that a change in the balance between telomere shortening and telomerase activity underlies cellular aging processes. Recently Jaskelioff et al. (2010) demonstrated that restoring telomere integrity reverses the aging process in mice.

The dysdifferentiation theory suggests that aging is governed by programmed differentiation that decreases normal gene activity and in turn activates genes that are disruptive to normal cellular function. The homeostasis of cellular metabolism and function is based on the balance between cell proliferation and cell death. Reactive oxygen species (ROS) are produced during mitochondrial action that underlies the cellular metabolism, as well as via auto-oxidation of chemical and biological molecules. These ROS are chemically reactive and extremely toxic to cellular and subcellular structures (Halliwell et al., 1992). The Bcl-2 protein appears to prevent oxidative damage to cellular organelles and lipid membranes, and protects cells, in a dose-dependent manner, from the toxicity of hyperoxides (e.g. H_2O_2). Another protein, Bax, operates as an accessory to Bcl-2. The ratio between Bcl-2 and Bax determines the survival or death following an apoptotic stimulus, i.e. a stimulus that starts programmed cell death. Specifically, an elevated expression of Bcl-2 appears to be preventative, while an elevated expression of Bax favors the apoptotic process. The balance of these two proteins thus determines the fate of a cell.

The membrane hypothesis of aging (MHA) connects aging with a decreasing effectiveness of cellular protective and repair mechanisms. The MHA further postulates that cellular senescence is caused by a cross-linking action of free oxygen radicals within the cellular membrane. This results in biochemical and metabolic errors, which progressively accumulate and end in cell death. Therefore, the MHA suggests that ROS-induced cell membrane structural damage is the primary mediator in cellular aging.

These three proposed mechanisms clearly overlap. The generation of ROS damages cellular integrity, which may lead to alterations in gene expression including telomere shortening and activation of the *Bax* gene resulting in aging and presbycusis. The term presbycusis will be reserved here for human age-related hearing loss, and ARHL will be used in relation to animal work.

10.1.2 **Presbycusis**

Gates et al. (2008) studied 241 volunteer members of a dementia surveillance cohort aged 71–96 years that presented average hearing losses as shown in Figure 10.1. The hearing loss was about 10 dB larger for frequencies above 2 kHz in males compared to females.

Deafness, which often leads to tinnitus, is clearly associated with mitochondrial DNA (mtDNA) mutations. Moderate to severe presbycusis in people 50 years and over is typically strongly associated with family history (McMahon et al., 2008). Specifically, an association between moderate to severe hearing loss in women and their maternal family history of hearing loss was observed. This would fit with hereditary changes in mtDNA as an important cause for presbycusis. Paternal family history of hearing loss was also significantly associated with moderate–severe hearing loss in men, but less strongly.

Common deletion levels in mtDNA appear to be related to the severity of hearing loss in individuals with presbycusis (Markaryan et al., 2009). One specific mtDNA deletion, mtDNA4834 has been linked to ARHL in rodents (Seidman et al., 1997). The equivalent mtDNA4977 deletion in humans has also been identified, using archived temporal bones from patients with presbycusis (Bai et al., 1997). However, tinnitus overall appears to

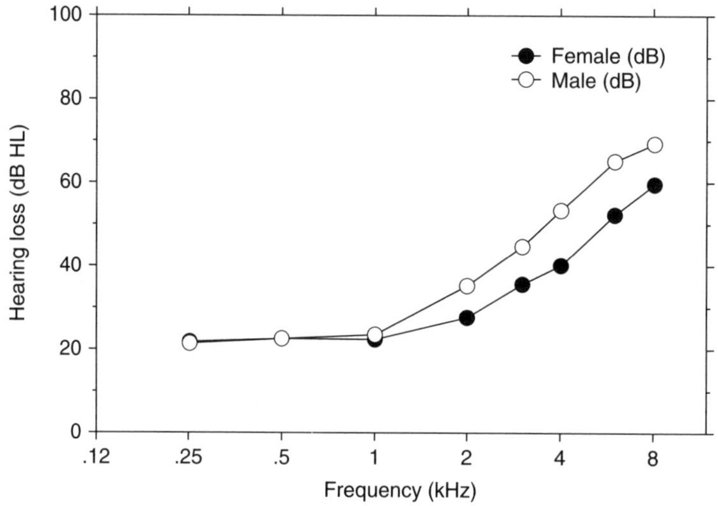

Figure 10.1 Mean hearing losses for men and women are displayed. Data from Gates et al. (2008).

have a low heritability according to a large study from Nord-Trøndelag County in Norway (Kvestad et al., 2010; Chapter 2).

10.1.3 Animal models for age-related hearing loss

Similar to humans, animals experience increasing hearing loss with aging (Syka, 2002). However, the extent of the deterioration is different not only among animal species but also within a particular species, i.e. between strains. A mouse or a rat strain is a group of animals that is genetically uniform. Mouse strains can be inbred, mutated, or genetically engineered, while rat strains are usually only inbred. Particularly in mice, there are many inbred strains that exhibit rapid progress in ARHL as well as many mutants that show progressive hearing loss with aging. One of the best-characterized animal models of human presbycusis is the Mongolian gerbil (Schmiedt et al., 1996; 2002). It is an outbred species, and it is therefore genetically diverse, like humans. As a result, the amount of hearing loss that gerbils develop over their lifespan is extremely variable. The F344 rat or F344 × Norway-F1 (FBN) inbred rat strains are used particularly in studies of the biochemical aspects of ARHL (see section 10.4 and section 5.3, Chapter 5).

One of the most commonly used animal models of human presbycusis is represented by the C57BL/6J (C57) strain of mouse (Willott, 1986). Jim Willott and his many students have nearly "single-handedly" advanced the use of mouse models of age-related hearing loss (Willott, 1991) as will be clear from the following. In contrast to the Mongolian gerbil, all individuals of the C57 mouse strain are genetically identical. The C57 mouse strain, which demonstrates a progressive auditory decline with onset at an early age, is usually compared with another model, the CBA/J (CBA) mouse strain (Willott et al., 1988ab; 1997), which displays a moderate auditory impairment with onset late in life. These two inbred mouse strains have become the favorite models of "peripheral" presbycusis (C57 strain) and "central" presbycusis (CBA strain) respectively. This difference can be used to separate central aging effects on hearing from cochlear ones. The peripheral mouse ARHL

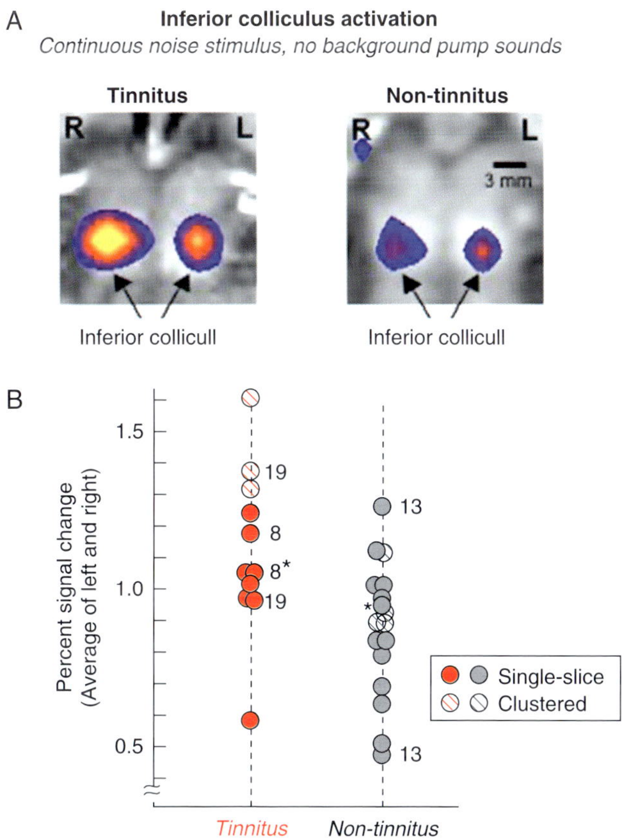

Color plate 1 (Also see Figure 4.2.) The inferior colliculi of tinnitus subjects showed abnormally high sound-evoked activation when the scanner coolant pump (and the acoustic noise it produces) was off. The sound stimulus was a continuous, broadband noise (binaural, 50–55 dB SL). A) Enlarged images of the inferior colliculi in one tinnitus and one non-tinnitus subject (corresponding to circles near asterisks in (B)). A map of activation (color) produced by the continuous noise stimulus is overlaid on a T1- weighted anatomical image (grayscale) obtained in the same imaging session. The color scale in the activation maps indicates the significance of the difference in image signal between stimulus on and off periods according to a t-test (uncorrected for multiple comparisons) (blue: $p = 0.01$; yellow: 2×10^{-9}). B) Percent signal change in the inferior colliculi of each tinnitus and non-tinnitus subject studied during the "pump off" condition. Each circle indicates % change averaged between the left and right inferior colliculi of a given subject. Reprinted from Melcher, J.R., Levine, R.A., Bergevin. C., and Norris, B. (2009) The auditory midbrain of people with tinnitus: abnormal sound-evoked activity revisited. *Hearing Research*, **257**, 63–74, with permission from Elsevier.

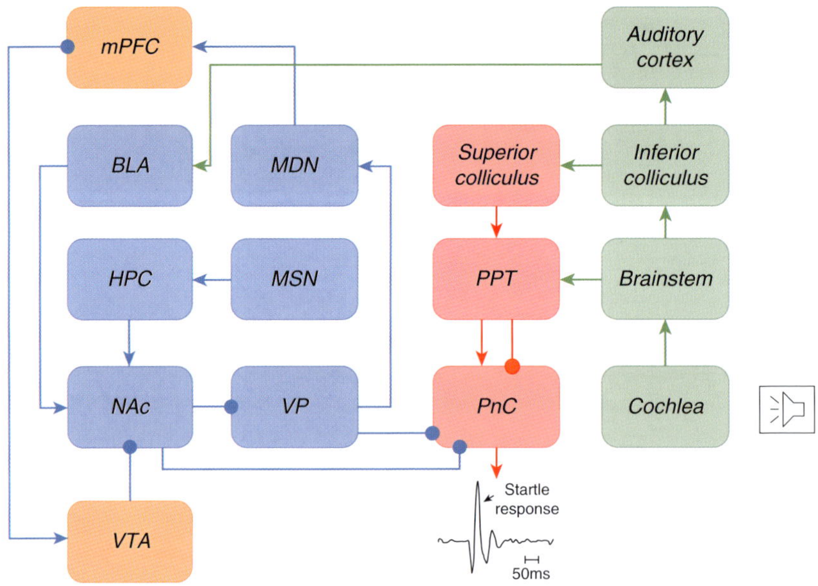

Color plate 2 (Also see Figure 5.5.) Startle response circuit with pre-pulse inhibition (PPI) circuit. The auditory pathway is indicated with olive colored boxes and connections. The startle circuit is indicated by pink boxes and red connections. The PPI modulating circuit is indicated in blue, it inhibits the PnC and is potentially affected by auditory cortex via the amygdala (BLA). Modulations from the mPFC via the VTA are indicated in orange-brown color. The arrowheads indicate excitatory connections. Round-dotted endings indicate inhibitory connections. BLA, basolateral amygdale; HPC, hippocampus; MDN, mediodorsal thalamic nucleus; MPFC, medial prefrontal cortex; MSN, medial septal nucleus; NAc, nucleus accumbens; PPT, pedunculopontine tegmental nucleus; PnC, nucleus reticularis pontis caudalis; VP, ventral pallidum; VTA, ventral tegmental area. After Swerdlow et al. (2001).

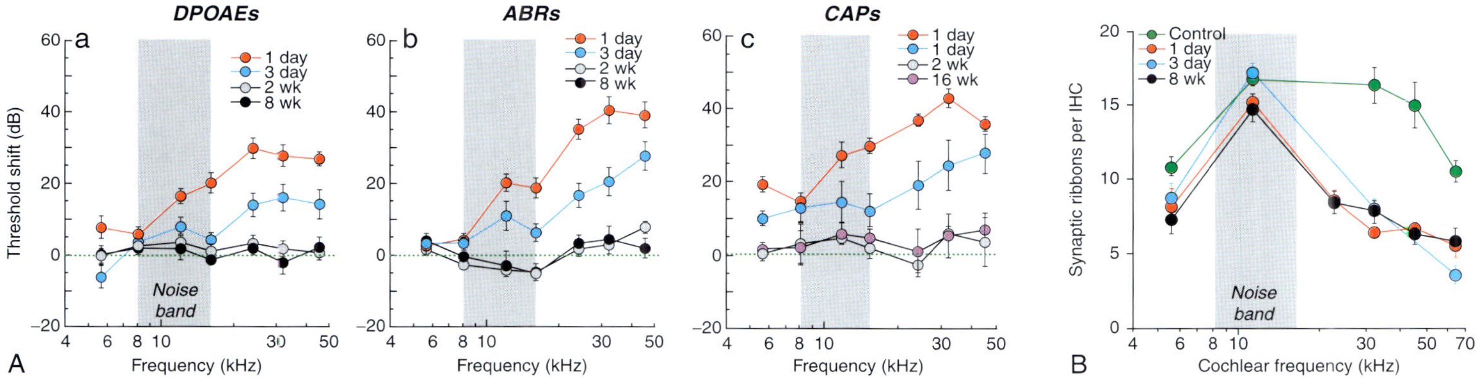

Color plate 3 (Also see Figure 7.1.) Noise trauma that only evokes temporary thresholds shifts and temporary changes in otoacoustic emissions (A) can still result in loss of synaptic ribbons (B). Compiled from Kujawa, S.G. and Liberman, M.C. (2009) Adding insult to injury: cochlear nerve degeneration after "temporary" noise-induced hearing loss. *Journal of Neuroscience*, **29**, 14077–85. © 2009, The Society for Neuroscience, with permission.

(continued)

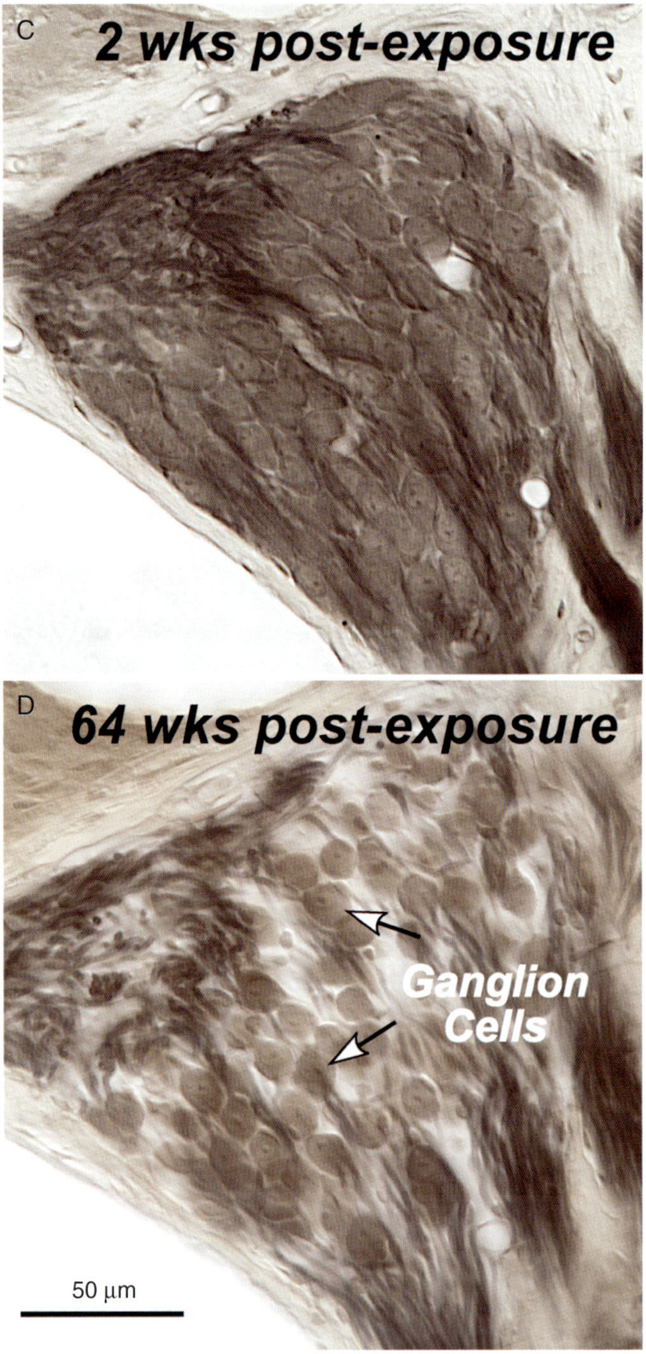

Color plate 3 (*cont'd*) (Also see Figure 7.1.) Delayed loss of ganglion cells (C and D). Compiled from Kujawa, S.G. and Liberman, M.C. (2009) Adding insult to injury: cochlear nerve degeneration after "temporary" noise-induced hearing loss. *Journal of Neuroscience*, **29**, 14077–85. © 2009, The Society for Neuroscience, with permission.

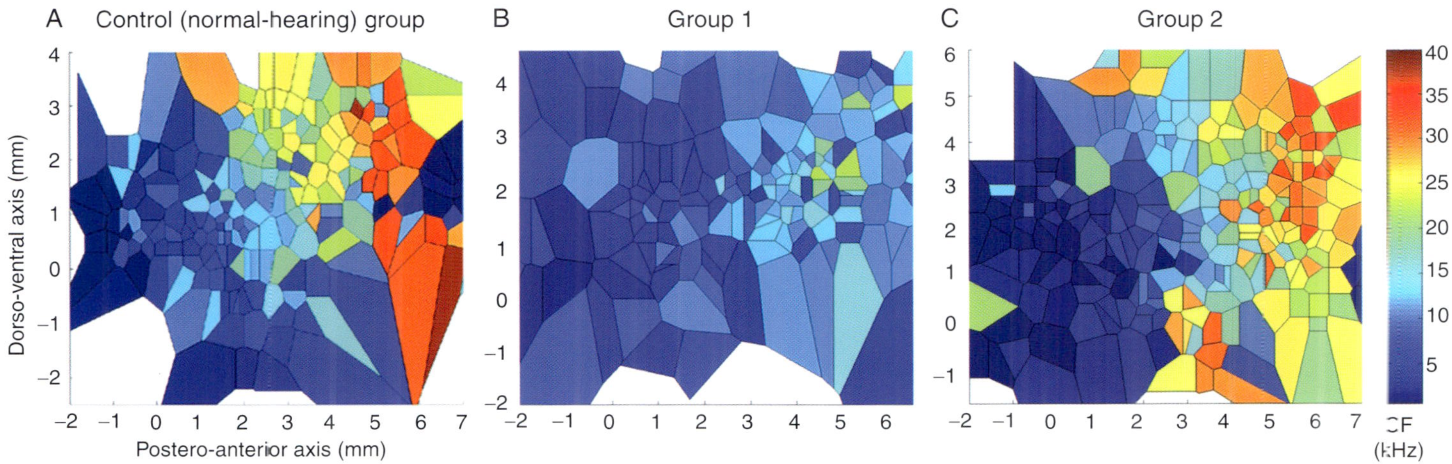

Color plate 4 (Also se Figure 7.6.) Compound CF maps in AI in control cats (A), cats recovered in quiet (B), and cats recovered in a high-frequency EAE (C). The center of each polygon,
constructed using the tessellation method (MatLab), corresponds to the coordinates of a
recording site in auditory cortex along the anteroposterior axis (abscises) and the ventrodorsal axis (ordinates). The tip of the posterior ecto-Sylvian sulcus was taken as the (0,0) coordinate. The CF is represented by color as indicated by the color bar. Reproduced from Noreña, A.J. and Eggermont, J.J. (2005) Enriched acoustic environment after noise trauma reduces hearing loss and prevents cortical map reorganization. *Journal of Neuroscience*, **25**, 699–705. © 2005,
The Society for Neuroscience, with permission.

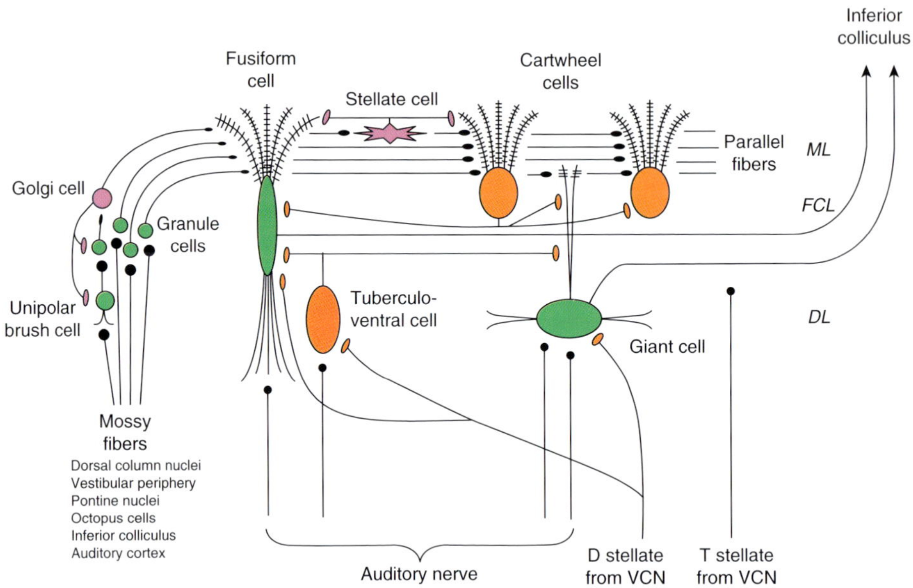

Color plate 5 (Also see Figure 8.2.) Schematic structure of the DCN. Granule cells lie in the fusiform cell layer (FCL) of the DCN, and are excited through mossy fibers originating from a large number of other regions of the brain (listed at lower left of the figure). The granule cell axons (parallel fibers) contact the spiny dendrites of cartwheel cells in the molecular layer (ML) and other inhibitory interneurons, i.e. superficial stellate cells and Golgi cells. Cartwheel cell axons contact fusiform cells, which are themselves contacted by parallel fibers on spiny apical dendrites. Auditory nerve fibers innervate the deep layer (DL), contacting the smooth basal dendrites of fusiform and giant cells, as well as tuberculoventral interneurons. Two groups of cells from the VCN, the D- and T-stellate cells, provide auditory inputs to the deep layer. Fusiform and giant cells, the principal cells of the DCN, project to the inferior colliculus. Glutamatergic neurons are shown in green and their terminals in black, glycinergic neurons and their terminals are shown in orange, and GABAergic neurons are shown in pink. Reprinted from Oertel, D. and Young, E.D. (2004) What's a cerebellar circuit doing in the auditory system? *Trends in Neurosciences*, **27**(2), 104–10. Copyright (2004), with permission from Elsevier.

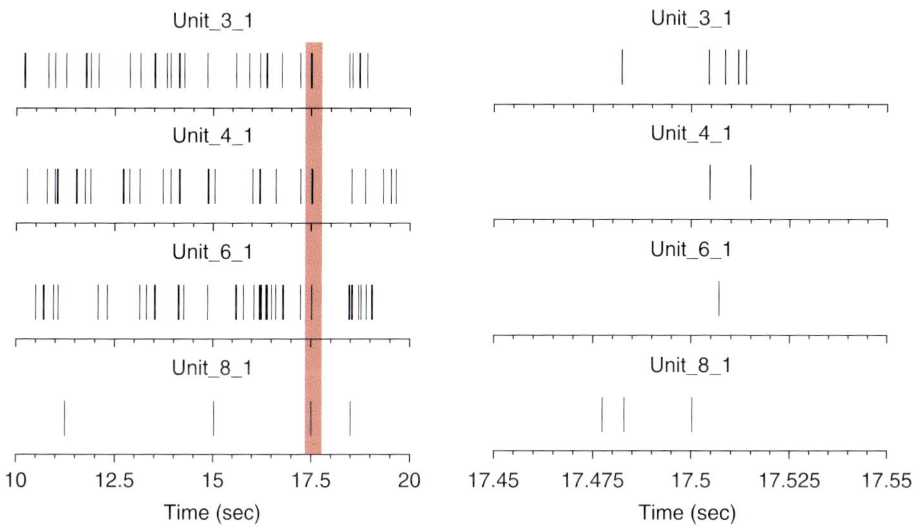

Color plate 6 (Also see Figure 9.1.) A 10-s section of spontaneous activity for four simultaneously recorded neurons from cat primary auditory cortex (left). On the right a 100-ms part around the time mark of 17.5 s (corresponding to the highlighted region on the left) shows synchronous firing within a 5–10-ms window.

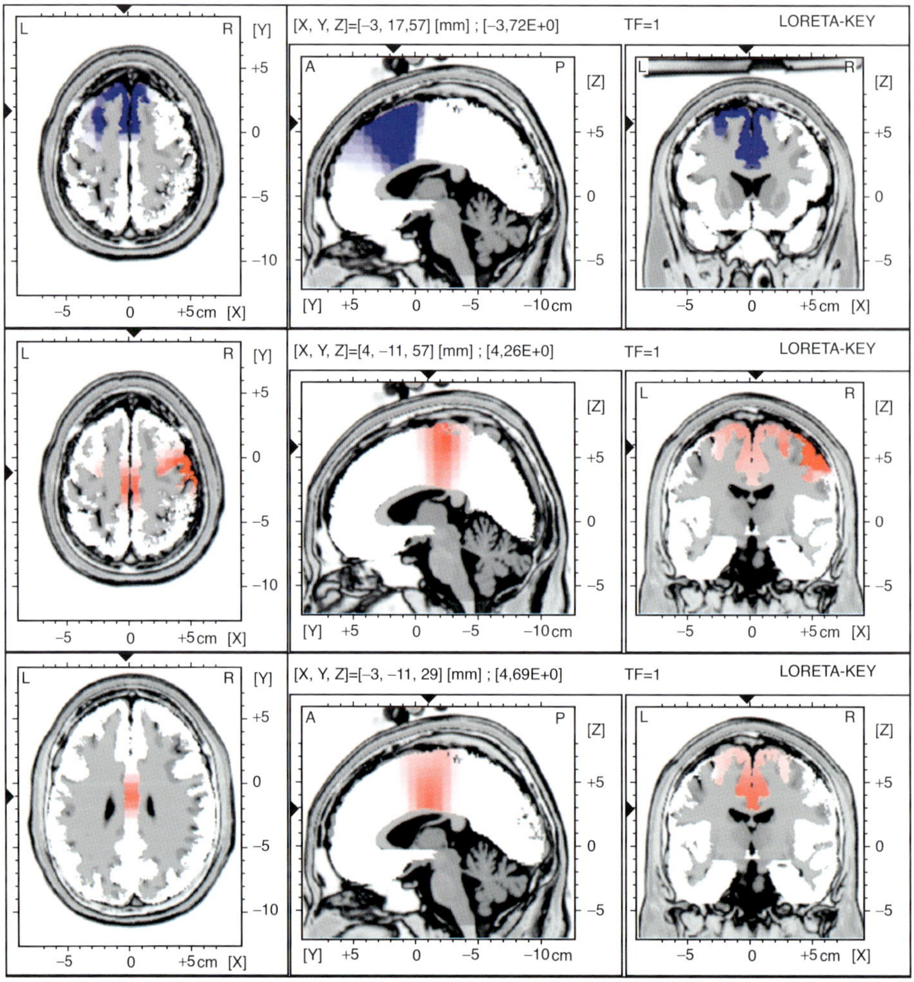

Color plate 7 (Also see Figure 11.4.) Significant results for current density amplitude analysis in respectively delta (2–3.5 Hz), theta (4–7.5 Hz), alpha (8–12.5 Hz), and beta (21–32.5 Hz) bands. Reprinted from Vanneste, S., Plazier, M., van der Loo, E., Van de Heyning, P., Congedo, M., and De Ridder, D. (2010) The neural correlates of tinnitus-related distress. *NeuroImage*, **52**(2), 470–80. Copyright (2010), with permission from Elsevier.

models show degeneration of the organ of Corti and also invariably include degeneration of afferent neurons, stria vascularis, and spiral ligament. The endolymphatic potential appears normal in these models even up to ages at which hearing loss is pronounced, and changes in the organ of Corti, typically hair cell loss, can account for most hearing loss. Thus, they best fit the pattern of sensory-based ARHL (Ohlemiller, 2006).

The recessive adult hearing loss gene (*Ahl*) that is mapped to chromosome 10, has been identified in the C57BL/6J and DBA/2J inbred strains of mice, and is the presumed cause of this species' progressive hearing loss. More recently, the *Ahl* gene has also been implicated in NIHL (Davis et al., 2001). This already suggests that interactions between ARHL and NIHL are very likely. The *Ahl* gene product was recently determined to be cadherin 23 (Noben-Trauth et al., 2003). The *Cdh23* gene is a member of the cadherin superfamily of genes encoding calcium-dependent cell–cell adhesion glycoproteins. The protein encoded by the *Cdh23* gene is a large, single-pass transmembrane protein composed of an extracellular domain containing 27 repeats that show significant homology to the cadherin ectodomain. An ectodomain is that part of a membrane protein that extends into extracellular space. Expressed in the neurosensory epithelium, the protein is thought to be involved in stereocilia organization and hair bundle formation. The gene is located in a region containing the human deafness loci DFNB12. The Cdh23^{Ahl} allele is common to C57 mice strains. Allele is a short form of allelomorph ("other form"), which was used in the early days of genetics to describe variant forms of a gene detected as different phenotypes.

Mouse mutations that promote both NIHL and apparent sensory ARHL, such as Cdh23^{Ahl}, suggest that there is often no useful distinction between the sensory and neural substrates of NIHL and ARHL. In these mice, environmental noise levels that would normally be harmless, may cause permanent hearing loss. A non-linear interaction between ARHL and NIHL was observed in gerbils born and raised in a quiet environment (Mills et al., 1997). They were exposed monaurally at 18 months of age to a 3.5-kHz pure tone for 1 h at 113 dB SPL. Six weeks after the exposure permanent threshold shifts in the exposed ear were approximately 20 dB in the 4- to 8-kHz region. Thresholds in the non-exposed ear were unaffected. The non-exposed ear would then reflect the pure ARHL, whereas in the exposed ear it would be combined with NIHL. This of course assumes that there is no central interaction between the activity from the exposed ear and the other ear. And, as we have seen in Chapter 7, this is generally not the case. Animals were then allowed to age in quiet until 36 months of age when thresholds were assessed again. The effect of NIHL and ARHL were non-additive, i.e. the resulting hearing loss in the exposed ear was larger than expected on basis of the loss in the pure NIHL and pure ARHL groups. Thus, sensory ARHL may represent cumulative damage, and alleles that promote this condition may make affected individuals prone to damage from otherwise benign exposures. Noise exposure early in life may also trigger progressive neuronal loss, the hallmark of neural ARHL (Kujawa and Liberman, 2006; Ohlemiller, 2006).

10.2 **Structural changes**

10.2.1 **Hair cells and auditory nerve**

ARHL is characterized by a loss of hair cells. For example, in the C57 mouse strain, the genetically programmed pattern of SNHL starts in early adulthood, i.e. around 2 months of age, and progresses to near total deafness by the second year of life (Willott, 1986; Li and Borg, 1991). The hair-cell loss starts in the high-frequency region and progresses to increasingly lower frequencies (Spongr et al., 1997). Both OHC and IHC losses are observed, although the percentage of OHC loss is always higher (Figure 10.2 top). In contrast, the CBA mouse shows little evidence of hair cell damage until late in life (Figure 10.2 bottom), i.e. until 18 months of age.

The hair cell loss in the C57 mouse is accompanied by a loss of spiral ganglion cells with nearly complete loss in the basal cochlea during the second year of life (Willott, 1991). Hearing threshold shifts based on evoked potential measurements during the lifespan of C57 mice reflect this hair cell loss. Threshold shifts are most obvious at high frequencies, later progressively spreading to low frequencies.

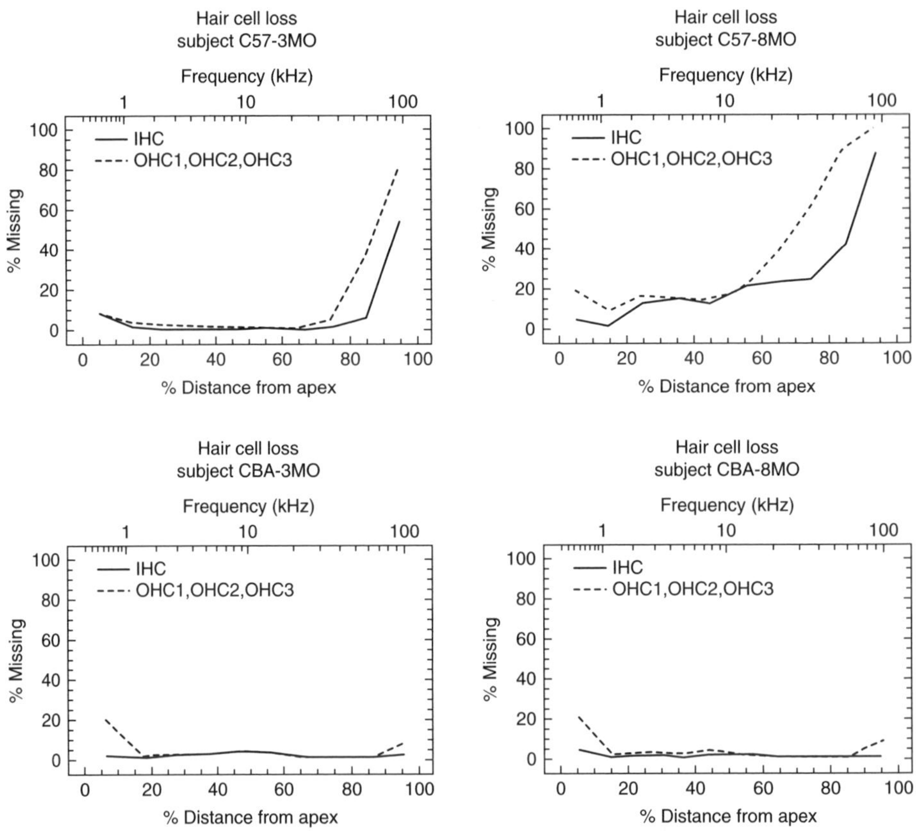

Figure 10.2 Mean cytocochleograms for C57BL/6 mice (top row) and CBA mice (bottom row) at 3 and 8 months of age. Cytocochleograms show percentage of missing OHCs and IHCs in 10% intervals as a function of percent total distance from the apex of the cochlea; cochlear position (bottom axis) was transformed to frequency (top axis) using a generalized cochlear frequency-place map conversion equation for the mouse. Reprinted with permission from Spongr, V.P., Flood, D.G., Frisina, R.D. and Salvi, R.J. (1997) Quantitative measures of hair cell loss in CBA and C57BL/6 mice throughout their life spans. *Journal of the Acoustical Society of America*, **101**, 3546–53. Copyright 1997, Acoustic Society of America.

Species with lifespans longer than mice or rats, i.e. guinea pigs and chinchillas, provide more extensive data about the anatomic and functional consequences of aging (Syka, 2002). In guinea pigs, cochlear pathology has been studied in animals up to an age of 5 years; however, no significant loss of hair cells was seen in the basal or middle turn of the cochlea in these aged animals (Ingham et al., 1999). In the apical turn, there was a significant loss of hair cells in all rows of the outer hair cells (up to 20%), most severe in the third row. There was no loss of apical inner hair cells in the aged animals. The apical loss of hair cells corresponded with the decrease in the numbers of spiral ganglion neurons, near the apex (Covell and Rogers, 1957). Detailed histopathology in chinchillas, at ages up to 19 years (!), demonstrated small but progressive losses of OHC and IHC (with the outer row of OHC most severely affected), more expressed at the apex and base. (This is similar to the audiograms of music-induced hearing loss in, typically young, DJs or music technicians; Chapter 2). In many cases IHC losses were accompanied by damage to spiral ganglion cells (SGCs; Bohne et al., 1990). A later study by McFadden et al. (1997) confirmed the losses of OHC and IHC in the basal and apical regions of the cochlea in aged chinchillas. The losses were paralleled by decreases in distortion-product OAEs and small but significant declines in auditory sensitivity, indicated by measurements of evoked potentials with high-frequency losses exceeding low-frequency losses.

Examination of the inner ears of animals raised in quiet environments, thereby avoiding NIHL, showed that degeneration of the stria vascularis might be the most prominent element in ARHL (Gates and Mills, 2005). Histopathological studies of ageing gerbils provided strong evidence for vascular involvement in ARHL. Analyses of lateral wall preparations stained to contrast blood vessels (Figure 10.3) have shown losses of the strial capillary area in aged animals. The vascular pathological changes first occurred as small focal lesions mainly in the apical and lower basal turns and progressed with age to encompass large regions at both ends of the cochlea. Not surprisingly, areas of complete capillary loss invariably correlated with regions of strial atrophy. Thus, considerable support exists for the major involvement of strial microvasculature in age-related degeneration of stria vascularis. This contrasts with the absence of these findings in NIHL (Chapter 7).

Loss of SGCs without associated loss of hair cells is common among aging mammals and called primary degeneration (Keithley and Feldman, 1979; Keithley et al., 1989; White et al., 2000). Moreover, apparent primary and secondary degeneration, i.e. following the loss of IHC, of SGCs may occur in the same cochlea (Hequembourg and Liberman, 2001). It is thus possible that age-related SGC loss and hair cell loss operate in parallel by independent mechanisms. It is difficult, however, to rule out the role of hair cell loss or dysfunction in SGC loss. Primary degeneration of SGCs has been observed in the cochlea of CBA/CaJ mice after mild noise exposure of juvenile animals (Kujawa and Liberman, 2006). In this study, CBA/CaJ mice were exposed to an 8–16-kHz noise band at 100 dB SPL for 2 h at ages from 4–124 weeks and held with unexposed cohorts for post-exposure times from 2–96 weeks. When evaluated 2 weeks after exposure, maximum threshold shifts in young-exposed animals (4–8 weeks) were 40–50 dB. Animals exposed at ≥16 weeks of age showed essentially no shift at the same post-exposure time. However, when held for long post-exposure times, these animals with previous exposure showed

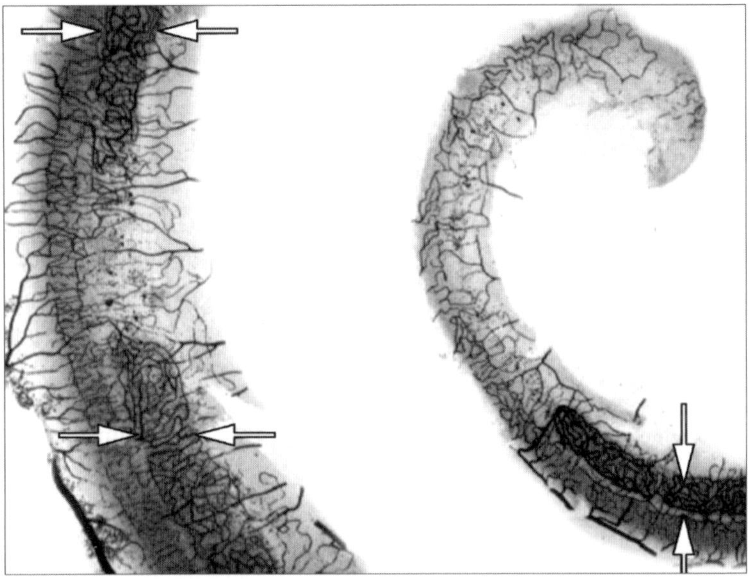

Figure 10.3 Surface preparations of lateral wall dissections from an old gerbil stained to contrast blood vessels. The strial capillary bed (between arrows) overlies vessels of the spiral ligament. A) Focal area of strial capillary atrophy. B) Complete loss of strial capillaries throughout the apical turn. Reprinted from Gates, G.A. and Mills, J.H. (2005) Presbycusis. *Lancet*, **366**, 1111–20. Copyright (2005), with permission from Elsevier.

substantial ongoing deterioration of cochlear neural responses and corresponding primary neural degeneration throughout the cochlea without changes in hair cell responses (DPOAEs). Delayed SGC loss was observed in all noise-exposed animals held 96 weeks after exposure, even those that showed no NIHL 2 weeks after exposure. This suggested a link between noise exposure—especially early exposure—and later apparent neural ARHL. Noise may produce a slight pathology of hair cells that interferes with the critical trophic support that they normally provide. Even if age-related SGC loss occurs in the presence of normal hair cells, other cell types in the cochlea may contribute. For example, as we have seen, C57BL/6 mice carry the $Cdh23^{Ahl}$ mutation that promotes progressive hair cell loss (Ohlemiller, 2004). Ultrastructural signs of synaptic pathology—a likely precursor to neuronal loss—can be found in these mice prior to overt hair cell loss. In addition, the number of radial fiber afferent synapses was reduced in excess of SGC loss and preceded IHC loss (Stamataki et al., 2006). This may reflect an early and subtle aspect of $Cdh23^{Ahl}$-related hair cell degeneration in ARHL.

10.2.2 Central nervous system

Frisina and Walton (2006) reviewed structural changes in the CNS of aging animals and made a distinction between pure age-related central changes and induced central effects resulting from age-induced sensory lesions. Typically these periphery-induced central changes were found in regions with high-frequency neurons, whereas the pure central aging-related changes were largely independent of the neural CFs, and potentially also independent of the auditory nucleus or brain area studied. The following survey was inspired by this review.

Cochlear nucleus Different types of neurons in the CN receive different types of synaptic input and project to different auditory nuclei (see Figure 3.3, Chapter 3). Thus, if the effects of aging and/or ARHL differ among these neuron types then central coding of various aspects of audition may be differentially affected as well. In the AVCN, bushy cells receive strong single axon input from the cochlea via the end-bulbs of Held, whereas stellate cells receive multiple inputs from auditory nerve fibers. Bushy cells project to the medial nucleus of the trapezoid body and are involved in binaural spatial localization, but stellate cells do not. These two cell types in the young (2 months) and old (2 years) C57BL/6J mice were investigated by transmission electron microscopy (Briner and Willott, 1989). Statistically significant age differences were found irrespective of neuron type and CF and thus likely reflect general cellular responses to aging. A greater percentage of the neurons was occupied by lipofuscin, one of the aging or "wear and tear" pigments typically found around the nucleus.

Willott and Bross (1990) subsequently evaluated the influence of aging and age-related cochlear impairment on the VCN by measuring morphological properties of the octopus cell area (OCA) of the posteroventral cochlear nucleus (PVCN) in five age groups of inbred C57BL/6J and CBNJ mice (young adult to very old). Aging, regardless of strain, was associated with a decrease in volume of the OCA, a loss of neurons, slight decrease in neuron size, increased packing density of glial cells, and changes in dendrites ranging from minor to total loss of primary branches. The greatest changes occurred beyond the median lifespan. This was followed by similar investigations in the DCN of C57 and CBA mice (Willott et al., 1992). DCN cell volume, cell number and size declined (large and small multipolar neurons) with age for C57s. For CBA mice, cell volume increased in the first year, and declined in old age, without major changes in cell size or number of cells. This suggests that the primary aging changes in the DCN are driven by the rapid loss of cochlear inputs experienced by C57s, and that these DCN changes do not occur for the most part in mouse strains with a slowly progressing loss (e.g. CBA).

Helfert et al. (2003) made electron micrographs from the mid-frequency area of the right AVCN from five 3-month-, four 19-month-, and five 28-month-old F344 rats. They found (using similar techniques as in Helfert et al., 1999; see inferior colliculus section which follows) significant age-related decreases in the size of terminals contacting small-diameter (<2 μm) dendrites. Dendrites of this size comprised the largest percentage of dendrites in the AVCN. On these targets, vesicle terminals were reduced in volume by nearly 44% and 24%, respectively, in 28-month-olds when compared to the 3-month-olds. The decrease in terminal size may be related to age-associated reductions in neurotransmitter levels found in the F344 CN (see section 10.4.2). The observations presented here contrast with those previously described by the same authors in the IC, in which there were significant age-related losses of synaptic terminals and dendrites, but no change in the size of synaptic terminals.

Morphological measurements were also made for the AVCN in the DBA/2J and C57BL/6J mice strains to determine the effects of cochlear pathology on the number, packing density, and size of neurons and on AVCN volume (Willott and Bross, 1996). Both strains possess alleles that cause progressive cochlear pathology initially affecting the organ of Corti: in DBA mice, hearing loss is evident at 4 weeks of age and progresses rapidly; in C57 mice, hearing loss begins after 2 months of age and progresses more slowly. In both strains AVCN volume decreased, some loss of neurons occurred, and these changes paralleled the progression of peripheral hearing loss. Central changes were rapid in DBA mice, but the ultimate magnitude of the changes in 1-year-old mice did not differ between strains. Both strains differed from well-hearing CBA/J mice, which exhibited no changes in the AVCN measures.

Interaction with NIHL The effects of chronic cochlear impairment on morphological features of the adult CN were assessed in CBA/J mice with severe cochlear damage resulting from noise trauma (Willott et al., 1994). Noise was presented for 5 or 10 min at a level of 135 dB SPL. ABRs obtained 1 day and/or 1-week after exposure found minimal or no response at 80 dB SPL) in all cases. Similarly, the acoustic startle response could not be elicited by noise pips of 80–90 dB SPL. Thus, the noise exposure had resulted in profound hearing loss that was present shortly after the exposure and persisted beyond the usual time range for TTS. Similarly aged but non-exposed CBA mice were used as controls, so the effects of peripheral damage and aging could be compared. Cochlear damage produced significant changes in CN subdivisions that receive the heaviest input from cochlear afferents (AVCN, OCA, DCN layer III). These changes included a reduction of the volume of the neuropil, the dense tangle of axon terminal, dendrites and glial processes in the space between neuronal cell bodies. Reductions in neuron size, and increases in neuronal packing density were complementary to reduced volume in these subdivisions. The age at onset and duration of damage had little to do with the severity of central effects of cochlear damage. The effects of cochlear damage were not additive with age-related changes seen in the old controls.

By making HRP injections in the center of functionally characterized ICC regions with neurons having larger than normal minimum gap-detection responses, Frisina and Walton (2001, 2006) could examine how CN inputs to this area changed with age. They observed an age-related decline in the number of input neurons from all three divisions of the contralateral CN. The decrease was roughly similar for all three areas of the CN and was reduced to about one-third at mid age and to about 10% at old age. These reduced inputs from the contralateral CN, are possibly a result of the reduction in axon terminals in CN principal cells with age, and could partly underlie the age-related temporal processing deficits in this ICC region.

Inferior colliculus The number of GABA-immunopositive neurons in the ventrolateral portion of the ICC of aged (18- to 29-month-old) F344 rats was significantly reduced (36%) compared to their matched young adult (2- to 7-month-old) cohorts (Caspary et al., 1990).

Helfert et al. (1999) compared changes in the synaptic organization of the ICC among three age groups (3, 19, and 28 months) of F344 rats. Their data suggest similar losses of excitatory and inhibitory synapses in the ICC (see also Kazee et al., 1995 in C57 mice). The reduction in the number of synapses was related to a similar reduction in the number of dendrites, in particular those with diameters between 0.5–1.5 μm. Thus, the decrease in GABA and excitatory amino acids identified in the IC may be attributable to synaptic and dendritic declines, rather than cell loss. As we have seen the number of projecting CN neurons is also greatly reduced (Frisina and Walton, 2006) and this could potentially result in fewer dendrites in the IC.

Auditory cortex Ling et al. (2005) found no age-related changes in neuron number across layers in AI of FBN rats.

Thus, the main structural finding in the CNS related to ARHL is the reduction in cell number and/or cell density in subcortical structures, potentially propagated by reduced input leading to a reduction in the number of dendrites.

10.3 **Electrophysiological changes**

10.3.1 **Cochlea and auditory nerve**

Audiograms in elderly humans without a history of noise exposure generally show a flat loss of ~20 dB at low frequencies and an increasing loss with frequency above 2 kHz (Figure 10.1). This profile and the relatively robust DPOAES that are found in elderly subjects (Gates et al., 2002) challenge the common belief that presbycusis is based primarily on hair-cell disorders. We already distinguished between peripheral and central ARHL as reflected in the particular mice strains (see section 10.1). CFs, SFRs, and thresholds were recorded from single ANFs of gerbils aged for 36 months in quiet (Schmiedt et al., 1996). The data were compared with those from young controls. Fibers were classified as "low-SFR" if their SFR were 5–18 spikes/s and "high SFR" for higher rates. For CFs >6 kHz, the percentage of low-SFR fibers decreased significantly from 57% of the population in young gerbils to 29% in aged gerbils. Their data suggest that some of these initially low-SFR fibers now have shifted to the high-SFR group. At CFs <6 kHz, the SFR distribution did not change significantly with age, with the low-SFR fibers comprising 30 and 39% of the population, respectively, for the young and aged animals.

10.3.2 **Central nervous system**

The cochlear nucleus Fusiform cells in aged FBN rats displayed significantly higher maximum discharge rates to CF tones than those recorded from young-adult animals (Caspary et al., 2005), potentially a hyperacusis like phenomenon. The mean SFR, however, was only slightly, and not significantly, higher in older (mean 36.8 sp/s) compared to younger (mean 33.8 sp/s) animals. Fusiform cells of aged rats displayed significantly fewer non-monotonic rate-level functions for tones at CF, consistent with an age-related loss of glycinergic input. This age-related downregulation of glycinergic inhibition alters the fusiform-cell output of the DCN but likely also alters the AVCN output. Thus, the age-related loss of ANF activity seems to result in a compensatory downregulation in the function of the inhibitory system of the CN. This suggests a selective loss of inhibitory input from vertical cells (labeled tuberculo-ventral cells in Figure 8.2, Chapter 8), providing narrow-band inhibition, and a less selective loss of inhibition from D-multipolar cells, providing wide-band inhibition, in the DCN of old animals.

Cartwheel cells comprise another large population of DCN inhibitory neurons. They receive excitatory inputs from granule cell parallel fibers (Chapter 8) and provide a source of glycinergic inhibitory input onto apical dendrites of DCN fusiform cells. Young (4–5 months) FBN rat single-unit thresholds ranged from 0–30 dB SPL while aged (31–33 months) animals had thresholds that ranged from 25–75 dB SPL. Single-unit recordings from aged cartwheel cells revealed also significantly increased SFR (mean 9.9 sp/s in young rats and 13.9 sp/s in aged rats) and rate-level functions characterized by hyperexcitability at higher intensities (Caspary et al., 2006). Collectively, these findings suggest a loss of tonic and perhaps also transient feedforward inhibition onto aged DCN cartwheel neurons. These changes may reflect a compensatory downregulation of synaptic inhibition in response to a loss of excitatory drive from auditory and non-auditory excitatory inputs via granule cells. Caspary et al. (2006) suggested that in spite of their increased SFR and stimulus-driven firing rates, cartwheel cell inhibition could ultimately be weakened by an age-related decrease in release of glycine or loss of receptor efficacy. Both fusiform and cartwheel cells appear to feature signs of hyperexcitability potentially underlying hyperacusis, depending on the balance of their combined actions.

The inferior colliculus Frequency tuning curves (FTCs) were obtained from ICC neurons across much of the approximately 2–2.5 year lifespan of two inbred mouse strains, the C57BL/6, and the CBA/J. Tonotopic organization was disrupted in the ICC of aging C57 mice (Willott, 1986). Low-CF FTCs from neurons in the dorsal part of the ICC changed little during the first year of life, but in more ventral regions the high-frequency portions of FTCs were eliminated, CFs became lower, and low-frequency thresholds were reduced. These changes made the FTCs more similar to one another along the tonotopic axis. During the second year of life, all thresholds became greatly elevated with neurons throughout the ICC responding only to middle frequencies at very high intensities. The frequency range of C57 FTCs begins to decrease at 14 months of age, when hearing loss is quite severe at all frequencies. In CBA mice, the above changes are minimal or do not occur even in 22 month olds, which consequently only have a moderate loss of sensitivity across all frequencies.

Differences in central response properties between young and old CBA mice should reflect aging effects only and not those of ARHL. The old mice had only a moderate hearing loss. The percentage of neurons that were spontaneously active, defined as SFR >1sp/s, increased with age in the ventral IC (the area most sensitive to high frequencies, including most of the central nucleus) and decreased in the dorsal IC (the area most sensitive to lower frequencies, including much of the dorsal cortex). Thus, there appears to be a positive correlation between the increase in SFR and loss of input to the ventral IC (Willott et al., 1988a). Recordings from IC neurons in young adult (2-month) and middle-aged (7-month and 12–15-month) C57BL/6J mice (Willott et al., 1988b) showed that the percentage of spontaneously active neurons increased significantly in the ICC, but not in other IC subnuclei. Combining the CBA/J and C57 results suggests that chronic loss of functional and/or anatomical input to the ICC is likely the cause of increased SFRs in surviving auditory neurons.

Extracellular single-unit recordings were made from ~300 IC neurons of 24-month-old F344 rats under ketamine–xylazine anesthesia (Palombi and Caspary, 1996), and compared with those obtained from young adults (3 months). Average thresholds increased from ~25 dB SPL in young rats to ~56 dB SPL in aged rats. The CF range and mean were similar for the two groups. For the IC as a whole, there was no change in SFR, which is somewhat hard to interpret. If SFR in the IC actually follow the SFRs lower in the auditory system, the age-related pathology in the lower system would result in a reduction in SFR with age. If SFR at the level of the IC reflects responses of neurons not driven at all by the lower auditory system, then a reduction in GABA-mediated inhibition with age in the IC could result in increased SFR. The lack of change in SFR found by Palombi and Caspary (1996) may reflect some combination of age-related changes within the IC neurons themselves as well as lower in the auditory system.

The auditory cortex Tonotopic maps were obtained in the A1 of C57 mice during young adulthood (1.5–2 months) when their hearing is optimal, and at 3, 6, and 12 months of age, a period during which progressive, high frequency SNHL occurs. Maps were also obtained from CBA/CaJ mice, which retain good hearing as they age (Willott et al., 1993). Following progressive loss of high frequency sensitivity in the periphery, virtually the entire auditory cortex became devoted to the mid frequencies (especially 10–13 kHz), which retained high sensitivity. Similar age-related changes were absent in normal-hearing CBA mice. These findings indicate that tonotopic-map plasticity in A1 is associated with the high frequency hearing loss in C57 mice. Comparison of collicular and cortical changes may have relevance regarding mechanisms of cortical plasticity. The over-representation of middle frequencies in the IC of 6-month-old C57 mice suggests that the changes in the tonotopic map of A1 may just reflect changes in IC. On the other hand, the lower mean CF thresholds of mid-frequency neurons in A1 compared to those of the IC suggest additional plasticity beyond the IC. It is thus likely that both cortical and subcortical mechanisms are involved in the tonotopic map changes in A1.

Recordings from ~100 neurons in both layer V from aged FBN rats (29–33 months) and layer V in young-adult rats (4–6 months) showed that single-unit SFRs were not significantly different between young (mean 6.7 sp/s) and aged rats (mean 6.4 sp/s) (Turner et al., 2005). This result reflects the observed constancy in DCN fusiform cell SFRs in young and aged FBN rats, and in IC of young and aged F344 rats (see earlier). A recent study by Hughes et al. (2010) confirms that there is no change in SFR for lower cortical layers in FBN rats with age, but a significant increase (~44%) in upper cortical layers. In this study, unit responses to noise bursts and SFRs were obtained from simultaneously recordings across A1 layers using a single shank, 16-channel electrode. Aged FBN rat AI units displayed increased peak (24%) and steady state response rates (38%) compared to young AI units. This suggests that corticocollicular output cells (layer V) show no change in SFR, whereas in the input and corticocortical output cell layers (layers II–IV) there is a substantial increased SFR.

10.4 Molecular biology of aging in relation to tinnitus

10.4.1 Calcium channels and homeostasis

Sound-increased basilar membrane motion results in coordinated movements of inner and outer hair cells that displaces the stereocilia from their resting position. This causes ion-conducting channels in these stereocilia to open. The resulting potassium influx from the endolymph that surrounds the stereocilia depolarizes the hair cells, which in turn, causes an opening of voltage-gated calcium channels in the basal part of the hair cell. Repolarization of the hair cell is partially accomplished by an outward calcium-activated potassium current. Much of the calcium entering the cell diffuses away from the membrane and will be buffered to proteins within a few milliseconds. This is an intermediate step to sequestering the calcium in the endoplasmic reticulum which happens in a time span of about 10 ms. Subsequently any excess free calcium as well as most of the sequestered calcium will slowly exit the cell through an adenosine-triphosphate (ATP)-driven pump modulated by calmodulin (Blaustein, 1988).

Ca^{2+} -pump mutations in cochlear hair cells lead to hearing problems. Parvalbumin, calbindin (CB) and calretinin (CR) are neuron-specific, high-affinity cytosolic calcium-binding proteins involved in the regulation of intracellular calcium levels. Parvalbumin and calbindin are co-localized with the inhibitory transmitter GABA dominantly in cortical chandelier and basket cell interneurons. Decreases in the synthesis of the inhibitory transmitter GABA and subsequent decreases in inhibitory inputs in auditory structures during aging may be one of the factors responsible for the increase in calcium-binding protein immunoreactivity.

In CBA/CaJ mice 5.5% of DCN neurons were parvalbumin positive (PV+) in the very old (30–39 months) mice, versus 2.2% in the 1-month-old mice. In the DCN, 3% of the neurons were CB immunopositive in the 30–39 months mice compared to 1.9% in the 1-month-old group. In the PVCN, 20% of the neurons in the very old mice were PV+, compared to 12% in the young mice. CB did not show any significant age-related differences in the PVCN (Idrizbegovic et al., 2001). In the DCN of C57BL/6J mice an age-related increase in the total number of PV+ neurons was found, while no changes in the total number of CB or calretinin (CR) positive neurons were demonstrated. In the PVCN, the total number of PV, CB, or CR positive neurons remained stable with increasing age. The percentage of PV, CB, or CR positive neurons significantly increased in the DCN, and the percentage of PV and CB-positive neurons increased in the PVCN (Idrizbegovic et al., 2004). Because the neuronal number of calcium binding protein positive neurons in the DCN and PVCN was decreased or unchanged, this implies a relative up-regulation of calcium-binding proteins in neurons that had not previously expressed these proteins. This plastic response in the profoundly hearing impaired C57 mouse may be a survival strategy for CN neurons.

10.4.2 Receptors and transmitter systems

The most studied auditory nucleus with neurochemistry techniques in relation to aging is undoubtedly the IC, and nearly all of this work was done by Don Caspary and coworkers.

Glutamate Glutamate plays a major role in neuronal plasticity and neurotoxicity, and has been implicated in a number of pathological states. One of these states could be ARHL-presbycusis. More than one mechanism of glutamate toxicity has been suggested. The most accepted mechanism is that mediated by NMDA glutamate receptors. It causes increased toxic intracellular calcium ion concentrations and activation of calcium-dependent proteases (that break down proteins) and apoptotic protein caspases, which damage mitochondria among others. As a result, excess glutamate may lead to either apoptotic (in lower concentrations) or necrotic (in higher concentrations) neuronal death. Though there is no evidence of typical apoptotic changes in the IC with age, the general synaptic loss in the ICC and its correlation with peripheral ARHL in the C57 mice (Kazee et al., 1995) and in F344 rats (Helfert et al., 1999) may be a sign of cellular dysfunction in cases where inputs from the cochlea decline with age.

GABA There are several extrinsic and intrinsic GABAergic inputs to neurons in the ICC. The dorsal nucleus of the lateral lemniscus projects bilaterally to the ICC and is comprised almost entirely of neurons that immunolabel for GABA and its precursor glutamic acid decarboxylase (GAD). In the SOC, the LSO contains neurons that co-immunolabel for GABA and glycine and that project to the ipsilateral ICC. Two additional SOC nuclei, the ventral nucleus of the trapezoid body and the dorsomedial periolivary nucleus, project to the ipsilateral IC, and both contain large populations of GABA-immunoreactive neurons (Helfert et al., 1999). Twenty to 40% of all neurons in the ICC are considered GABAergic, a much higher percentage than in auditory cortex, where estimates run from 15–20% (Winer, 1992). Many of these neurons send projections that terminate in the ipsilateral and/or contralateral IC. Commissural fibers from the contralateral ICC, which continue beyond the ipsilateral ICC to the external cortex of the IC, have recently been shown to immunolabel for GABA.

Aging results in a major GABA deficit in the ICC. Considering the importance of inhibitory functions in auditory processing and in controlling SFR, the selective loss of GABA in the ICC may have a significant influence on hearing and tinnitus in the elderly. Neurons in the ICC were immunolabeled in young adult (2–7-month-old) and aged (18–29-month-old) F344 rats. A significant age-related reduction in both basal (–35%) and K^+-stimulated (–42%) efflux of GABA from the ICC was observed. In addition, there was a corresponding significant decrease (–30%) in tissue content of GABA in ICC of aged rats (Caspary et al., 1990).

Raza et al. (1994) showed age-related changes in GAD, in the degradative enzyme GABA-transaminase (GABA-T), and in the uptake system for GABA in the central nucleus of the ICC, the CN, and nuclei of the lateral lemniscus (NLL) of F344 rats. In young adults GAD activity was highest in the ICC intermediate in NLL, and lowest in CN. Reductions in GAD activity were seen in the ICC of mature (–31%) and aged (–30%) rats when compared to young adult. The CN, however, did not show any loss of GAD activity. In contrast to the loss of GAD in aged ICC, high affinity uptake processes for [^{14}C]GABA were not significantly altered. Similar to the ICC, the NLL showed remarkable age-related deficits, GAD activity: –35% aged versus mature. None of these areas showed a significant loss of GABA-T activity with aging. These findings suggest that age-related loss of GABA-mediated inhibition in the ICC of F344 rats is not attributable to changes in uptake or degradation of GABA, but may be related to a loss of biosynthetic capacity of the GAD present. In ICC and NLL, but not CN there are age-related neurochemical deficits.

Quantitative receptor autoradiography was used to assess $GABA_B$ receptor binding in three primary subdivisions of the IC: dorsal cortex (ICD), ICX, and ICC of 3-, 18–20-, and 26-month-old F344 rats (Milbrandt et al., 1994). In three IC subdivisions, $GABA_B$ receptor binding was significantly reduced in 26-month-old rats when compared to 3-month-old rats (ICD, −44%; ICX, −36%; ICC, −32%). A significant decrease in $GABA_B$ receptor binding in the IC of aged versus young F344 rats was demonstrated. In contrast, $GABA_A$ receptor binding was unchanged in the IC in F344 rats.

Milbrandt et al. (1996) found that [^{3}H]GABA binding to $GABA_A$ receptors was significantly reduced in the IC of young adult (3 months) and aged (18–26 months) F344 rats when compared to 2-month animals. However, no significant changes were observed after 3 months of age. GABA modulation of the picrotoxin binding site was examined and showed an increased sensitivity of the receptor to GABA modulation in aged rats and, thus, may serve as a compensatory mechanism to enhance $GABA_A$ receptor function in response to a presynaptic loss of inhibition. Subsequently it was found that: (1) $GABA_A$ receptor subunit mRNA levels were significantly altered in the IC of old F344 rats, and (2) age-related changes in subunit levels appeared to be regionally selective and subunit specific. A highly significant increase in the level of the γ1 subunit mRNA was observed with little change in the levels of the α1, β2, and γ2 subunit mRNAs (Milbrandt et al., 1997). Taken together, these data suggest the potential for an alteration in the balance between excitation and inhibition. However, compensatory mechanisms, such as up-regulation of the α2 and γ1 $GABA_A$ receptor subunits, may increase the sensitivity of the IC receptors to GABA and assist in maintaining GABA inhibitory levels.

Caspary et al. (1999) assessed age-related changes of the $GABA_A$ receptor composition and function in the IC of young-adult, middle-aged, and aged rats. Western blotting was used to measure protein levels of selected $GABA_A$ receptor subunits in preparations obtained from the inferior colliculus of F344 and FBN rats. In both strains, the aged group exhibited significant increases in γ1 subunit protein and a decrease in α1 subunit protein. Age-related changes in $GABA_B$ receptors and presynaptic markers of GABA neurotransmission in the IC of F344 rats include a significant loss of GAD activity and GABA release, and are summarized in Figure 10.4.

Glycine Milbrandt and Caspary (1995) used quantitative receptor autoradiography to examine the effects of aging on the binding profile of the strychnine-sensitive glycine receptor in the F344 rat. Glycine receptor binding sites were localized using [^{3}H]strychnine, a glycine receptor (GlyR) blocker, in two principal subdivisions of the cochlear nucleus; the DCN and AVCN. In young rats, single concentrations of [^{3}H]strychnine showed significantly higher binding levels in the DCN than the AVCN (+ 38%). Little binding was detected in regions of the PVCN, and no specific binding was apparent in the cerebellum. A significant age-related decrease in [^{3}H]strychnine (8-nM) binding was observed in the AVCN (−37%) and DCN (−23%) of 26-month-old rats compared with 3-month-old rats.

Willott et al. (1997) studied glycinergic neurons in the CN of C57 and CBA mice. In the CN of old C57 mice (18 months) with severe hearing loss, the number of glycine-immunoreactive neurons decreased significantly. The number of strychnine sensitive GlyRs decreased significantly in the DCN of old C57 mice. Significant effects were not

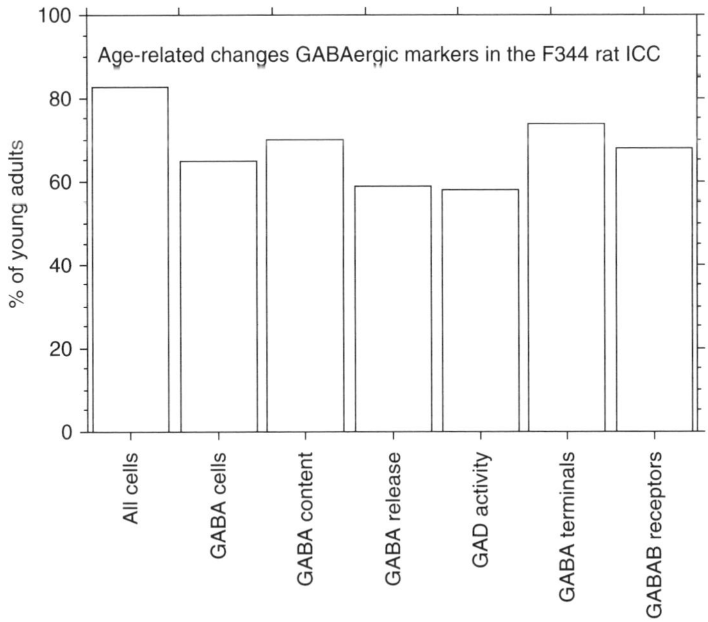

Figure 10.4 Graphic summary of significant age-related changes in markers for GABA neurotransmission across various studies. This suggests a loss of presynaptic GABA in the aged IC of F344 rats. The IC of old rats showed decreased GABA levels, number of GABA immunoreactive neurons, GABA release, tissue content of GABA, GAD activity, number of GABAergic terminals, and $GABA_B$ receptor binding. Data from Caspary, D.M., Holder, T.M., Hughes, L.F., Milbrandt, J.C., McKernan, R.M., and Naritoku, D.K. Age-related changes in GABAA receptor subunit composition and function in rat auditory system. *Neuroscience*, **93**(1), 307–12. Copyright (1999) Elsevier.

observed in the CN of middle-aged C57 mice (with less-severe hearing loss) or in very old CBA mice (which do not exhibit severe hearing loss). The findings of the present study indicate that neither cochlear pathology prior to old age (12-month-old C57 mice) nor extreme age in the absence of severe cochlear pathology (CBA mice) resulted in significant changes in glycine immunolabeling or GlyR binding. However, the combination of severe hearing loss and age, as occurs in old C57 mice, was accompanied by substantial decline in measures of glycinergic function.

Caspary et al. (2008) in reviewing their findings over nearly two decades of research, posited that aging can be thought of, in part, as comprising a slow peripheral deafferentation, which in turn can result in compensatory changes throughout the specific auditory nervous system. Homeostatic plasticity describes how, in response to activity-dependent input changes in development and deafferentation, neural systems undergo pre- and postsynaptic compensatory changes to stay within a relatively narrow operating range of excitation and inhibition. Changes in chronic neuronal activity (over a period of days) trigger compensatory changes in synaptic activity, which in turn, contribute to a return toward original levels of neuronal activity (see Chapter 1, section 1.6).

10.4.3 Hormones and neuromodulators

Hormones In humans, hearing loss is more profound in elderly males than females (Figure 10.1). In that context, Hultcrantz et al. (2006) asked: "Is the female sex steroid estrogen the key to preserved hearing in the aging human?" The female sex hormone estrogen contributes to roles as diverse as the differentiation and function of the reproductive tract, memory storage, and bone growth (Petterson and Gustafsson, 2001). Estrogens work through two intracellular receptors, estrogen receptor (ER)-α and -β. Receptors of both kinds have been demonstrated in the inner ear of fish, rodents, and humans. The effects of estrogens are primary believed to occur through mRNA transcription and protein synthesis via the two ERs, but there are also non-genomic effects. The exact mechanism by which estrogen affects hearing function is not known, but a study of the brain of an ERβ knockout mouse revealed several morphological abnormalities with increasing age (Wang et al., 2001). At 2 months of age neural loss of neurons was apparent, especially in the somatosensory cortex, and at 2 years of age degeneration was found throughout the brain. This indicates that ERβ is essential for neural survival and that it may play an important role in the processing of sensory stimuli in the brain.

Contrasting all this are findings that male and female C57BL/6J (B6) mice exhibit similar degrees of hearing loss until about 3 months of age, after which, the loss accelerates in females (Willott, 2009). Several findings indicate a negative effect of ovarian hormones on the female B6 auditory system. Whereas the sex difference in high-frequency hearing loss was not significantly affected by castration, the female disadvantage in hearing thresholds at lower frequencies was erased by ovariectomy. Finally, intact females had more severe loss of neurons in the low-frequency region of the AVCN than other groups. In contrast, the presence of androgens had beneficial effects. So the story of B6 mice and men shows a new twist.

Neuromodulators Serotonin may modulate afferent fiber discharges in the cochlea, IC, and auditory cortex. Specific functions of serotonin are exerted upon its interaction with specific receptors; one of those many receptors is the serotonin 2B receptor. Biochemical studies confirmed the presence of serotonin in the cochlea (Vicente-Torres et al., 2003). Serotonin may regulate the function of stria vascularis and may cause decreases in cochlear blood flow and result in cochlear microcirculation dysfunction. Recently, serotonin-containing fibers have been found accompanying the olivocochlear efferent system (Gil-Loyzaga et al., 2000). These fibers form glomerulus-like structures under the IHCs and send collaterals to the first row of the OHCs. Serotonin may function as a neuromodulator of the efferent cholinergic and GABA innervation to the inner ear.

Tadros et al. (2007) investigated the differences in gene expression of serotonin 2B receptors with age in cochlea and IC, and the possible correlation between gene expression and functional hearing measurements in CBA/CaJ mice. The serotonin 2B receptor gene was upregulated with age in cochlea and IC. A significant correlation between gene expression and functional hearing results was established. Immunohistochemical protein expression studies of IC showed more serotonin 2B receptor cells in old mice relative to young adult mice, particularly in the external nucleus. The upregulation of 5-HT 2B receptor gene expression with age in cochlear and IC tissues and correlations with age and hearing measurements indicate that this upregulation may contribute to ARHL pathogenesis. The 5-HT 2B receptor in the cochlea may also mediate endothelial regulation of blood flow, possibly in the vessels of the basilar membrane of the organ of Corti and in the contractile actin filaments of basal cells that control the vessels entering and leaving the stria vascularis. Through these vascular effects of serotonin and 5-HT 2B receptor on cochlear or brain blood supply, the upregulation of the receptor in old age may be related to ischemia and physiological dysfunction.

Noradrenaline (NA), dopamine (DA), and serotonin were determined using HPLC in AVCN, DCN + PVCN, locus coeruleus (LC) and dorsal raphe of dark Agouti-Hanovre rats aged 4, 21, and 24 months (Cransac et al., 1996). In older rats, the main noradrenergic changes were a selective NA increase in AVCN, and 5-HT levels were also increased in all the brainstem nuclei except the dorsal raphe. DA remained unchanged. These data show that noradrenergic neurons in sensory nuclei are differently affected by aging whereas serotonergic activation occurs in most of them possibly as a compensatory response to dysfunction of sensory input and processing. The increase of NA stores in the AVCN of aged rats is in line with the elevated auditory brainstem threshold reported in old rats and could improve the signal-to-noise ratio. When considering age-dependent serotonin metabolism, the major finding of Cransac et al. (1996) was a general increase of 5-HT levels in locus coeruleus, medial vestibular nucleus and cochlear nuclei. Since all these nuclei receive a serotonergic input from the raphe dorsalis, the results suggest a general activation of raphe dorsalis serotonergic projections in old rats.

10.4.4 Genes involved in ARHL

Genetic factors determine the ageing process but are under the influence of intrinsic and environmental factors. Liu and Yan (2007) reviewed heritability, and allelism and modifier genes of ARHL, and evaluated the genetic analyses to identify the genetic factors that are involved. I summarize parts of that review.

The cochlea requires a broad range of proteins with different functions, including maintenance of structural and mechano-electrical transduction integrity and neuronal innervation. The interaction of genes and proteins at different levels influences among others, hearing thresholds, audible frequency range, and age of hearing onset. To date, genes that underlie approximately 40 forms of non-syndromic hearing loss (NSHL), and even more for syndromic HL, are cloned. These genes belong to different gene families with various functions, including transcription factors, extracellular matrix molecules, cytoskeletal components, ion channels, and transporters. Potentially, moderate functional variants in any of the numerous genes involved in cochlear function could have impact on an individual's risk of ARHL. All known monogenic hearing loss genes are potential candidates for susceptibility to ARHL. Genes that protect against oxidative stress and mitochondrial genes can also be considered important candidates.

Impaired function of antioxidant enzymes caused by genetic variation has been postulated as leading to failure of cellular responses against the toxic effects of ROS and subsequent peroxidative cell injury. Studies of knock-out models of two antioxidant genes *Gpx1* and *Sod1* have shown that deletions therein can lead to both ARHL and NIHL (Ohlemiller et al., 2000). Glutathione-S-transferases (GST) play a role in antioxidant pathways and in detoxification, and thus might help in the protection of the cochlea. GST consists of several gene classes, including *GSTM* and *GSTT*, coding for cytosolic enzymes (Strange et al., 2001). Up to 50% of the Caucasian population do not carry the *GSTM1* gene (Pemble et al., 1994). These individuals are more prone to damage caused by oxidative stress, have lower amplitudes of high-frequency OAEs compared to individuals possessing the gene, and are possibly more susceptible to ARHL (Rabinowitz et al., 2002).

The *Ahl* gene As we already described in section 10.1.3, the *Ahl* gene product is cadherin23 (Cdh23). The protein encoded by this gene is a large transmembrane protein that is expressed in the neurosensory epithelium. The protein is involved in stereocilia organization and hair bundle formation. The gene is located in a region containing the human deafness loci DFNB12. The Cdh23^{Ahl} allele is common to C57 mice strains. The *Ahl* gene is implicated in both ARHL and NIHL. The C57BL/6J (B6) and DBA/2J (D2) inbred strains of mice exhibit an age-related hearing loss (ARHL) due to a recessive gene (*Ahl*) that maps to Chromosome 10. The *Ahl* gene is also implicated in the susceptibility to NIHL. The B6 mice (*Ahl/Ahl*) are more susceptible to NIHL than the CBA/CaJ (CB) mice (+ *Ahl*) (Erway et al., 1996). These genetic effects implicate the *Ahl* gene as contributing to NIHL susceptibility. Davis et al. (2001) demonstrated the segregation for the putative *Ahl* gene and mapping of such a gene to chromosome 10, consistent with other independent mapping of *Ahl* for ARHL in 10 strains of mice (Johnson et al., 2000).

10.5 Comparison of ARHL with NIHL

We have seen that the Cdh23^{Ahl} gene drives both NIHL and ARHL. Furthermore there is an interaction; age makes animals more susceptible to NIHL. Progressive hearing loss associated with aging and NIHL may result from an increasing mutational load expansion toward the apex in IHCs and SPGs. The observed structural, electrophysiological, and molecular changes, relevant for tinnitus are compared in Table 10.1.

Table 10.1 Comparison of NIHL and peripheral loss ARHL animals

Structure; function	NIHL	ARHL
Hair cells	⇓	⇓
Ganglion cells	⇓	⇓
Cell density CN	⇓	⇓
Cell density in IC		⇓
Cell density in AC	≈	≈
SFR ganglion cells	⇓	≈ (LF) ⇑ (HF)
SFR DCN	⇑	≈ (ff) ⇑ (cw)
SFR ICC	⇑	≈ (F344) ⇑ (C57)
SFR AI	⇑ (upper)	≈ (lower) ⇑ (upper)

(Continued)

Table 10.1 Comparison of NIHL and peripheral loss ARHL animals)

Structure/ function	NIHL	ARHL
Tonotopic map	Reorg.	Reorg. (C57)
Glutamate CN	⇑	⇓
Glutamate IC	⇑	⇑
Glutamate AC	⇑→⇓	
Glycine CN	⇓	⇓
GABA IC	⇓→⇑	⇓ ($GABA_A$) ⇑ ($GABA_B$)
5-HT CN		⇑ (CBA)
5-HT IC	⇑	⇑ (CBA)
5-HT AC	⇑	

⇑ upregulation; ⇓ downregulation; ⇑→⇓ first up- then downregulation; ⇓→⇑ first down- then upregulation; ≈ no significant change; cw, cartwheel cells; ff, fusiform cells; HF, high-frequency; LF, low-frequency; lower and upper refer to layers in AI.

10.5.1 Structural changes

The structural changes observed in NIHL (Chapter 7) consist of hair cell loss in basal regions progressing apically with continued exposure. This is followed by degeneration of ganglion cells, loss of synaptic inputs to the CN followed by new synaptogenesis, but with permanent loss of inhibitory terminals. The cell density in thalamus and auditory cortex was reduced. In addition proliferation of trigeminal ganglion synapses on DCN granule cells is described (similar to those following NIHL, Chapter 7). Structural changes in ARHL are dependent on the strain: in C57 mice one obtains very similar findings as in NIHL, i.e. hair cell loss in the basal turn and progressing to the lower-frequency region. This is followed by ganglion cell degeneration. In contrast, CBA mice only present significant hair cell loss in old age and then mostly affecting the OHCs in the apical turn. Primary degeneration of ganglion cells is observed. Reduced cell density in the CN resulting in reduced inputs from CN to IC so that one-third of inputs remain in mid age and only 10% in old age. There is a loss of GABA+ cells and GABAergic terminals in IC. No cell loss was seen in auditory cortex.

10.5.2 Physiological and neural changes

The physiological substrates of NIHL (Chapter 7) are a reduction in DPOAE amplitude, a reduction in SFR and stimulus evoked firing rate in auditory nerve fibers, increased SFR in DCN superficial layers, including fusiform cells, and increased SFR in ICC and AI. This is combined with increased serial (IC) and parallel (AC) neural synchrony (Chapter 9), and with a reorganization of the tonotopic map in AI, and likely also in other tonotopic regions in auditory cortex. Age-related changes in ARHL were not found for DPOAEs, the endolymphatic potential is likely reduced with aging, and there is a loss of low-SFR auditory nerve fibers for CF >6 kHz (gerbils). There was no SFR change for fusiform cells in the DCN, but an increased SFR was found in cartwheel cells. An increase in the number of spontaneously active (>1 sp/s) neurons in the high-frequency part of the IC was found, but was reduced in the low-frequency part of the IC. No change in SFR was found in the IC of F344 rats. In AI no SFR changes were seen in layer V, but SFR was increased in the upper layers in FBN rats. Tonotopic map changes were seen in IC (but with elevated thresholds leaving the possibility of residual responses and not a real reorganization) and AI accompanied by low thresholds at CF (required for genuine reorganization) such that with age the entire AI appears to be tuned to mid frequencies in C57 and very old CBA mice.

10.5.3 Changes in neurotransmitters and neuromodulators

In NIHL, transmitter-related changes (Chapter 7) appeared transient for GABA in IC, first downregulated and then, after a while, upregulated to nearly normal levels. A complementary change was found for glutamate in AC, which started with upregulation followed by downregulation. Others reported downregulation of GABA in CN and IC, and upregulation of glutamate in CN and IC. Upregulation of the neuromodulators ACh in CN and 5-HT in IC and AC was seen. Transmitter changes in ARHL, suggest glutamate decrease in CN and upregulation in IC (using PCR). Glycine binding was down in the entire CN of F344 rats and down in the DCN of C57 mice but not in CBA mice suggesting that changes are dependent on the loss of input from the cochlea (as in NIHL). In the IC of F344 rats, GABA concentration and GABA release were down. So was GAD activity, i.e. synthesis of GABA. $GABA_B$ binding was down but no change in $GABA_A$ binding was found. 5-HT was up regulated in the CN and IC of CBA mice.

BDNF protein levels were significantly upregulated by 9% in exposed young (7–8 months) FBN rats in fusiform cells ipsilateral to the exposure side. BDNF levels in aged (28–29 months) sound-exposed fusiform cells increased 31% ipsilateral to the exposure to the 1-h, 17-kHz-centered octave-band noise (116 dB SPL). Protein levels of the BDNF receptor, TrkB, were also significantly increased in aged but not in young sound-exposed DCN fusiform cells. This suggests a relationship between the upregulation of BDNF/TrkB and the increase in spontaneous and driven activity previously observed for aged and sound-exposed fusiform cells (Wang et al., 2011). Following sound-exposure, young FBN rats demonstrated a significantly greater capacity for threshold recovery than aged FBN rats, which suggests the sound exposure in young rats induced minimal functional damage as demonstrated using ABR thresholds. In Wang et al.'s study, the elevated BDNF protein level 80 days following the present PTS induced by sound exposure was similar to that observed in a DCN unilateral cochlear ablation model (Suneja et al., 2005). In response to the shown decrease in glycinergic inhibition in aged and sound-exposed DCN, BDNF could enhance glutamate release at glutamatergic synapses and directly cause neuronal hyperactivity in the short term following sound exposure.

The comparison between NIHL and ARHL suggests that aging may only play a minimal role in tinnitus. Nearly all changes in SFR are the result of peripheral hearing loss. That aging in itself does not induce tinnitus may have been glanced from the leveling off of the tinnitus prevalence for ages >65 years, i.e. when most people stop working and occupational noise effects do no longer play a role. The overall decrease in GABA and related inhibitory activity appears to be compensated by other mechanisms.

Chapter 11

Hyperactivity and hypersynchrony in neural networks as substrates for tinnitus?

11.1 Summary of bottom-up mechanisms: salicylate, noise trauma, aging

11.1.1 SFR, transmitters, neuromodulators, and behavioral tests

Some of the effects from salicylate, somatic stimuli, noise exposure, and aging on SFR and changes in transmitter and neuromodulators as compiled in previous chapters are shown in Table 11.1 (expanded from Table 10.1).

The most obvious structural difference resulting from salicylate application and the effects of noise exposure and aging is the only minor structural effect that salicylate has on the cochlear hair cells. There are, however, transient changes in the functioning of the OHCs as witnessed by the initial decrease and later increase of OAEs. Chronic salicylate application has similar effects on the SGCs as the other tinnitus-inducing agents: loss of ganglion cells. The effects on SFR are also quite different; whereas acute salicylate in high dose increases the SFR in ANFs, NIHL is characterized by decreases in the SFR of ANFs. A somewhat paradoxical result is obtained in quietly-aged gerbil ANFs with high CFs showing an increase in SFR. This is, however, entirely due to the loss of low-SFR fibers and not to an actual increase of the high SFRs compared to controls. The findings in the brainstem, midbrain, and primary auditory cortex show typically reduced SFRs for salicylate, and increased SFRs for NIHL and ARHL. Neural synchrony appears to be increased after NIHL both in IC and AI, but not following salicylate in AI. Results for neurotransmitters and modulators are generally similar for all three tinnitus-inducing agents. This somewhat surprising result, different SFR and neural synchrony profiles and similar neurotransmitter profiles, indicates that it is not clear if there is a unique neural substrate of tinnitus in the auditory system. The effects of trigeminal nerve stimulation do lead to excitatory as well as inhibitory influences on SFR in the DCN and IC.

It is apparent that putative electrophysiological correlates of tinnitus and putative behavioral indicators of tinnitus in animals are not consistent with each other. In Table 11.2 I gathered comparisons for electrophysiological responses from auditory cortex in rat and cat and gap-startle responses and the schedule-induced polydipsia avoidance-conditioning test (from Chapters 6 and 7). I also show the results of the EAE (covering the region of hearing loss) applied immediately after the noise exposure and present for

Table 11.1 Comparison of salicylate, somatic stimuli, NIHL, and ARHL animals

Structure	Salicylate	Somatic	NIHL	ARHL
Hair cells			⇓	⇓
OAEs	⇓→⇑	⇑	⇓	⇓
Ganglion cells	⇓(chronic)		⇓	⇓
Cell density CN			⇓	⇓
Cell density in IC				⇓
Cell density in AC			≈	≈
SFR ganglion cells	≈ ⇑		⇓	≈ (LF) ⇑ (HF)
SFR VCN			⇑	
SFR DCN	⇓ (ff) ≈ (cw)	⇑⇓	⇑→⇓ (ff)	≈ (ff) ⇑ (cw)
SFR ICC	⇓	⇑⇓	⇑	≈ (F344) ⇑ (C57)
SFR AI	≈ ⇓		⇑ (upper)	≈ (lower) ⇑ (upper)
SFR AII	⇑			
Synchrony IC			⇑	
Synchrony AI	≈		⇑	
Tonotopic map	Reo		Reo	Reo (C57)
Glutamate CN	⇑	⇑	⇑	⇓
Glutamate IC			⇑	⇑
Glutamate AC			⇑→⇓	
Glycine CN	⇓		⇓	⇓
GABA IC			⇓→⇑	⇓ ($GABA_A$) ⇑ ($GABA_B$)
5-HT CN				⇑ (CBA)
5-HT IC	⇑		⇑	⇑ (CBA)
5-HT AC	⇑		⇑	

cw, cartwheel cells; ff, fusiform cells; Reo reorganization.

at least 3 weeks. In addition the effects of vagus nerve stimulation paired with tone responses that were outside the frequency region of the behavioral indications for tinnitus are presented.

The results for salicylate indicate increased stimulus-evoked responses measured as LFP or spike activity. Concommitant with this is the finding of decreased SFR but also a finding of increased spontaneous FDG-PET responses. Because the FDG is likely more related to spontaneous LFPs than to spikes it is still surprising that increased LFPs and reduced SFRs co-occur. Rat (Yang et al., 2007; Paul et al., 2009; Sun et al., 2009) and cat data (Ochi et al., 1996; Zhang et al., 2011) agree well. The two behavioral tests also do correlate: increased startle (Sun et al., 2009) and decreased licking in silence (SIPAC; Paul et al., 2009). However, the decreased SFR in auditory cortex does not fit with the

Table 11.2 Comparison of electrophysiological and behavioral responses

Application	SFR	Stim-LFP	Xcor	Tonotopy	FDG	Behavioral
Salicylate rat (250 mg/kg) systemic	⇓	⇑			Spont ⇑	⇑ (startle) ⇓ (SIPAC)
Salicylate cat (200 mg/kg) systemic	⇓	⇑	≈	Reo		
Salicylate (round window)		⇓				
Noise trauma cat (6 kHz, 2h, 115 dB)	⇑		⇑	Reo		
Noise trauma rat (16 kHz, 1h, 115 dB)	⇑	⇑	⇑	Reo		⇑ (startle)
EAE cat 0 weeks after trauma	≈		≈	≈		
VNS-tone rat >4 weeks after trauma	⇑		≈	≈		≈ (startle)

⇓,≈,⇑: decreased, no change, increased, compared to controls. Reo: reorganized tonotopic map.

standard assumption of "tinnitus equals increased cortical SFR." Salicylate does appear to produce reorganization of auditory cortex (Stolzberg et al., 2011).

The findings following noise trauma are similar but differing in exposure time: 2 h (Seki et al., 2003; Noreña and Eggermont, 2005) vs. 1 h (Engineer et al., 2011) and this has an effect on the expected PTS. Most notably the triad of reorganized tonotopic maps, increased SFR, and increased neural synchrony was found. The gap-startle response was increased (a positive test for tinnitus) as expected from the hypothesis that tinnitus fills the gap in the continuous sound preceding the startle stimulus. Noise trauma also has a tendency to increase the central gain and may lead to hyperacusis. A correlation of these three electrophysiological markers with the strength of the gap-startle response (Engineer et al., 2011) revealed that:

> tonotopic map reorganization and tuning curve broadening, but not increased SFR or synchronization, were significantly correlated with the degree of gap-startle impairment in untreated noise-exposed rats. In addition, hearing loss and potential hyperacusis (as assessed from steeper rate level functions) were not correlated with gap-startle response impairment.

Following long-term exposure to an enhanced acoustic environment (EAE) after noise trauma, all electrophysiological indices of tinnitus were back to normal values (Noreña and Eggermont, 2005). Following 18 days of VNS-tone pairing after noise trauma the tonotopic maps were back to normal, the neural synchrony decreased to control levels but surprisingly the SFR further increased. The startle response strength returned to the control level. In addition the cortical thresholds were back to control levels (whereas the ABR threshold surprisingly remained elevated); this return to control levels also followed sham treatment (Engineer et al., 2011). I assume that this is just the result of the 1-h stimulation that typically does not result in a PTS. In that light one has to evaluate the conclusion of Engineer et al. (2011): "that noise-induced increases in cortical SFR and local synchronization were not significantly correlated with behavioral correlates of tinnitus in individual rats."

This again poses the question that we already raised in Chapter 5: is an increased gap-startle response an indication of tinnitus? Sun et al. (2009) finds it a sign of hyperacusis that could amplify the loudness of tinnitus following salicylate application. Engineer et al. (2011) finds the same but the decrease in gap-startle strength was not correlated with the increased SFR and synchrony in auditory cortex following noise trauma. The latter is surprising as in the post-exposure results the average synchrony was increased and the average startle strength was decreased and the reverse happened after VNS-multitone pairing. The control-value gap-startle response obtained after VNS-multitone pairing surprisingly was accompanied by increased SFR, so obviously this could even better fill in the gap in noise preceding the startle stimulus. Given all the problems with evaluating spontaneous firings and their synchrony during changing anesthesia levels these findings still do not rule out a model that relates increased neural synchrony in primary auditory cortex that accompanies reorganization of the tonotopic map as the most likely electrophysiological correlate of tinnitus (Noreña and Eggermont, 2003).

The gap-startle model of tinnitus may only depend on increased SFR in the gap-startle pathway, which does not include cortex. It is widely assumed that the gap activates the ascending auditory pathway in the IC and thereafter the colliculus superior, which in turn affects the startle pathway by activating an inhibitory cholinergic pathway from the pedunculopontine tegmental nucleus to the caudal pontine reticular nucleus (Koch, 1999). Thus increased SFR in the pathway leading up to the IC may be sufficient, however, it is till surprising that decreased SFR after salicylate is accompanied by a decreased gap-startle response compared to control.

11.1.2 Spontaneous versus driven transmitter release in central synapses

Current models of tinnitus, mostly based on homeostatic mechanisms (Schaette and Kempter, 2006), predict gain changes in the central auditory system that should similarly affect SFR and stimulus evoked measures such as LFPs. It is clear that cortical SFR decreases whereas LFP amplitude increases following salicylate application. The potential reason for this difference may be sought in different synaptic release mechanisms for spontaneous and stimulus-driven activity. Action potential-evoked neurotransmitter release from central neuron synapses requires Ca^{2+} influx through voltage-gated Ca^{2+} channels. Increasing or decreasing the extracellular Ca^{2+} concentration increases respectively decreases spontaneous neurotransmitter release. Blocking voltage-gated Ca^{2+} channels with non-specific blockers such as Cd^{2+} and Ni^{2+} does not affect spontaneous transmitter release, suggesting that these channels are not involved in the spontaneous release process (Glitsch, 2008).

Mechanisms regulating spontaneous transmitter release remain poorly understood. At central synapses, evoked and spontaneous release have been distinguished among others by differences in vesicle pools used, intracellular Ca^{2+} sensors for exocytosis, and by the spatial separation of the postsynaptic receptors that they target (Vyleta and Smith, 2011). Atasoy et al. (2008) found in 1-day-old hippocampal cultures that spontaneous and evoked neurotransmissions activate distinct sets of NMDA receptors. Another recent study (Fredj and Burrone, 2009) suggested that spontaneous release originates from a resting pool of synaptic vesicles that is normally not mobilized by stimulus-driven neuronal activity. However, Hua et al. (2010) found that spontaneously recycling synaptic vesicles were depleted by subsequent stimulation, excluding the existence of two different synaptic vesicle pools, recycling and resting, for evoked and spontaneous activity, respectively.

GABA's principal action, mediated by ionotropic $GABA_A$ receptors, is to increase membrane permeability to chloride ions. In most mature neurons, this leads to an inhibitory postsynaptic potential. This event occurs when postsynaptic $GABA_A$ receptors are activated following brief exposure to a high concentration of GABA, which is released from presynaptic vesicles. The resultant increase in membrane conductance underlies what is known as "phasic" inhibition. Low GABA concentration in the extracellular space can result in the persistent or "tonic" activation of $GABA_A$ receptors, in a manner that is temporally dissociated from phasic inhibition. Tonic activation of $GABA_A$ receptors results in a persistent increase in the cell's input conductance. Thus, for a given excitatory post synaptic current, the size and duration of the EPSP will be reduced, and the temporal and spatial window over which signal integration can occur will be shortened, reducing the probability of action potential generation (Farrant and Nusser, 2005). It is tempting to equate loss in tonic inhibition with increased SFR, and loss of phasic inhibition with increased gain reflected in LFP-amplitude increase. If tonic and phasic inhibition are differentially affected by salicylate the opposite effect on LFP amplitude and SFR in auditory cortex can be explained.

11.1.3 Immediate early genes and ion channels

A recent review (Knipper et al., 2010) provides the basis for the following comparison between the molecular level effects of noise trauma and salicylate on the auditory periphery and cortex. A diversity of signaling cascades for these two tinnitus-inducing agents became evident from microarray studies after impulse noise exposure (Kirkegaard et al., 2006), or salicylate treatment (Im et al., 2007). Both studies analyzed altered gene expression in the cochlea using a differential oligonucleotide microarray (3–24 h) in rodents. With few exceptions, no overlap of altered gene expression after either salicylate or impulse noise was observed, emphasizing the complexity of signaling cascades activated upon these two traumatizing interferences.

A compilation of data from the groups of Walhaüsser-Franke and Knipper is presented in Table 11.3. BDNF in spiral ganglion cells was upregulated after both acoustic trauma and salicylate (Tan et al., 2007; Panford-Walsh et al., 2008), and likely resulted from altered glutamate/NMDA receptor activation including involvement of L-type Ca^{2+} channels, all of which were shown in various studies to lead to recruitment of transcription

Table 11.3 IEGs: salicylate versus NIHL

Activity re: sound	**Salicylate**			**Noise trauma**		
	c-fos	**Arc/ Arg1.3**	**BDNF**	**c-fos**	**Arc/ Arg1.3**	**BDNF**
ANF		≈	⇑	⇑		⇑
DCN	⇓	≈				
IC	⇓	≈				
AC	⇑	⇑⇓	⇓		⇓	⇑⇓
Amygdala	⇑	⇑				

≈: no change; ⇑⇓: up respectively down.

factors that act on BDNF. At the other end of the auditory pathway, in auditory cortex, a decline of the *Arc/Arg3.1* expression was described after tinnitus-inducing acoustic trauma (Tan et al., 2007) as well as tinnitus-inducing salicylate application (Panford-Walsh et al., 2008). Neurons in Arc/Arg3.1 knock-out animals exhibit markedly reduced endocytosis, increased steady-state AMPA receptor levels and increased amplitude of spontaneously evoked mEPSCs. Reduced AMPA receptor trafficking is a direct molecular correlate of deprivation phenomena, suggesting that reduced Arc/Arg3.1 levels leads to increased AMPA responses, and increased SFR. In contrast to this, Mahlke and Wallhässer-Franke (2004) found an increase in the number of Arc/Arg3.1 immunoreactive neurons after both acoustic trauma and salicylate application (red arrows in Table 11.2). According to the authors, the latter data may have been affected by high background noise levels in the saline controls.

So whereas very different signaling cascades are evident for salicylate and NIHL at the level of the cochlea and auditory nerve as well as in the auditory cortex very similar IEG expressions are suggested. The findings by the two different groups do not agree at all, but they do predict similar effects on SFRs for salicylate and NIHL in auditory cortex, increases (Tan et al., 2007; Panford-Walsh et al., 2008) and decreases (Mahlke and Wallhässer-Franke, 2004) respectively. In contrast to the similar results in the IEG expression, NIHL results in increased SFRs and salicylate in decreased SFRs in cortex (Chapters 6 and 7). Clearly, more research is needed here.

11.2 **Top-down aspects of perception**

It is time to investigate what the role of the auditory cortex is in the tinnitus percept. This section presents my personal view of auditory cortical function. Auditory cortex is likely necessary for perceiving tinnitus; without auditory cortex there is usually not a conscious auditory percept, and potentially not an annoyance aspect although there is a bypass from the MGB to the amygdala, with return to the auditory cortex. Secondly, the auditory cortex does more than just relay information from the MGB to higher-order cortical areas. A case in point is that >99% of neural inputs to a cortical neuron are from other cortical cells. Even in the input layers of auditory cortex at most 20% of the inputs are of thalamic origin (Abeles, 1991). It is thus likely that the various auditory cortical areas work mostly on their own reentrant activity. Furthermore, the output of the cortex to the thalamus also far outweighs the input it receives from the auditory midbrain, suggesting that the cortex exhibits a control function on subcortical structures. This control function of auditory cortex may reach as far down as the cochlear hair cells (Xiao and Suga, 2002).

I also view the cortex more as a representational system than an afferent information processing and transmission system. By this I mean that the information in the afferent signals is used first and foremost for a comparison with the "world-view" stored in the brain. This view of the world can be changed whenever the input from the outside world (i.e. from subcortical auditory areas) violates its expectations, as an old and trusted learning rule expresses (Rescorla and Wagner, 1972). The flow of this comparison process through the cortex is also reflected in the large series of event-related potentials that are generated by such violations. One has only to think about the mismatch negativity and the P300 as oddball or deviance signaling components. Furthermore, in language studies there are the additional semantic (N400) and syntactic (P600) violation-indicating components (Chapter 4). An alternative description of these phenomena is that the cortex performs a Bayesian integration of sensory information, where the prior probability of an event is updated by the sensory stimulation to a posterior probability, which then functions as the new prior etc.

Tinnitus, as a potential consequence of changes in the cortical tonotopic maps, may be a result of maladaptive auditory plasticity. In this respect, it is useful to briefly summarize again what properties remain plastic in the adult auditory system. Cortical receptive fields of individual neurons are pliable by learning (Fritz et al., 2003), and so are cortical tonotopic maps (Polley et al., 2006). Peripheral hearing loss causes changes of tonotopic maps in AI (Rajan et al., 1993) and auditory thalamus (Kamke et al., 2003), but not in the ICC (Irvine et al., 2003) or CN (Rajan and Irvine, 1998). We have previously shown the correlations between tonotopic maps changes, increased SFR and increased neural synchrony (Noreña and Eggermont, 2005, 2006; Engineer et al., 2011; Chapter 7). These findings clearly point to an important role for the thalamocortical complex in the generation of tinnitus through maladaptive plasticity, whereas other mechanisms (gain change) or systems (e.g. somatic) may be responsible for the changes observed in the DCN and IC (Mulders and Robertson, 2009).

11.2.1 Feedback loops in the auditory system

An aspect of the "learning by violation" rule mentioned earlier may be that the cortex also tries to adjust the output of subcortical structures through its corticofugal feedback activity. In this way, increased activity at a particular cortical site can, for instance, change the representation of frequency in the IC (Suga et al., 2000). We will explore this aspect further down and derive from it the potential importance of subcortical structures in the generation of tinnitus. The auditory system is not just an afferent projection system but has a myriad of efferent connections that make it a reentrant system characterized by multiple, loosely interconnected, regional feedback loops (Spangler and Warr, 1991; Winer, 2006; Schofield, 2010).

The first loop is thalamocortical and comprises both the interaction between the thalamus with a given cortical area and between cortical areas. A reflection of these interactive loops is found in the various oscillatory brain rhythms; for instance, the gamma-band oscillation with its frequency in the 40–60 Hz range relies on connections that produce delays of 15–25 ms comprising conduction times between cells and synaptic integration times. This is a purely cortical rhythm, whereby each cortical area generates its own frequency. The rhythms in the 8–14 Hz range (delays of about 100 ms required) likely are all dependent on the thalamus, where the interplay between the reticular nucleus and the thalamic projection cells can generate rhythmic bursting with long delays caused by the duration of inhibition or hyperpolarizations, or both. Disturbances in that rhythm have been implicated in various positive syndromes including tinnitus (Llinás, et al., 2005). The cortex feeds back to the thalamus from pyramidal cells in layers V and VI (Lee and Sherman, 2011).

The amygdala, the fear center of the brain, gets two inputs from the auditory system, a fast one via the MGB and a slower one via the secondary auditory cortex (Farb and Ledoux, 1999). This also constitutes a loop, as the amygdala feeds back on the auditory cortex. This integration of the limbic system and the thalamo-cortical complex is involved in emotional aspects of tinnitus (see section 11.3). A potentially important loop from MGB to amygdala, nucleus acumbens (NAc), thalamic reticular nucleus and back to the thalamus may function as a gate to filter out unwanted sound such as tinnitus (Rauschecker et al., 2010).

The corticofugal connections from layer V also affect the auditory midbrain and have been demonstrated to affect its response properties. The midbrain is subsequently involved in a loop comprising the dorsal cochlear nucleus. The cochlear nucleus is also directly affected by corticofugal fibers (Schofield and Coomes, 2005) as is the olivary complex (Coomes and Schofield, 2004). The olivo-cochlear bundle in turn connects the hair cells with the brainstem.

Feedback loops tend to stabilize system's activity. It may well be that this feedback over time also stabilize tinnitus that originates from increased SFR at more peripheral sites such as the DCN or at more central ones such as auditory cortex. In the long run, peripheral and central activity may enhance each other, and the result is that there is no particular site in the central auditory system that can be held solely responsible for tinnitus. Opening the loop by blocking connections is likely the only way to successfully probe for a cure of tinnitus. This can be done, for instance, by using drugs such as lidocaine (Baguley et al., 2005b) or by desynchronizing the activity of the nested loops, i.e. by stimulation the auditory system through a cochlear implant (Quaranta et al., 2004) or by direct electrical (De Ridder et al., 2006b) or transcranial magnetic stimulation (Plewnia et al., 2003) of the auditory cortex. Frontal cortical areas may become part of the reentrant loops establishing the tinnitus sensation (section 11.3.2).

11.2.2 Corticofugal action

The investigation into the role of descending connections from cortex to the ICC of the big brown bat (Yan and Suga, 1998) was the start of a large series of investigations into cortical control of the auditory system. They found that electrical stimulation of cortex paired with a tone, with a frequency equal to the BF of the stimulation site, altered the frequency map in the ICC such that it enhanced the extent of the paired-tone frequency representation (Figure 11.1). Moreover, the plastic changes were asymmetric across the BF range in the ICC: ICC neurons with best frequencies higher than that of the stimulated cortical site showed downward shifts in their best frequencies and toward that of the stimulated neurons in cortex, whereas neurons tuned to lower frequencies were relatively unaffected. Surprisingly 30 min of stimulation with tone bursts at 50 dB SPL alone also induced a shift in the ICC frequency map that was smaller than but similar to that observed after electrical stimulation of the corresponding BF site in the auditory cortex (Figure 11.1D). Thus, biologically irrelevant tone bursts and/or direct cortical electrical stimulation can augment the midbrain representation of the stimulus tone frequency or the electrically stimulated cortical neurons' BF.

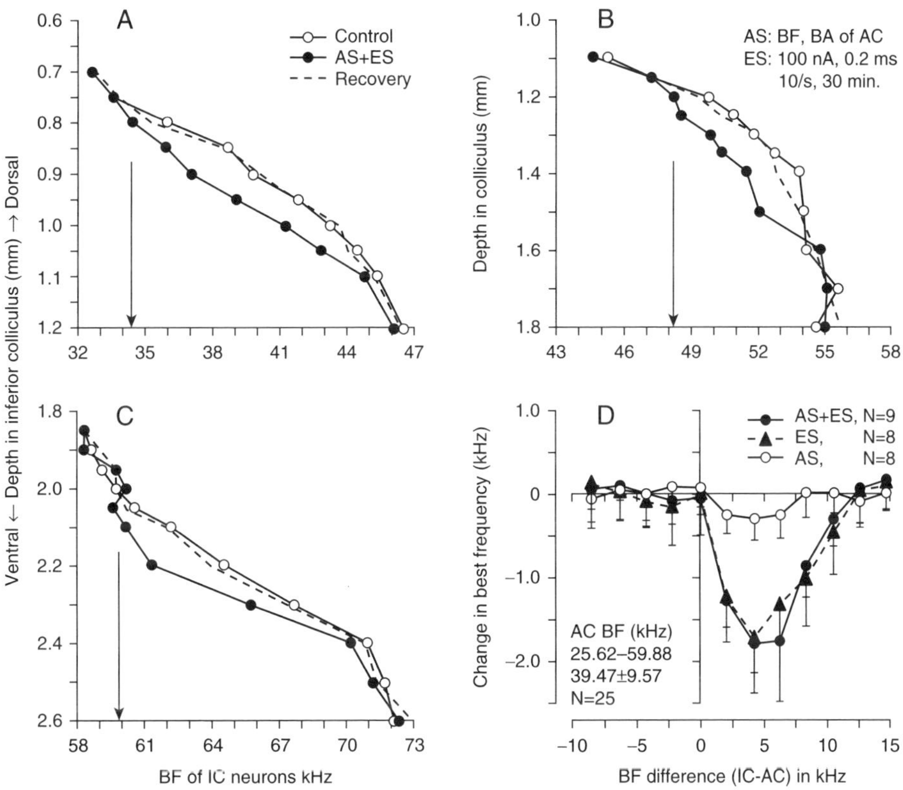

Figure 11.1 Shifts in the frequency map of the IC evoked by focal electrical stimulation of the auditory cortex paired with an acoustic stimulus (A–C). Arrows indicate the BF of the stimulation site in auditory cortex. In (D), the amount of shift in BF is shown as a function of BF difference between the collicular and electrically stimulated cortical neurons. Changes in BF of IC neurons are shown for: electrical stimulus alone (ES;▲), acoustic stimulus alone (AS; o), and both (●). Each symbol and vertical bar indicate a mean and standard deviation. Ns are the numbers of the BF-depth curves used for averaging. Reproduced by permission from Macmillan Publishers Ltd. from Yan, W. and Suga, N. (2002). Corticofugal modulation of the midbrain frequency map in the bat auditory system. *Nature Neuroscience*, **1**(1), 54–58. Copyright (1998).

This effect was replicated and extended in mice (Yan and Ehret, 2001, 2002; Yan et al., 2005). Yan and colleagues observed that bipolar electrical stimulation of primary auditory cortex, one electrode at the surface, the other one in layer VI did not affect best frequencies in ICC when the BFs of stimulated cortical neurons and recorded collicular neurons were similar (Figure 11.2C). However, BFs in ICC shifted toward the BF of the cortical stimulation site when cortical and collicular frequencies were different (Figure 11.2A,B). In addition to frequency-specific shifts in collicular BFs, cortical stimulation reduced firing rates if cortical and collicular BFs were different (Figure 11.2D).

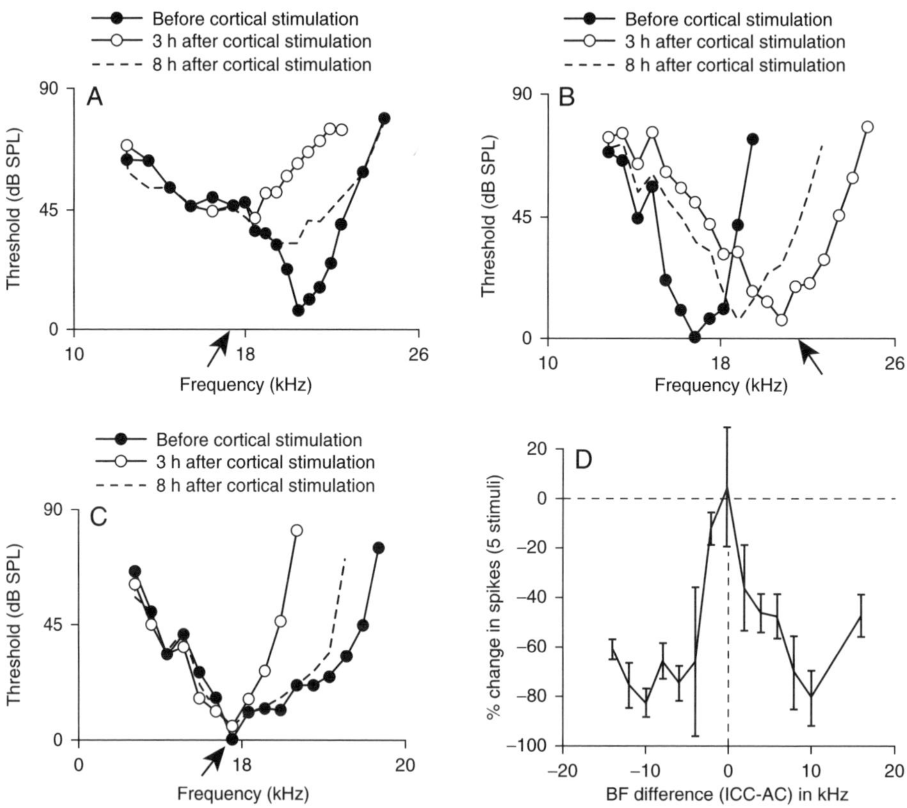

Figure 11.2 A) Changes in frequency-tuning curve, amplitude-response function and response pattern of an unmatched collicular neuron of which the BF was higher than the stimulated cortical BF. The cortical activation abolished all responses at the sharp portion and at the high-frequency tail of the original tuning curve while the low-frequency tail was less or not at all affected. B) Changes in an unmatched collicular neuron of which the BF was lower than the stimulated cortical BF. The cortical activation shifted the entire frequency-tuning curve upward toward the cortical BF with little effect on the sharpness of tuning. C) Changes in a collicular neuron of which the BF was similar to the stimulated cortical BF (matched neuron). The cortical activation did not shift its BF but sharpened its frequency-tuning curve. D). Effects of the cortical activation on the average collicular response magnitude at the original BF. Response rates were strongly reduced by the cortical activation when cortical and collicular BFs were different, but they were unaffected when cortical and collicular BFs were similar (zero BF difference). Means ± SEM are shown. Reproduced from Yan, J. and Ehret, G. (2002). Corticofugal modulation of midbrain sound processing in the house mouse. *European Journal of Neuroscience*, **16**, 119–28. © 2002, John Wiley and Sons with permission.

In exploring the corticofugal effects to levels below the midbrain Luo et al. (2008) found that cortical activation increased the response magnitudes and shortened response latencies of CN neurons with BFs matched to the cortical stimulation site, whereas it decreased response magnitudes and lengthened response latencies of unmatched CN neurons. In addition, cortical activation shifted the frequency tunings of unmatched CN neurons toward those of the activated cortical neurons. The auditory cortex apparently implements a long-range feedback mechanism to select or filter incoming signals from the ear. Yan and Suga (1998) suggested that the corticofugal system is involved in the long-term improvement and adjustment of subcortical auditory information processing, largely because the corticofugal effects slowly disappeared over 2–3h after the cessation of the asymptotic 30-min cortical electrical stimulation (Ma and Suga, 2001).

Corticofugal feedback may be an important factor in the manifestation of tinnitus (Jastreboff, 1990, Eggermont, 2008). Brain magnetic field source imaging suggests that tinnitus is accompanied by a reorganization of the auditory cortical tonotopy (Muhlnickel et al., 1998; Wienbruch et al., 2006). The putative pattern of reorganization correlates with the subjective tinnitus strength and with the shift in the representation of tinnitus frequencies in the auditory cortex. Corticofugal feedback, induced by the tinnitus to which a person directs her/his attention, could enhance the processing of tinnitus-related frequencies and suppress the processing of surround frequencies in the brainstem and auditory midbrain. Therefore, this frequency-specific amplification by corticofugal feedback in subcortical areas might contribute to the stabilization of the tinnitus percept, leading to a chronic form of tinnitus. This is, however, pure speculation and requires further investigation.

11.3 The resting brain networks

Tinnitus is aberrant spontaneous or resting state brain activity interpreted as sound. An exploration of recent findings on resting state brain activity and its spatial correlations may shed light on some findings in tinnitus patients.

11.3.1 Correlated spontaneous activity in the brain

The observation that spontaneous BOLD activity is not random noise, but is specifically organized in the resting human brain has created a new avenue of neuroimaging research (Fox and Raichle, 2007). Fox et al. (2005) identified two widely distributed brain networks on the basis of either spontaneous correlations within each network or anticorrelations between networks. One network consisted of regions that are also involved in task-related activations and the other of regions corresponding also to task-related deactivations. Thus, the same type of correlation between brain areas is present during spontaneous activity as well as task-related activity. This resting-state functional connectivity reflect correlations in slow (<0.1-Hz) spontaneous fluctuations in the BOLD signal.

A promising, and potentially related, electrophysiological correlate of spontaneous BOLD fluctuations is the slow (<0.1-Hz) voltage fluctuation that has been observed with DC-coupled EEG recording (Steriade et al., 1993). These slow electrical fluctuations are themselves correlated with changes in the power of higher frequency bands (Young and Eggermont, 2009). The idea that slow fluctuations may modulate the power of higher frequency activity has been referred to as "nested" frequencies and may be an important principle in brain organization. Rapid fluctuations such as alpha or gamma activity might coordinate neuronal activity across small spatial scales, whereas slower fluctuations in the power of these frequency bands may be required for long-range coordination. There is evidence for such a frequency–space trade-off in the brain (Buzsaki and Draguhn, 2004; Steriade, 2006). Whether spontaneous BOLD activity is directly related to slow electrical fluctuations previously seen with DC-coupled EEG or to changes in the power of higher frequency electrical activity remains to be determined. Finally, much animal work has focused on slow (<1-Hz) fluctuations in electrical membrane potential, referred to as up and down states (Luczak et al., 2007). Whether this phenomenon is related to the slow fluctuations in the BOLD signal also remains a topic for future work.

If spontaneous neural activity reflects ordered brain states it should show highly specific patterns, and these should reflect the functional architecture of the networks. Eggermont (1994) was the first to establish that spike pair-correlations in anesthetized cat auditory cortex during spontaneous activity were predictive of those during stimulus-evoked activity. I showed a comparison of the cross-correlation coefficients for 71 dual-electrode pairs with clearly visible correlation peaks in both conditions between the stimulus (after shift-predictor correction) and spontaneous conditions. A paired t-test showed that the spontaneous cross-correlation coefficient values were on average 0.004 larger ($P < 0.0001$) than during stimulation. Under stimulus conditions the peak of the neural correlogram was generally narrower than under spontaneous firing conditions. This was the case in 62–87% of correlograms, and depended on the stimulus type. The adequacy of a particular stimulus played a role as well; in general the narrower correlograms were obtained for the more effective stimuli. Most likely the cause for this narrowing of the correlogram was a more synchronous arrival of thalamic afferent activity at the cortical cells, so that especially the delayed coincidences resulting from di- and polysynaptic interactions that make up the base of the correlogram peak were better synchronized. Thus, spontaneous functional connectivity is very similar to that under stimulus conditions. Nevertheless, a hierarchical cluster analysis showed that neuron clusters typically consisted of pairs with ≤1 mm distance, and changed their borders somewhat between spontaneous and stimulus conditions (Eggermont, 2006).

Tsodyks and coworkers (Tsodyks et al., 1999) also showed that, in the anaesthetized cat, spontaneous activity is endowed with specific patterning that reflects the functional architecture of the underlying network. Using optical imaging, largely reflecting LFPs, they demonstrated that spontaneous activity in the cat visual cortex is highly coordinated across large assemblies. This coordination is feature specific, as the spontaneous discharges of individual neurons were temporally locked to the activation of other cells with similar orientation preferences. The patterns of local field potential coherence in spontaneous activity can span several millimeters of cortical surface (Destexhe et al., 1999; Eggermont et al., 2011), indicating that specific long-range interactions are possible even in the absence of visual input. Moreover, these patterns are highly variable and their fluctuations can largely account for the variability of the responses evoked by a subsequent visual stimulus. These data make it likely that spontaneous activity may contain structured information and therefore plays an important role in cortical function. Pathological changes in these network correlations may be underlying tinnitus.

11.3.2 Network correlations in tinnitus

A reciprocal relationship between reduced alpha-band activity and increased gamma-band activity in tinnitus patients was found (Lorenz et al., 2009). Depending on the frequency band of the EEG oscillations, local or more global neural synchrony is implicated. At the single subject level there was a strong inverse relationship between the alpha- and gamma-coupling both in tinnitus patients and in controls (Figure 11.3). In tinnitus patients especially the long-range synchrony coupling appears to be affected. In particular long-range coupling in the 9–12-Hz frequency band was reduced and that in the long-range 45–52-Hz frequency band was increased (Schlee et al., 2009a). In patients with a tinnitus history of <4 years, the left temporal cortex was predominant in the gamma network whereas in patients with tinnitus duration of >4 years, the gamma network was more widely distributed including more frontal and parietal regions.

One could generalize this by postulating that the hyperactivity of the temporal cortices in tinnitus is integrated in a global network of long-range cortical connectivity (Schlee et al., 2009b). They found that the inflow into the left and right temporal cortex, i.e. a top-down influence from the global network onto the temporal areas, relates to the subjective strength of the tinnitus distress. If and how these findings relate to positively or negatively correlated BOLD signals is presently unclear.

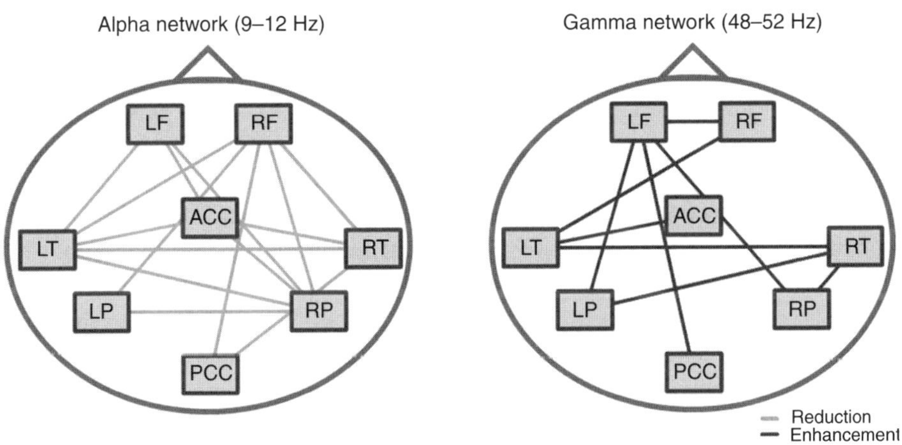

Figure 11.3 Schematic display of the alpha and the gamma network. Connections with a significant group difference were plotted as edges in the networks. The nodes were named by: LF = left frontal, RF = right frontal, LT = left temporal, RT = right temporal, LP = left parietal, RP = right parietal, ACC = anterior cingulate cortex, and PCC = posterior cingulate cortex. Reproduced from Schlee, W., Hartmann, T., Langguth, B., and Weisz, N. (2009). Abnormal resting-state cortical coupling in chronic tinnitus, *BMC Neuroscience*, **10**(11). © 2009, BioMed Central, with permission.

Kahlbrock and Weisz (2008) found that changes of tinnitus loudness induced by residual inhibition (Chapter 3) were mediated by alterations in the pathological patterns of spontaneous brain activity, specifically by the reduction of delta activity. Delta activity is a characteristic oscillatory activity generated by deafferented/deprived cortical neuronal networks (Llinás et al., 2005). This implies that residual inhibition effects might reflect the transient reestablishment of balance between excitatory and inhibitory neuronal assemblies that has been perturbed (in most tinnitus individuals) by hearing loss.

11.4 **Imagery, hallucinations, and tinnitus**

Tinnitus is an internal sound, just like imagined sounds or hallucinations (where the patient however thinks it is an actual external stimulus). It is instructive to compare brain activation under these varied endogenous generated sounds. A crucial difference appears to be the absence/presence of the involvement of the primary auditory cortex. From the experiments using lidocaine to suppress tinnitus (Chapter 4) the suggestion arose that major blood flow changes during tinnitus occur in the parabelt area of auditory cortex (see Figure 4.1, Chapter 4). Similar conclusions were drawn for gaze-induced tinnitus that appears to localize tinnitus activity to the parabelt area of auditory cortex.

Using fMRI, Bunzeck et al. (2005) investigated the neural basis of mental auditory imagery of familiar complex sounds that did not contain language or music. In the first condition (perception), the subjects watched familiar scenes and listened to the corresponding sounds that were presented simultaneously. In the second condition (imagery), the same scenes were presented silently and the subjects had to mentally imagine the appropriate sounds. During the third condition (control), the participants watched a scrambled version of the scenes without sound. To overcome the disadvantages of the acoustic scanner noise a sparse temporal sampling technique was used with acquisition at the end of each movie presentation. Compared to the (visual) control condition, bilateral activations in the PAC (including Heschl's gyrus) and secondary auditory cortices (SAC) (including planum temporale) were found during perception of complex sounds. In contrast, the imagery condition elicited bilateral BOLD only in SAC, whereas no significant activity was observed in the PAC. The results suggest that imagery and perception of complex sounds rely on overlapping neural correlates of the SAC but not PAC.

The finding that only perception but not imagery of the utilized stimuli was associated with activation of the PAC is consistent with theories featuring modular psychological mechanisms. According to these theories, perceptual inputs like sounds are processed bottom-up by several perceptual mechanisms that operate in a sequential and hierarchical mode. The cortical processing of the presented auditory stimuli starts at the first elementary level where basic acoustic features are extracted. The PAC and SAC are supposed to be the neural basis of these elementary acoustic analyses. However, mental auditory imagery is a top-down process that is characterized by reversed-order analysis. This process is initiated by neural networks specialized for higher-order cognition like the inferior frontal gyrus and insular regions (Griffiths, 2001; Hoshiyama et al., 2001; Jäncke and Shah, 2004).

Only three studies reported activations in PAC during auditory imagery (Dierks et al., 1999; Hoshiyama et al., 2001; Yoo et al., 2001). In contrast to Bunzeck et al. (2005) and most other studies cited therein, Hoshiyama et al. (2001) and Yoo et al. (2001) used rather simple sounds like computer-generated tones. Irrespective of the underlying mechanisms of hallucinations, most other studies did not find activations in primary auditory cortex during verbal (Silbersweig et al., 1995; Lennox et al., 1999; Shergill et al., 2000) or musical hallucinations (Griffiths, 2000). Therefore it is possible that the observed activation in different subdivisions of the auditory cortex varies as a function of the complexity of the imagined sound. Maybe this would also apply to the complexity of the tinnitus sound; tonal tinnitus may involve PAC and more complex tinnitus only SAC. This could relate to the finding that the preferred TMS stimulation type is different for tonal and noise-like tinnitus (de Ridder et al., 2007b).

A clinical condition in which musical imagery is not voluntary is referred to as musical hallucinosis. Those patients "attributed the experience to a problem with the brain or the ears" (Griffiths, 2000) Findings from brain imaging studies are consistent with the hypothesis that musical hallucinosis might result from abnormal spontaneous activity in neural substrates normally underlying musical perception and imagery (e.g. Griffiths, 2000). Additionally, there is evidence that, e.g. abrupt bilateral hearing loss following injection of gentamycin (Tanriverdi et al., 2001), and desynchronization of the right auditory cortex (Kasai et al., 1999) are linked to musical hallucinosis. Thus, whereas tinnitus originates in a different way than hallucinations and voluntarily imagined sounds, the underlying activation of auditory cortex appears to be similar.

11.5 Similarities between tinnitus and epilepsy mechanisms

Loss of peripheral input to central neurons can promote the development of synchronous spiking activity (Noreña and Eggermont, 2003; Chapter 9) by prolonging postsynaptic depolarization and increasing the likelihood of temporally coincident inputs converging on synapses. In the normal central auditory system, inhibition is more or less co-tuned with the excitatory part of the receptive field of a neuron produced by thalamocortical input (Wu et al., 2008). This would restrict synchronous activity to neurons tuned to properties of the acoustic stimulus thereby leading to normal auditory perception. However, when intracortical inhibition is weakened, widely distributed synchronous spike-firing activity can develop and lead to the perception of sounds that are physically absent, i.e. tinnitus. This is again speculation, but the following observations suggest its plausibility.

A highly persistent modification of brain functioning occurred after repeated application of electrical stimulation (kindling) to auditory cortex and resulted in the development and spread of epileptiform activity (Valentine et al., 2004). Hearing sensitivity measured by the auditory brainstem response, and reflected in the thresholds at the CF of cortical neurons, was not affected by kindling. Recordings were obtained from primary auditory cortex contralateral to the kindled site on the day following the last kindling session. Spontaneous neural synchrony, as measured by spike cross-correlation, was enhanced by 40%

compared to sham controls. This was accompanied by a profound alteration of the tonotopic map in primary auditory cortex with a large area becoming tuned to a narrow frequency range that was related to the location of the kindling electrodes.

Noise trauma acting at the periphery and kindling applied at a central level, both cause profound changes in the tonotopic map in primary auditory cortex and both induce a strong increase in spontaneous neural synchrony. These correspondences suggest a role of neural synchrony in tinnitus and epilepsy indicating that tinnitus could result from an epileptiform-like pattern of activity in cortex.

As reviewed in Valentine et al. (2004) the changes occurring in GABA and glutamate receptors after kindling and noise-induced hearing loss are also comparable. Kindling has been associated with a loss in GABAergic inhibition. It has also been hypothesized that kindling stimulation in cortex produces an accumulation of glutamate that triggers altered NMDA channel functions and long-lasting changes in synaptic efficacy of long-range horizontal connections. In cat primary auditory cortex horizontal connections generally follow a course parallel to the isofrequency contours (Read et al., 2001), but 10–15% project in directions orthogonal to the isofrequency contours and may even extend as far as the anterior and posterior auditory fields (Wallace et al., 1991; Lee et al., 2004). Strengthening of the horizontal connections after kindling may explain the larger percentage of double-tuned frequency tuning curves that we observed. If these horizontal connections become stronger than those of the specific thalamic inputs, they could elicit an outward migration of glutamate hyperactivity and subsequent synaptic changes. This process would likely stop when all the neurons in the stimulated cortical area (and its projection region) were combined into one large neural assembly.

11.6 The limbic system, prefrontal cortex, and tinnitus

11.6.1 Role of stress, and its physiological effects

Hébert et al. (2004) hypothesized that tinnitus, a permanent internal sound, should behave as a stressor and should be accompanied by chronically elevated cortisol levels in individuals with severe tinnitus. Basal cortisol levels were measured using saliva samples (five saliva samples per day for 3 days within a week) taken in the participant's natural environment. Secretion of cortisol (a naturally occurring glucocorticoid) is under the control of the hypothalamic–pituitary–adrenal (HPA) axis, a closed-loop endocrine system. Cortisol levels follow a circadian rhythm with elevated cortisol levels in the morning (AM) phase, and decreased levels in the afternoon (PM) phase. Following perception of a stressor, there is a rapid release of corticotropin-releasing factor (CRF) by the hypothalamus, which triggers subsequent release of adrenocorticotropin hormone (ACTH) by the pituitary, and glucocorticoids (cortisol) by the adrenal gland. Intolerance to external sounds was assessed psychometrically. The high tinnitus-related distress group had chronic cortisol levels greater than both the low tinnitus-related distress and control groups, and also displayed greater intolerance to external sounds. The low tinnitus-related distress and control groups did not differ from each other on either of these measures, which is compatible with the clinical observation that severe tinnitus is associated with high stress levels (Hébert and Lupien, 2009). Long-term exposure to high levels of cortisol lead to a down-regulation of glucocorticoid receptors, and to an increased responsivity of the HPA axis to stress, thus leading to an endocrine feedback loop of increased cortisol levels and increased responsiveness to stress (Hébert and Lupien, 2007).

Using fMRI Wolf (2009) showed the association of higher cortisol levels with a stronger amygdala response to emotional stimuli. This emphasizes the correlated activation of the HPA axis and amygdala activity during stress responses. Activation of the HPA axis likely influences lateral efferent feedback projections, which feature a co-distribution of dynorphin and enkephalin-like opioids (Altschuler et al., 1988). Thus, a stress-induced release of opioid peptides, specifically dynorphins from lateral efferent terminals in the auditory periphery, may be involved in tinnitus induction, due to its potentiation of excitatory effects on NMDA receptors (Sahley and Nodar, 2001). These findings justify a direct impact of stress responses on the physiology of auditory nerve activity of hearing-impaired animals and humans with tinnitus.

Mazurek et al. (2010) found that 24-h stress in Wistar rats decreased the thresholds and increased the amplitudes of auditory brainstem responses and distortion product OAEs. The resultant auditory hypersensitivity was transient and most pronounced between 3–6 h post-stress, returning to control levels 1 week later. The concentration of corticosterone and tumor necrosis factor alpha was systemically elevated in stressed animals between 3 and 6 h post-stress, confirming the activation of the HPA axis. In addition, expression of the HPA-axis associated genes: glucocorticoid receptor (GR) and hypoxia-inducible factor 1 alpha (Hif1α) was modulated in the auditory tissues. In the spiral ganglion no differences in gene expression between stressed and control animals were found. In the organ of Corti, expression of GR mRNA remained stable, whereas that of Hif1α was significantly down-regulated 1 week after stress. In addition, the expression of an OHC marker prestin was significantly upregulated 6 h post-stress. This indicates that stress influences the auditory periphery. How this interacts with, for example, existing NIHL and tinnitus remains to be seen.

11.6.2 Stress- and emotion-related areas in the brain

As we have seen in section 11.2, the amygdala is involved in fear and other aversive emotions, as well as evaluative processes associated with these emotions, whereas the ventral striatum mediates processes associated with reward and motivation/approach behavior. The amygdala and hippocampus both receive inhibitory presynaptic input from cholinergic neurons intrinsic to the NAc, suggesting a possible mechanism for decreased activity in these regions as a consequence of activity increases in ventral striatum. Cells in the human hippocampus and parahippocampal areas respond to novel stimuli with an increase in firing. However, already on the second presentation of a stimulus, neurons in these regions show very different firing patterns. In the parahippocampal region there is dramatic decrease in the number of cells responding to the stimuli (Viskontas et al., 2006), suggesting a rapid habituation. Thus a novel stimulus normally is associated with parahippocampal habituation and active hippocampal inhibitory activity. It is thus possible that in tinnitus this mechanism is disrupted (likely because tinnitus ceases to be a novel sound) resulting in persistent parahippocampal activity and thus preventing habituation (De Ridder et al., 2006a). The parahippocampal area may also play a role in memory recollection, and a dysfunction in this mechanism could explain auditory hallucinations. In line with this hypothetical mechanism tinnitus can result from continuous sending stored auditory information from the hippocampus to the auditory association areas. This is compatible with electrophysiological studies demonstrating that auditory habituation is disrupted after amygdalo/hippocampal resections in humans (Hamalainen et al., 2007).

In distressed tinnitus patients low frequency activity (delta and theta) in the dorsal anterior cingulate cortex (ACC; BA24, BA32) is decreased, and alpha and beta activity is increased (Figure 11.4). In ACC of highly distressed versus lowly distressed patients alpha is even more increased, suggesting that the amount of alpha activity correlates with the amount of distress (Vanneste et al., 2010c). This contrasts, with the finding (Figure 11.3) that the long-range coupling in the alpha band was reduced and that in the long-range gamma band was increased in tinnitus patients (Schlee et al., 2009a).

11.6.3 Is there a neural gate to block tinnitus?

The first model of tinnitus that included the limbic system (Jasteboff, 1990) has also been its longest lasting and is influential to date. Limbic and auditory brain areas are thought to interact at the thalamic level. While a tinnitus signal originates from lesion-induced plasticity of the auditory pathways, it may be tuned out by feedback connections from limbic regions, which block the tinnitus signal from reaching auditory cortex. If the limbic regions are compromised, this "noise-cancellation" mechanism breaks down, and chronic tinnitus results (Rauschecker et al., 2010). In this model, efferents from structures in the subcallosal area, which includes the NAc of the ventral striatum and the ventral medial prefrontal cortex (vmPFC), are involved in the blocking of the tinnitus signal at the thalamic level. Rauschecker et al. argue that reorganization of auditory cortex and thalamus in response to peripheral deafferentation may be necessary but not sufficient for the tinnitus perception to arise. The NAc contains dopaminergic and serotonergic neurons among other types. Whereas the dopaminergic system within the NAc is well known for its involvement in reward behavior and avoidance learning (Blood and Zatorre, 2001), serotonergic neurons play a modulatory role in various emotion-related systems. Importantly, the NAc receives glutamatergic input from the amygdala, as well as projections from the hippocampus and the raphe nuclei.

In addition to corticofugal and transcortical glutamatergic projections from vmPFC, anatomical data indicate that serotonergic axons (from the dorsal raphe nucleus, the NAc, and other paralimbic regions) innervate the thalamic reticular nucleus (TRN) and the dorsal thalamus. Serotonin excites the GABAergic neurons of the TRN (McCormick and Wang, 1991), which in turn have a powerful inhibitory influence on sensory thalamic relay cells (Guillery and Sherman, 2002). TRN-mediated inhibition can also cause thalamic relay neurons to shift between tonic and burst-firing modes, the latter of which requires cells to be in a hyperpolarized state (Llinás and Jahnsen, 1982).

According to the model laid out by Rauschecker et al. (2010), tinnitus most likely results from the following factors: (1) in most, if not all, cases, the process leading to

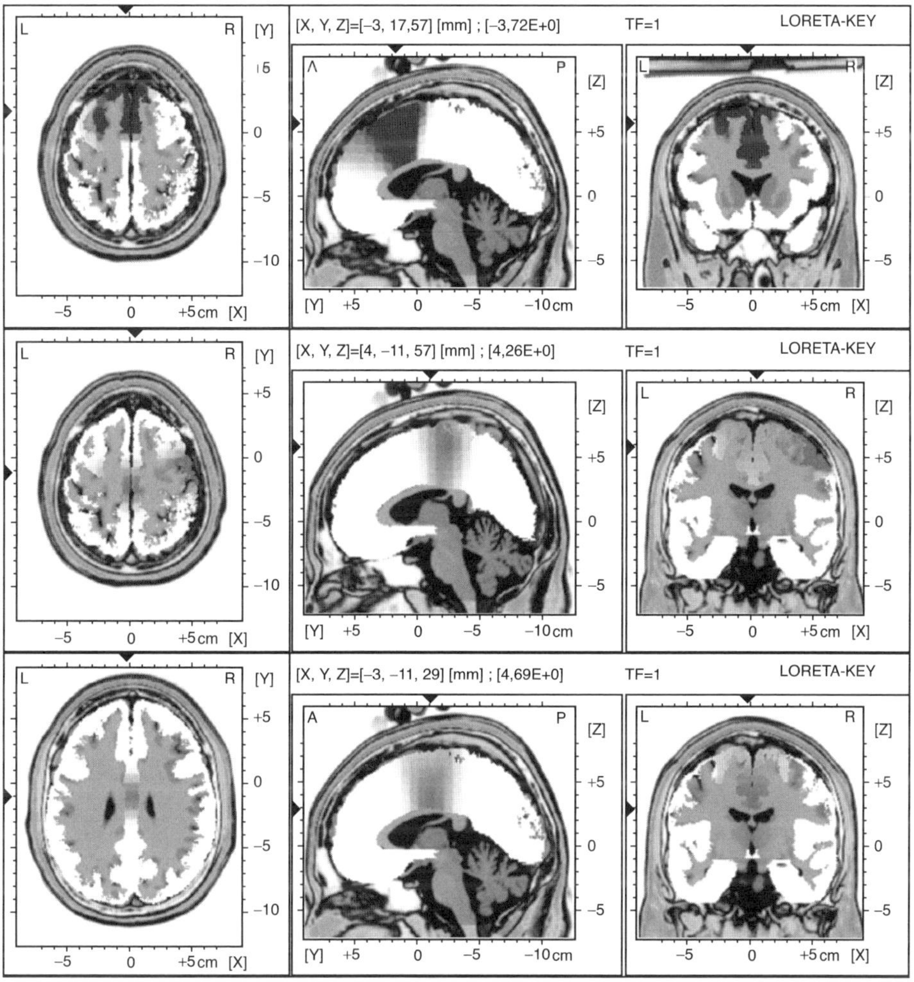

Figure 11.4 (Also see Color plate 7.) Significant results for current density amplitude analysis in respectively delta (2–3.5 Hz), theta (4–7.5 Hz), alpha (8–12.5 Hz), and beta (21–32.5 Hz) bands. Reprinted from Vanneste, S., Plazier, M., van der Loo, E., Van de Heyning, P., Congedo, M., and De Ridder, D. (2010). The neural correlates of tinnitus-related distress. *NeuroImage*, **52**(2), 470–80. Copyright (2010), with permission from Elsevier.

tinnitus is triggered by a lesion to the auditory periphery, e.g. a loss of hair cells in the inner ear resulting from acoustic trauma or aging; (2) loss of input in the lesioned frequency range leads to an overrepresentation of lesion-edge frequencies, which causes hyperactivity and possible burst firing in central auditory pathways, constituting the initial tinnitus signal; (3) under normal circumstances, the tinnitus signal is cancelled out at the level of the thalamus by an inhibitory feedback loop originating in paralimbic structures: activity from these structures reaches the TRN, which in turn inhibits the MGN. If, however, paralimbic regions are compromised, inhibition of the tinnitus signal at the thalamic gate is lost, and the signal is relayed all the way to the auditory cortex, where it leads to permanent reorganization and chronic tinnitus.

Using fMRI and voxel-based morphometry, Leaver et al. (2011) assessed tinnitus-related functional and anatomical anomalies in auditory and limbic networks and provided some quantitative data to back up the previous hypothesis. Within auditory cortex, hyperactivity was found in medial Hexsch's gyrus (mHG), the likely location of primary auditory cortex and in posterior superior temporal cortex (pSTC), a secondary auditory region. This stimulus-evoked increased activity in tinnitus patients was present for all stimuli in pSTC; however, hyperactivity in mHG was restricted to tinnitus-frequency-matched stimuli. However, the NAc exhibited the greatest degree of hyperactivity, specifically to sounds frequency-matched to patients' tinnitus. Complementary structural differences were identified in vmPFC, another limbic structure heavily connected to the NAc. These are areas that have been shown to evaluate stimulus significance. Hyperactivity in NAc correlated with the magnitude of structural changes in the vmPFC in these same patients. Overall, Leaver et al. (2011) suggest that both auditory and limbic regions are involved in tinnitus. My personal impression is that they only measured BOLD response correlates for hyperacusis rather than for tinnitus, albeit that the patients "reported no history of *severe* hyperacusis or phonophobia and in a short survey reported limited or no sensitivity to noise."

Other evidence for a modulatory role of the striatum comes from the finding that deep brain stimulation to caudate neurons (area LC) in the body of the nucleus, a subsite of the striatum, was found to decrease or increase tinnitus loudness (Cheung and Larson, 2010). The abnormal functioning of the NAc or the caudate suggests the involvement of genetic factors and it is hard to explain how these changes could be produced by, for example, NIHL.

11.7 **Can hyperacusis and tinnitus both result from an auditory system gain change?**

The concept of auditory gain change is illustrated in Figure 11.5. In tinnitus without hearing loss, increased neural gain (upward pointing arrow) results in a leftward shift of the loudness-level function. This results in a factual decrease in hearing threshold, and sound levels below the normal threshold (here arbitrarily set at 15 dB, vertical dashed line) may become audible. This could explain the emergence of sound sensations experienced by people staying in a silent soundproof room for a few minutes (Chapter 1), provided that in such a short time the gain can increase. One could call this tinnitus. The subjective tinnitus loudness would likely be low, e.g. equivalent to that for a 15 dB HL sound. For normal suprathreshold sound levels the loudness thereof is considerably increased by the increased central gain, resulting in hyperacusis. In case of hearing loss, in this example set at 45 dB, with recruitment, the loudness function is shifted to the right and becomes steeper. If this is combined with an increased gain, hyperacusis (or over-recruitment) may be present at higher sound levels. In case the subjective tinnitus loudness would be 10 dB SL, this then converts to the equivalent loudness of 60 dB HL and is quite substantial.

This scenario suggests that central gain change alone can account for both hyperacusis and tinnitus. The finding that in patients with a primary complaint of hyperacusis the prevalence of tinnitus is about 86% supports this. However, in patients attending tinnitus

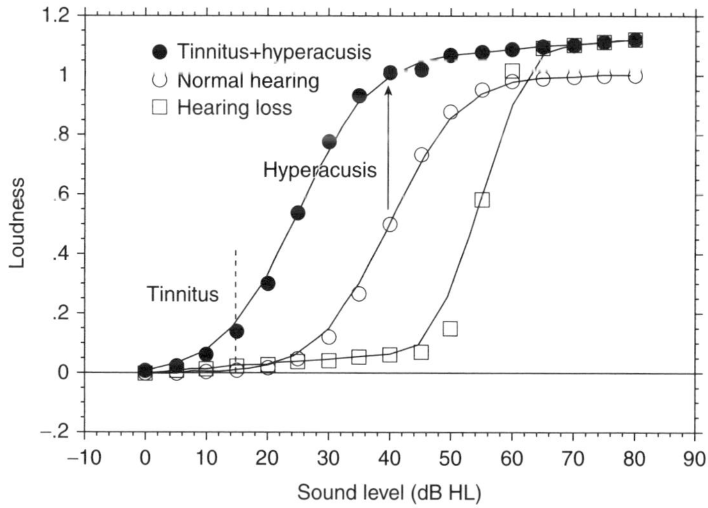

Figure 11.5 Hypothetical loudness functions in normal hearing with hyperacusis and for NIHL.

clinics with a primary complaint of tinnitus the prevalence of hyperacusis is only about 40% (Baguley, 2003). This suggests a strong link with hyperacusis only for a particular subset of tinnitus patients and it is tempting to suggest that in that group hyperacusis is causal to the tinnitus. More detailed typing of tinnitus etiology is obviously needed to find the reasons why there are so many more patients with tinnitus without accompanying hyperacusis (Chapter 3).

Noreña (2011) considered the effects of a homeostatic gain change covering the entire auditory system from auditory nerve to cortex. The effect of this mechanism is that ANF activity, spontaneous as well as driven, is amplified by the CNS. Noreña's model suggests that reducing the SFR in ANFs could ameliorate tinnitus. Following this, one would assume that cochlear nerve resection in humans would also alleviate tinnitus, but this is typically only partially the case (House and Brackman, 1981). An interesting consequence of this homeostatic plasticity model is the distinction made between so-called "VCN-tinnitus" and "DCN-tinnitus." VCN-tinnitus results of increased SFR in the central auditory system because of amplification of ANF SFR activity and this requires a large number of functionally intact IHCs. The required gain change may result from diffuse loss of ribbon synapses (Kujawa and Liberman, 2009) and subsequent diffuse reduction in ANF SFR, which initiates the central gain increase. DCN-tinnitus would occur in case the ANF activity is zero or very low, implying damaged IHCs, and where the increased spontaneous activity in the DCN is supplied by the dorsal column and/or trigeminal nerve activity (Shore et al., 2008). This is the result of potentially the formation of new synapses or the enhancement of the efficacy of existing synapses between dorsal column and/or trigeminal nerve on the granule cells in the DCN, potentially also a result of homeostatic mechanisms. The SFR in the DCN apparently does not depend strongly on the SFR of ANFs; after cochlear ablation the increased SFR as a result of noise trauma was reduced but still stayed well above the control levels (Koerber et al., 1966; Zacharek et al., 2002).

It is thus tempting to state that VCN-tinnitus is a consequence of central hyperacusis, whereas DCN-tinnitus will occur without hyperacusis. The electrophysiological measure for hyperacusis would potentially be reduced ABR amplitudes in the presence of enhanced MLR or late obligatory cortical potentials.

This model provides a solution for the dependence of the SFR in the IC on spontaneous ANF activity (Mulders and Robertson, 2009; Chapter 7). Ablation of the cochlea in normal hearing animals results in silencing the VCN, whereas there are no changes in the SFR of the DCN neurons. In the DCN, the dorsal column and/or trigeminal nerve become more active and increases the SFR (this takes some time). In the VCN and IC, after only OHC loss, the increased neural gain will amplify the incoming SFR from the cochlea. In partial hearing loss, both effects will be present. If there is only a TTS (IHCs intact, but nerve fiber degeneration may follow) as in Mulders and Robertson (2009) the VCN tinnitus prevails, subsequent ablation immediately thereafter then reduces the SFR to baseline because the dorsal column/trigeminal factor did not have enough time to kick in. The time needed was estimated as about 2 weeks (Shore et al., 2008). However, Mulders and Robertson (2011) observed that there is a much longer delay before the hyperactivity in the IC becomes independent of afferent input. After 8–12 weeks following the trauma, cochlear ablation no longer abolishes the SFR increase in IC. So at short durations following noise trauma, hyperacusis may cause the increase in firing rate, but after 8–12 weeks intrinsic changes in the IC are sufficient to maintain the increased SFR. After ablation a sizable but non-significant decrease remains suggesting that the hyperacusis effect remains and that amplified peripheral SFR still contribute to the overall SFR in IC.

The model also needs to explain why in BOLD responses tinnitus-patients without hyperacusis (Gu et al., 2010; Chapter 4) show only increased stimulus-evoked responses in auditory cortex, whereas those with hyperacusis but no tinnitus also show this in the inferior colliculus. As suggested in Chapter 4, this would require clearly reduced spontaneous activity in auditory cortex, which so far has only been shown for salicylate. Alternatively, it could be explained if the hyperacusis experience does not fall in the standard class of perceptions and is caused by subcortical gain change that is fed to the limbic system via the thalamus, and thus bypassing the primary auditory cortex.

A testable alternative is based on the presumption that there are two different sources of spontaneous activity in the central nervous system; peripheral ones whose outputs are not much affected by the state of the animal (awake, drowsy, attentive, etc.), and central ones that are affected by the state of the animal. The differences reflect itself in the shape of the interspike interval histograms. The two systems interact at the level of the midbrain (Eggermont, 1990a). It is likely that in the absence or reduction of peripheral spontaneous firing the corticofugal effects induce changes in the SFR of the midbrain neurons.

Chapter 12

Management of tinnitus

From the previous chapters it is clear that tinnitus is related to alterations of neural activity in the CNS. These alterations are most probably due to altered sensory input, for example, by hearing loss, or even in its (audiometric) absence by dead regions in the cochlea. The neural substrates of tinnitus suggest various approaches to modify neural processing and change the properties of tinnitus and so obtain some alleviation of it. These approaches include neurophysiologic, psychologic, and pharmacologic ones. Surgical approaches so far mostly have a bearing on objective tinnitus and I will not consider these here. The neurophysiologic-based interventions for tinnitus include substitution methods to compensate for missing activity in the output of the cochlea via specially tailored acoustic environments, and via amplification of environmental sounds in the hearing frequency range, i.e. by hearing aids. In deaf persons the missing sounds can be applied by a cochlear implant. Other approaches in this area comprise masking or suppression of the tinnitus, but these methods partially overlap with the hearing-aid approach. An exception is formed by direct stimulation of the auditory cortex using either electrode arrays on the cortical surface or depth electrodes in the auditory cortex or other brain areas. A non-invasive method that may be useful to suppress tinnitus is based on transcranial magnetic stimulation (TMS). I will introduce and briefly discuss these approaches to manage tinnitus.

Psychologic and counseling approaches may be based on neurophysiologic models of tinnitus (section 12.3) or derived from treatment paradigms of people with depression, but will only be briefly discussed here.

With respect to the pharmacologic approaches I will elucidate the rationale for the use of those drugs that have been and are part of clinical trials, and describe the so far disappointing results obtained in humans and the somewhat more promising findings in animals.

12.1 Sound therapy (passive)

12.1.1 In a quiet environment nearly everyone experiences tinnitus

It has been known for more than a century (Chapter 1) that people without tinnitus in normal day life experience sounds when they are in a soundproof room for a few minutes. In addition, they describe the sounds they hear in the same fashion as people describe their tinnitus. This suggests that environmental masking is sufficient to prohibit perception of normal spontaneous neural activity. Potentially, then, it may be helpful to mask or suppress the spontaneous activity underlying tinnitus.

12.1.2 Noise and other sounds

The central auditory system has a frequency-specific gain regulation that depends on the spectrum level of ANF output. Formby et al. (2003) showed that after continuous ear plugging for 2 weeks, listeners rated the loudness of sounds higher than before the ear plugging, whereas after an equal continuous period of exposure to low-level noise they rated the sound's loudness lower. This indicates that when there is less sound coming in, the gain of the auditory system is turned up, with the consequence that the spontaneous activity is also amplified, potentially leading to tinnitus. When sound is provided, e.g. by a hearing aid or a sound-generating device, the benefit is that the central gain, which works on both spontaneous and stimulus-driven activity, is turned down and amelioration of the tinnitus may result. This effect is likely not long lasting, as residual inhibition experiments (Chapter 3) have shown.

Electrophysiologic support for the potential of sound therapy, at least in preventing tinnitus to occur following noise trauma, was provided in animal studies by Noreña and Eggermont (2005, 2006) and reviewed in Chapters 7 and 9. Other evidence that acoustic stimulation in the frequency range of the hearing loss is better than using overall broadband noise masking comes from the residual inhibition studies of Roberts (2007) and Roberts et al. (2008), that demonstrated longer-lasting residual inhibition for sounds matched to the hearing loss.

The sound therapy method is likely not applicable in persons with normal hearing and tinnitus. However, provided that despite normal thresholds the ANF output is less in a certain frequency range than in the remainder, sound therapy may still prove to be a viable clinical procedure. The main therapeutic aspect of sound is based on the assumption that it balances the firing rate of ANFs across frequency. If the output is already balanced, as is likely the case in normal hearing persons without tinnitus, the sound would provide an imbalance and if applied for a long time be more harmful than beneficial (Noreña et al., 2006; Pienkowski and Eggermont, 2009).

The same objective of balancing the output of the cochlea has been implicitly promoted by a procedure that combines acoustic stimulation in the frequency range of the hearing loss with counseling, and this approach appears effective (Davis et al., 2007). In a sense, tinnitus-retraining therapy (TRT) also uses sound as part of the treatment in combination with counseling (section 12.3). Here, however, the sound is used to help the patient habituate to the emotional aspects of tinnitus, and subsequently to the perceptual aspects of tinnitus, and is somewhat subservient to the counseling aspects of the procedure (Jastreboff et al., 1994; Henry et al., 2002).

A quite contrasting approach to change putative maladaptive reorganization of the auditory system was presented by Okamoto et al. (2010) and intended for people with tonal tinnitus. They used an innovative approach based on "notched" music, intended to silence the sound only in a narrow frequency range surrounding the tonal tinnitus (Figure 12.1A). Another group of people with tonal tinnitus received a placebo treatment where the notch was at either side of the tinnitus frequency range (Figure 12.1B). The group that received the target treatment (n = 8) showed reduced subjective tinnitus

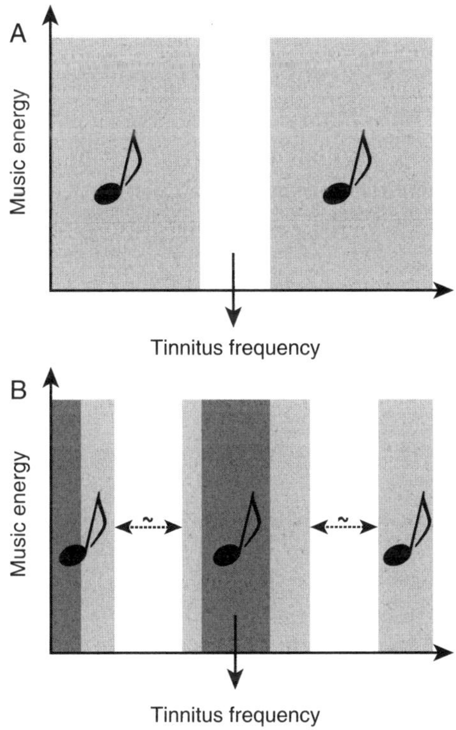

Figure 12.1 A) Target treatment. A frequency band of one octave width centered at the individual tinnitus frequency was removed from the music energy spectrum. B) Placebo treatment. Two frequency ranges of one octave width (dotted arrows) were removed from the music spectrum. The energy in the frequency band ranging from 0–707 Hz and the energy in the 1-octave frequency band (dark gray) surrounding the individual tinnitus frequency was not changed. The energy in the remaining frequency ranges was filtered (light gray). From Okamoto et al. (2010) with permission from the authors.

loudness and also showed reduced magnetic field 40-Hz auditory steady-state responses that are generated in Heschl's gyrus (Lütkenhöner et al., 2003a; Pantev et al., 1996), but no change in the N100 that is generated by neurons in the planum temporale (Lütkenhöner and Steinsträter, 1998; Lütkenhöner et al., 2003b). This suggests that the changes were occurring at the level of the primary auditory cortex. The placebo group did not report any changes nor did the subjects show changes in the evoked auditory fields.

These results are unexpected in the light of reported findings (Noreña and Eggermont, 2005; 2006) that an EAE covering the frequency range of the hearing loss eliminates the biological substrates attributed to tinnitus, i.e. reorganization of primary auditory cortex, increased SFR, and increased neural synchrony, in animal experiments whereas post-trauma exposure to a low-frequency EAE did not. The idea of Okamoto et al. (2010) was that the frequency regions above and below the frequency range of the tinnitus would "inject" increased inhibition in the tinnitus range and so suppress the SFR of those neurons. In the placebo group, the only expected phenomenon was the masking of the tinnitus as long as the sound was on.

The different findings may be better understood by stipulating that the EAE in the Noreña and Eggermont (2005; 2006) experiments was aimed at obtaining an equal output of the cochlea across frequency in cases where the tinnitus was not yet centralized or, if it was already established to reverse the imbalance in excitation that feeds the maladaptive plasticity. The Okamoto et al. (2010) exposure would not change the cortical reorganization, in effect would strengthen it, but would in some way recruit inhibitory effects on the tinnitus generating neurons. It is also not clear why the expected decrease in gain above and below the tinnitus frequency as a result of continuous exposure, and an increase in gain around the tinnitus frequency (Formby et al., 2003) would not happen here. On the other hand, the notched music results might fit with the findings by Moffat et al. (2009) discussed in the following section.

12.1.3 Hearing aids

Hearing aids, after first being proposed for the alleviation of tinnitus more than 60 years ago (Saltzmann and Ersner, 1947), were only considered as a viable option for relief of tinnitus from the late seventies on (Vernon et al., 1977; Roeser and Price, 1980; Surr et al., 1985; Folmer and Carroll, 2006; Del Bo and Ambrosetti, 2007). Amplification provides masking in the frequency range of tinnitus, especially if the hearing aids have a good high-frequency response, and in addition it allows brain plasticity to play a role as our previous discussion has indicated. It is important that the amplification in this high-frequency range (>4 kHz) is high enough to balance the output of the ANFs across frequency. There is, however, some doubt that hearing aids really work. Moffat et al. (2009) compared effects of conventional hearing-aid amplification and high-bandwidth amplification (Figure 12.2), the idea being that the high bandwidth would provide more amplification in the tinnitus frequency range and thus would induce more plastic changes in the tinnitus spectrum (Chapter 3). However, the tinnitus percept was affected only weakly in the conventional amplification group where the low-frequency components of the tinnitus (<2 kHz) were attenuated, and was not at all affected in the high-bandwidth group. This was explained as due to more robust changes in cortical reorganization in the high-frequency, i.e. the hearing loss, range, that would not be malleable over the relatively short (30-day) treatment period. It is likely that the reduction at lower frequencies is an effect of reducing the gain as a result of normalizing the input to the high-frequency range and therefore increasing its inhibitory effect on low frequencies.

A more positive outlook was provided by Schaette et al. (2010) who demonstrated that hearing aids (amplification up to 6 kHz) worn for 6 months had a significant effect (tinnitus loudness decrease, and decrease of annoyance as measured by the Tinnitus Questionnaire) on persons with tinnitus pitch <6 kHz but not for those with higher tinnitus pitch. They interpreted that as a definite effect of sound therapy where it is sufficiently loud to have an impact. Searchfield et al. (2010) observed improved scores on the THQ when counseling was accompanied by wearing a hearing aid compared to counseling only.

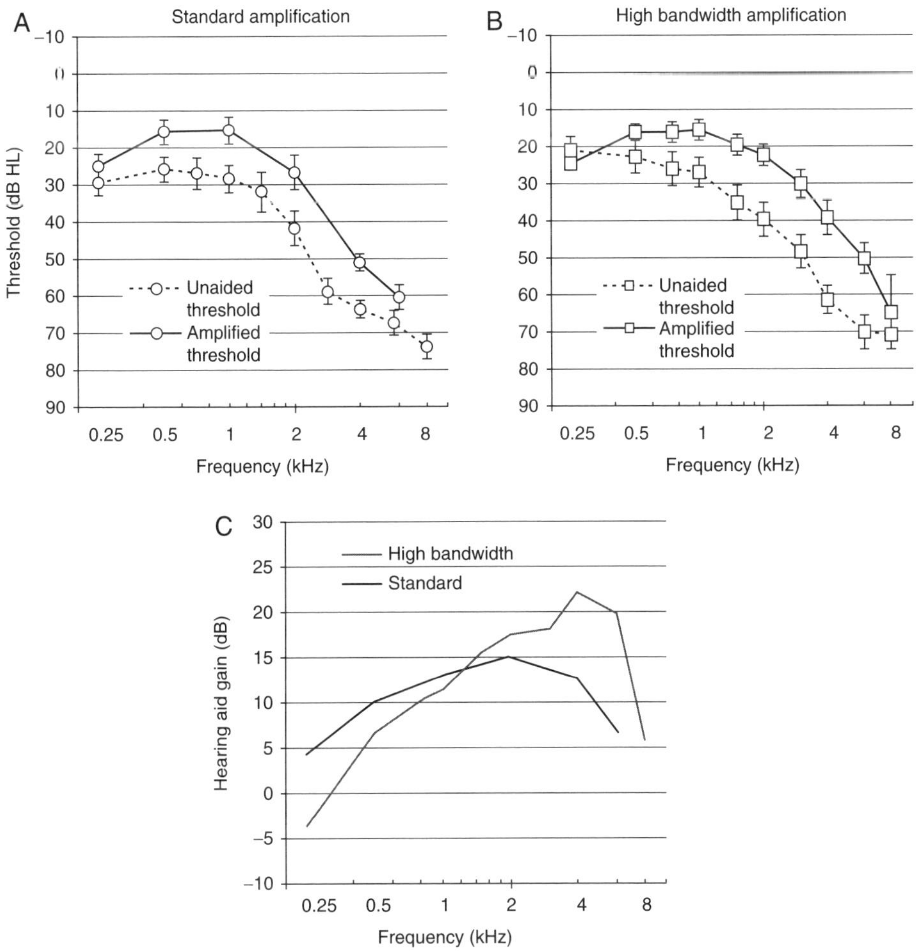

Figure 12.2 Hearing thresholds with and without amplification for (A) the standard amplification group (group 2) and (B) the high-bandwidth amplification group (group 3). Average gain profiles for the two groups are compared in (C). Reprinted from Moffat, G., Adjout, K., Gallego, S., Thai-Van, H., Collet, L., and Noreña, A.J. (2009). Effects of hearing aid fitting on the perceptual characteristics of tinnitus. *Hearing Research*, **254**(1–2), 82–91. Copyright (2009), with permission from Elsevier.

12.1.4 Cochlear implants

A cochlear implant (CI) is a surgically implanted electronic device with a multielectrode array that is placed into the cochlea and provides a sense of sound to a person who is profoundly deaf or severely hard of hearing. The basic parts of the CI processer include a microphone which picks up sound from the environment, and a speech processor which selectively filters sound to prioritize audible speech and sends the electrical sound signals through a thin cable to the transmitter. A coil held in position by a magnet placed in the petrous bone behind the external ear, transmits the processed sound signals to the internal part of the CI by electromagnetic induction. This internal part consists of a receiver and stimulator secured in bone beneath the skin, which converts these signals into electric impulses and sends them through an internal cable to the electrode array containing up to 24 electrodes inserted in the cochlea. This set of electrodes sends impulses to the auditory nerve fibers and subsequently to the brain. The electrode array typically reaches about 20 mm into the cochlea up to the region most sensitive to 1-kHz sounds. A signal-processing algorithm determines how the individual electrodes are activated. CIs are currently being used more and more to provide alleviation for tinnitus in those cases where hearing aids no longer provide enough amplification. For a general discussion on the use of CIs in persons with tinnitus, see Baguley and Atlas (2007).

One of the first studies suggesting that electrical stimulation of the cochlea could alleviate tinnitus came from Portmann et al. (1979). They found that stimulating the wall of the cochlea (the promontory), or even better the round window (Portmann et al., 1983), with negative pulses induced sensations of sound and enhanced tinnitus, whereas positive pulses reduced tinnitus. This polarity sensitivity will determine the performance of CIs, which typically use bipolar pulses, on potentially suppressing tinnitus.

All currently available CIs can provide effective and similar levels of tinnitus suppression. Ruckenstein et al. (2001) found a significant reduction for 35 of 38 patients (92%) in tinnitus intensity using CIs. More recently, a comparison between pre- and postimplantation THI scores (Chapter 3) showed a decreased score in 13 of 20 patients with preimplantation tinnitus, an unchanged score in six and increased score in one (Di Nardo et al., 2007). Pan et al. (2009) found in 153 implanted patients with tinnitus that 61% had total suppression and 39% reported a partial reduction. Of 91 people without tinnitus receiving a CI, only 12% reported tinnitus post-operatively.

Twenty-one unilaterally-deaf subjects who complained of severe intractable unilateral tinnitus confined to the deaf ear that was unresponsive to drug treatment, to cognitive therapy or to TRT, received a CI (Van de Heyning et al., 2008). Electrical stimulation via the CI resulted in a significant reduction in tinnitus loudness, as measured by the VAS, from 8.5 to about 2.4 at 1 and 2 years year after implantation. This delayed improvement suggests the induction of plastic changes in the central auditory system. The Tinnitus Questionnaire (Chapter 3) also revealed a significant positive effect of CI stimulation. The beneficial effect of a CI on tinnitus, reported by a majority of patients, could be due to a masking effect and above all to a possible CI stimulation-dependent reorganization of the central auditory pathways and cortex. In the context of the latter suggestion, it would be interesting to look at these effects over time following implantation as was done in the study by Van de Heyning et al. (2008). A more recent, but more restricted study covering only five patients (Buechner et al., 2010) concluded that improvements were found regarding the hearing and the tinnitus. However, not all participants benefited from the CI to the same degree and in the same situations.

A potential reason for the varying effects of CIs may be in the type of stimulation used. The standard use of a CI with a high pulse rate is to aid in speech perception, however, as Zeng et al. (2011) have shown, a low rate (<10-Hz) stimulus applied through the apical electrodes appears to suppress tinnitus. Other stimulus regimens or using the low rate stimulation with other electrodes were not effective. The suppression of the tinnitus was also reflected in the increase of alpha-band EEG power (compare to the reduced alpha power in tinnitus patients (Chapters 4 and 11)) and ended with the cessation of stimulation.

It may also be feasible to implant a partially deaf ear with useful hearing in the low frequencies but no hearing in the high frequencies, where amplification cannot be used to treat tinnitus. Special short electrodes have been developed for such cases (Dunn et al., 2010), and it is reported that they do not affect the low-frequency acoustic hearing. They may reduce tinnitus.

12.2 Direct stimulation of auditory cortex and deep brain structures

In the mid 20th century, Penfield and Rasmussen (1950) first reported that electrical pulses applied to the superior temporal gyrus (STG) of neurosurgical patients produced alterations in auditory perception. In this and subsequent reports (Penfield, 1958; Mullen and Penfield, 1959) it was observed that one form of auditory perceptual alteration was suppression of hearing. This hearing suppression effect was reported in only a small number of the more than 1100 patients studied, and the effective stimulation sites were located mainly on the posterolateral aspect of the STG and the anterior portion of Heschl's gyrus (HG). In a more recent study (Fenoy et al., 2006), sites where hearing suppression occurs were identified within anterior lateral HG and posterior lateral STG, in what may be auditory belt and parabelt fields. In contrast, stimulation of presumed core auditory cortex within posterior medial HG generally evoked sound perceptions, or in one case an increase in tinnitus intensity, that affected the contralateral ear and did not persist beyond the period of stimulation.

The first results in patients treated by extradural auditory cortex stimulation demonstrated statistically significant tinnitus suppression in cases of unilateral pure tone tinnitus without suppression of white or narrow band noise like tinnitus (De Ridder et al., 2007). For chronic extradural stimulation, two patients showed a persistent reduction of pure-tone tinnitus, and six patients had short periods of total tinnitus suppression. Significant improvements in the Beck Depression Inventory and tinnitus questionnaires were found, although objective measures of tinnitus loudness remained fairly stable. No surgical or stimulation-related complications were noted (Friedland et al., 2007). One of the major problems with extradural stimulation was the recurrence of tinnitus. A potential solution was found by changing the stimulation parameters to prevent presumed habituation of the auditory cortex to the stimulus in the presence of the not habitual tinnitus.

Seidman et al. (2008) implanted two patients with electrodes in and on HG. A patient with bilateral tinnitus for 2 years was implanted with an electrode in HG and experienced a sustained reduction to near elimination of tinnitus. The second patient with unilateral tinnitus for 7 years was implanted with an electrode on HG and had only a temporary reduction in her tinnitus. De Ridder et al. (2009) subsequently implanted five patients with an electrode on the auditory cortex. All patients had tinnitus consisting of a high-pitched pure tone and noise components. For the narrowband noise component of tinnitus a significantly better suppression was obtained with burst stimulation (bursts consisting of five pulses at a frequency of 500 Hz repeated at 40 burst/s) in comparison to tonic 40-Hz pulse stimulation. For the pure tone tinnitus component, no difference was found between tonic and burst stimulation.

Seven patients implanted with deep brain stimulation (DBS) systems for movement disorders who also reported having tinnitus were interviewed about their tinnitus conditions (Shi et al., 2009). Four were available for testing in a specialized tinnitus clinic with their DBS systems turned off or on. Three of the seven patients reported reduced tinnitus loudness when DBS was turned on. Of the four patients tested in the clinic, results indicated that DBS of the ventral intermediate nucleus of the thalamus caused decreases in tinnitus loudness in two patients with relatively prolonged residual inhibition. These results suggest that DBS of non-auditory thalamus structures may provide tinnitus relief for some patients. DBS therapy to a locus of caudate neurons (area LC) in the body of the nucleus, a subsite of the striatum that is not part of the classical auditory pathway, can decrease or increase tinnitus loudness perception. The DBS lead traversed through or was adjacent to area LC in six Parkinson's disease and essential tremor subjects with concomitant tinnitus who underwent implantation of the subthalamic or ventral intermediate nucleus. In five subjects where the DBS lead tip traversed area LC, tinnitus loudness in both ears was suppressed to a nadir of level 2 or lower on a 0–10 rating scale. In one subject where the DBS lead was outside area LC, tinnitus was not modulated (Cheung and Larson, 2010).

12.3 Transcranial magnetic stimulation

TMS is a technique for non-invasive stimulation of the human brain. For that purpose, a brief high-intensity magnetic field is generated that can reach up to about 2 Tesla (comparable to the standard MRI static magnetic fields of 1.5–3 Tesla) and typically lasts for about 100 ms. The field can excite or inhibit a small area of cortex below the coil. The magnetic coils may have different shapes. Round coils are relatively powerful. Figure-eight-shaped coils provide more focal stimulation by producing maximal current at the intersection of the two round components. These are the commonly used ones.

The orientation of the TMS coil has an effect because neurons are activated best by voltage gradients that run parallel to the axon. Neurons of the cortex are orientated either perpendicular or tangential (in fissures) to the surface and, depending on the folding of the cortex, have particular orientations with respect to the TMS-induced currents that will favor a particular neural population. The neurons in HG have mostly a tangential orientation that is particularly suited for TMS stimulation. The induced electrical stimulus activates a mixture of neurons beneath the coil (Ridding and Rothwell, 2007). Some are local to the area of cortex under the coil, others project axons to or from the site of stimulation; some are excitatory, others inhibitory. The final outcome of such stimulation might be complex and quite unlike the organized patterns of activity that occur in natural behaviors. However, some selectivity arises owing to the fact that different neurons have different thresholds to electrical stimulation. Low stimulation intensities therefore activate a much more limited selection of neurons than higher intensities. Repetitive TMS (rTMS) can produce effects that outlasts the period of stimulation; inhibition is obtained with tonic stimulation at about 1 Hz, and excitation with tonic stimulation at 5 Hz and higher (Hallett, 2007).

There are two important limitations of TMS. One is that the magnetic field falls off rapidly with distance from the coil surface, effectively limiting direct stimulation to the outer parts of the cerebral cortex under the skull. If large magnetic fields are used to induce currents deeper in the brain, then a large area of superficial cortex is also strongly activated, so that effects become difficult to interpret. The second limitation is that the site of stimulation is not focal.

For TMS to be of any help in alleviating tinnitus it needs to induce long duration after effects. The nature of the after-effects of rTMS depends on the number, intensity, and frequency of stimulation pulses. For example, stimulation at frequencies higher than 1 Hz tends to increase rather than decrease cortical excitability. After-effects also depend on the pattern of the pulses applied. For example, Huang et al. (2005a) used a pattern of stimulation termed theta burst stimulation (TBS) in which three 50-Hz pulses are applied regularly five times per second for 20–40 s. At low intensities this produced suppression of motor cortex excitability; however, if rather than using one long period (usually 40 s) of TBS each burst was applied only for 2 s followed by a pause of 8 s and then repeated the effect became facilitatory. Effects lasting for a while (more than a few minutes) after stopping the TMS stimulation are the most interesting from the point of view of tinnitus treatment and possible involve mechanisms such as LTD and LTP of synaptic connections. This is supported by evidence from pharmacological interventions in humans showing that the after-effects of rTMS can be blocked by NMDA-receptor antagonist such as dextromethorphan and memantine (Ridding and Rothwell, 2007). RTMS is known to produce a mixture of many excitatory and inhibitory effects. This makes its use problematic in restoring the putative imbalance between excitation and inhibition in tinnitus.

The problem in any clinical trial is to come up with a good placebo or sham stimulus. This is specifically difficult for TMS. During TMS of the temporoparietal region, peripheral activation of the ipsilateral facial nerve and trigeminal afferents can occur and this is easily detectable by the patient. The use of low concurrent peripheral stimulation during both sham and active rTMS is potentially a way to minimize this possible bias and to reduce behavioral differences between the two stimulation conditions. Using this procedure Rossi et al. (2007) found daily application of high-intensity 1-Hz rTMS to the left temporoparietal region an efficient strategy to transiently alleviate chronic tinnitus, independent of tinnitus lateralization or bilaterality and of eventual concurrent changes in mood. Plewnia et al. (2007) found that tinnitus loudness was reduced after temporoparietal, PET-guided, low-frequency rTMS. This reduction, lasting up to 30 min, was dependent on the number of stimuli applied, and was negatively correlated with the length of the medical history of tinnitus in our patients. These effects were related to stimulation of cortical association areas, not primary auditory cortex.

Burst rTMS at 5-, 10-, or 20-Hz intraburst frequency was found to suppress tinnitus better than tonic rTMS at the same rates (De Ridder et al., 2007a). Some patients showed a lasting benefit of tonic rTMS (1, 10 or 25 Hz) applied for 10 days at 1 year after stopping treatment (Khedr et al., 2009). Combined prefrontal and temporal rTMS suggested benefits 3 months after stimulation (Kleinjung et al., 2008). Veterans receiving rTMS at a relatively low rate of 0.5 Hz applied for 20 min for 5 consecutive days did not show improved tinnitus (Lee et al., 2008). Marcondes et al. (2010) reported improvement after 5 days of rTMS at 1 Hz that was still present after 6 months post-TMS. This was corroborated by comparison of the SPECT recordings before and 6 months after treatment. A single session of burst rTMS (5 Hz) did not improve tinnitus in a group of 20 patients (Poreisz et al., 2009). Vanneste et al. (2010b) found the effect of burst TMS (5, 10, and 20 Hz) to be more promising for bilateral tinnitus, compared to unilateral and that highly depressed patients failed to profit form the TMs application.

This brief summary suggests that the large differences in the results still defy TMS as a scientific procedure (which requires independent replication) and points to one that is extremely sensitive to the particular setting, and potentially different patient groups, in which the results were obtained.

12.4 **Tinnitus retraining therapy and cognitive behavioral therapy**

I base this summary largely on a comparative study by McKenna (2004). There are currently two main behavioral approaches to tinnitus management. They are the psychological approach that treats tinnitus within a cognitive behavioral model (Hallam et al., 1984), and the neurophysiologic approach (Jastreboff, 1990) that gave rise to tinnitus retraining therapy (TRT).

The cognitive behavioral approach assumes that the process of habituation to tinnitus is fundamentally similar to that to external stimuli. However, if this happened naturally nobody would experience tinnitus. It is believed that emotional aspects somehow interfere with this habituation process. Tolerance to tinnitus will be facilitated in the cognitive behavioral approach by reducing levels of autonomic nervous system arousal, changing the emotional meaning of the tinnitus, and reducing other stresses.

The main components of TRT are directive counseling and use of sound generators. TRT proposes that the process of increased auditory gain within the auditory system during periods of restricted sound input, or as a consequence of cochlear damage, may account for the emergence of tinnitus. The key assumption in TRT is that tinnitus becomes problematic because of the improper activation of the brain's limbic system, which plays a role in emotion, memory, and learning.

One of the important points in the therapy is to assure the patient of what tinnitus is and that it is not a forerunner of a drastic disease. Sexton (1880) said it very succinctly:

> As might be expected, this symptom often carries with it great terror to the patient. He fancies that he is pursued by warning voices of impending death, or that the sounds are forerunners of some grave cerebral affection, or that they at least prognosticate an ever-advancing aural disease, which

threatens total deafness. The patient can usually be promptly assured, as regards these matters, of the simplicity and harmlessness of tinnitus aurium. When the patient once has this symptom associated with chronic diseases of the ear affecting the conductive apparatus, it seldom if ever entirely disappears, but it does not necessarily indicate activity in the aural affection, and is not prognostic, therefore, of advancing aural disease. Tinnitus aurium is a symptom of frequent occurrence in old age, and may depend partly on trophic changes in the blood-vessels themselves or the tissues in which they are situated, thus altering the sounds of the circulation.

The cognitive behavioral approach and TRT have much in common. They both consider the emotional, i.e. limbic system, processing of tinnitus as the key factor in determining distress. Both approaches see habituation as the essential process involved in resolving the problem. One potentially important difference is that cognitive behavioral therapy is provided by psychologists, whereas TRT is provided by trained audiologists.

Kroener-Herwig et al. (2000) evaluated 10 clinical studies in TRT research and found methodological shortcomings such as a lack of controlled randomized group studies. Since no comparative studies had been conducted, there was no support for the assumption that TRT was superior to the cognitive behavioral approach. Even at the time of this writing, comparative studies are not available. Martinez Devesa et al. (2007, 2010) found in a meta-analysis of six randomized control trials that cognitive behavioral therapy did not result in significant changes in the subjective loudness of tinnitus, or in the associated depression. However there was a significant improvement in the quality of life (decrease of global tinnitus severity) of the participants, thus suggesting that cognitive-behavioral therapy has an effect on the qualitative aspects of tinnitus and contributes positively to the management of tinnitus.

A recent controlled trial (Bauer and Brozoski, 2011) found that both TRT and general counseling without additional sound therapy are effective in reducing the annoyance and impact of tinnitus. The global improvement in tinnitus handicap with TRT accrued over an 18-month period and appeared to be a robust and clinically significant effect.

12.5 Drugs, clinical trials

12.5.1 What should drugs ameliorate?

Loss of peripheral input to central neurons can promote the development of synchronous spiking activity (Noreña and Eggermont, 2003) by prolonging postsynaptic depolarization and increasing the likelihood of temporally coincident inputs converging on synapses. In the normal central auditory system, the inhibition surrounding the excitatory part of the receptive field of a neuron produced by thalamocortical input would be expected to restrict synchronous activity to neurons tuned to properties of the acoustic stimulus, thereby leading to normal auditory perception. However, when intra-cortical inhibition is weakened, widely distributed synchronous spike-firing activity can develop and lead to the perception of sounds that are physically absent, i.e. tinnitus.

Potential tinnitus-alleviating drugs are often selected from those used in treating the putative transmitter imbalances in the CNS, as occurring in epilepsy, neuropathic pain, and depression. Elgoyhen and Langguth (2010) provide a concise overview that in comparison with much earlier ones (Davies, 2004; Dobie, 2004) shows that little progress has been made. I give some examples illustrating this situation.

12.5.2 GABA enhancers and NMDA blockers

Hearing loss causes a downregulation of presynaptic GABA release, which reduces both phasic inhibitory synaptic transmission, and tonic extrasynaptic inhibition. Phasic inhibition corresponds to inhibitory synaptic transmission, tonic inhibition is caused by activation of extrasynaptic $GABA_A$ receptors by ambient GABA (Belelli, et al., 2009). Drugs (e.g. benzodiazepines) that enhance GABA-mediated phasic inhibition have not been found effective for treating tinnitus (Langguth et al., 2009). Enhancing tonic inhibition can be done by blocking GABA transaminase using vigabatrin, and in animals reversibly eliminated tinnitus (Brozoski et al., 2007). These findings suggest that drugs enhancing tonic inhibition are potential candidates for alleviating tinnitus. However, vigabatrin can cause serious irreversible visual field defects in humans. No clinical trials have been conducted.

Gabapentin increases the synthesis of GABA from glutamate by enhancing the action of the synthetic enzyme glutamic acid decarboxylase (GAD) thereby increasing the tonic GABA concentration in the brain. Gabapentin does not appear to bind to GABA receptors, nor does it modulate GABA receptor function. In animal experiments (Bauer and Brozoski, 2001), gabapentin was delivered in the animal's drinking water and significantly reduced the evidence of acoustic trauma-induced tinnitus in rats without affecting hearing or general psychophysical performance.

Two placebo-controlled human trials of the effect of gabapentin on tinnitus have been reported. The first study, a prospective placebo-controlled single-blind study (Bauer and Brozoski, 2006) showed some effectiveness in treatment of tinnitus in patients suffering tinnitus from noise trauma or other etiology. There was a significant improvement of the tinnitus annoyance for the trauma group only. High responders in both groups showed significant reduction in tinnitus loudness. The second study (Witsell et al., 2007) did not detect any improvement in tinnitus handicap, but did report a significant improvement in subjective tinnitus annoyance when compared with placebo.

Glutamate receptor antagonists have been tried in tinnitus sufferers. The rationale behind their use is that imbalance between inhibitory and excitatory neurotransmission is observed in several regions of the auditory pathway. Moreover, blockade of glutamatergic neurotransmission could also exert neuroprotectant effects because it is known that noise overexposure is followed by an excitotoxic injury of the hair cells. Therefore, local administration of glutamate antagonists might prevent inner ear damage and, possibly, tinnitus development in the acute phase (Guitton et al., 2004).

The putative non-selective NMDA receptor antagonist acamprosate has been tried in a double-blind study (Azevedo et al., 2007). Patients received placebo or acamprosate (333 mg, three times per day) and rated the loudness and annoyance of their tinnitus before and at monthly intervals during treatment. Acamprosate had no beneficial effects after 30 days of treatment, a modest benefit at 60 days, and a significant effect at 90 days.

Approximately 87% of the subjects in the acamprosate group showed some improvement, including three subjects in which tinnitus disappeared, compared with 44% in the placebo group.

Treatment with intravenous caroverine, an antagonist of non-NMDA and NMDA receptors, has also been analyzed, but with contradictory results. Denk et al. (1997) looked at 60 patients with tinnitus of which 30 were treated with caroverine and 30 with placebo. In the caroverine group, 63.3% responded to therapy immediately after the infusion. In the placebo group, none of the patients treated showed a significant response according to the defined success criteria. Domeisen et al. (1998) could not replicate these results using the same treatment protocol.

There are similarities in animal models regarding the neural mechanisms underlying epilepsy and central tinnitus (Eggermont, 2005a). Anticonvulsants therefore theoretically have the potential for relieving tinnitus distress, as their mode of action is to reduce central excitation and/or increase inhibition. However, the success has been minimal as the following examples show (more in Davies, 2004; Dobie, 2004 and Eggermont, 2005b). The effective anticonvulsant lamotrigine is ineffective for tinnitus relief (Simpson et al., 1999). In some individuals, gabapentin, a GABA analogue, is effective in reducing subjective and objective aspects of tinnitus, with the best therapeutic response when tinnitus was associated with acoustic trauma (Bauer and Brozoski, 2007; Witsell et al., 2007). There is also little evidence to recommend memantine, an NMDA blocker, for the treatment of tinnitus (Figueiredo et al., 2008).

12.5.3 Antidepressants

Tinnitus is often accompanied by some degree of depression and it is almost impossible to separate the etiology of the two conditions (Zöger et al., 2006). Typically, controlled studies did not show any influence on antidepressant drugs (e.g. nortriptyline, a tricyclic antidepressant) on the tinnitus severity, although nortriptyline improved the depressive state of the patients (Dobie et al., 1993; Dobie, 2004) and showed a reduction in tinnitus loudness (Sullivan et al., 1993). Nortriptyline appears to be uneffective in non-depressed tinnitus patients (Katon et al., 1993). Benzodiazepine anxiolytics have shown some effect in small size studies (Johnson et al., 1993; Kay 1981) but the effects on tinnitus loudness are small and emotional effects were not measured. Selective serotonin uptake inhibitors (SSRIs) such as paroxetine showed a significant improvement in tinnitus annoyance (Robinson, 2007), but there are only a few studies conducted so far.

Summarizing, much more needs to be done to evolve from tinnitus management to tinnitus treatment.

Chapter 13

Future directions

As shown in the preceding chapters, tinnitus research is making tremendous progress in both the understanding of mechanisms and in developing tinnitus management. Yet, we still do not know why only two-thirds of people with hearing loss develop tinnitus, or why tinnitus occurs in people without hearing loss. We also do not know the reason why both loudness and annoyance of tinnitus are enhanced by stress, whereas the two are typically only weakly correlated. Most importantly, we do not know how findings in animal models and those in humans can be integrated. Some other important questions that still need to be addressed in both basic and applied research areas are discussed in the following sections.

13.1 Molecular and cellular mechanisms

The study of molecular, genetic, and cellular aspects of tinnitus is still in its infancy albeit that exciting findings have emerged (Knipper et al., 2010). There appears to be no clear heritability of tinnitus (Kvestad et al., 2010), potentially limiting the way these molecular mechanisms can be probed in people with tinnitus.

13.2 Modeling and theoretical aspects

Mathematical modeling is important since it can quantify the results of putative mechanisms and clearly indicate the limitation of these mechanisms. Only an initial approach has been made to quantitatively explain some aspects of tinnitus and its neurobiological substrates. Parra and Pearlmutter (2007), based on insights from a gain adaptation model, experimentally demonstrated a direct link between the percept of a Zwicker tone and tinnitus, as previously suggested by Noreña and Eggermont (2003). The Zwicker tone is an auditory negative afterimage, perceived after the presentation of, e.g. a ½ octave wide notched noise. In this case this results in an audible Zwicker tone with a pitch in the frequency range of the notch. Trenado et al. (2009) proposed a multiscale model of neural correlates of auditory selective attention in the tinnitus decompensation. The application of homeostatic mechanisms to explain the increased SFR following hearing loss in the DCN has been demonstrated using a computational model (Schaette and Kempter, 2006, 2008). These modeling results are very encouraging and show that general mechanisms of brain plasticity might help our understanding of tinnitus. This field is likely to expand quickly in the near future.

13.3 Physiological mechanisms

Physiological data have been the main product of animal research as demonstrated in many previous chapters, but how to link these animal data to the non-invasive imaging and scalp-recorded data obtained in humans with tinnitus is far from clear. Specifically, the human equivalent of the triad of proposed tinnitus substrates in animals increased SFR, increased neural synchrony, and tonotopic map reorganization—has not yet been established. For instance, it is only by inference from MEG recordings that cortical reorganization is suggested, but the low spatial resolution of that technique combined with reversing tonotopic gradients at the border of auditory areas makes drawing conclusions difficult. We clearly need tonotopic maps in tinnitus patients based on high-resolution fMRI to assess this issue. FDG-PET scans can detect increased metabolic baseline activity in the auditory system, but the spatial resolution of this technique is also very low, so, for instance, discriminating between different auditory cortical areas will be difficult. Neural synchrony determines the amplitude of spontaneous EEG and evoked potentials. Synchrony changes in spontaneous activity in humans with tinnitus are different in the various frequency bands of the EEG. For instance, in temporal cortex alpha-band activity is reduced and gamma-band activity is enhanced. Here no clear equivalents from animal data are available. It also requires a leap of faith to extrapolate from local neuron spike-pair correlations to the macrosynchrony synchrony that results in scalp-recordable EEG activity.

A second aspect that needs to be addressed in animal models is the effect of cortical activity on SFR and neural synchrony in subcortical structures after noise trauma, salicylate application, etc. Ideally recordings should be made simultaneously from several key structures in the brain, for instance, DCN, IC, and AC. This would be very helpful in understanding the sequence of activation after, for example, noise trauma or salicylate application, and the steady state several months after a trauma or following chronic salicylate application. It would also allow the characterization of bottom up flow of increased SFR; is this flow possible without increased neural synchrony? By activating the cortex electrically, or silencing it with topical application of muscimol, a $GABA_A$ receptor agonist, one could, for instance, observe subcortical changes in SFR. The interplay between corticofugal effects and bottom-up hyperactivity could then be examined.

FMRI investigations of tinnitus also need to be adapted to resting state measurements recorded under low background noise conditions. This could result in a good measure of tinnitus metabolic activity (if distinguishable from the resting state) with good spatial resolution. Network activity across temporal and prefrontal cortical areas in animals (long-range correlation etc.) could help to explain the findings that suggested functional network changes in tinnitus patients (Schlee et al., 2009a,b). Multielectrode EEG spectra obtained before and a variable time after noise trauma could indicate the time span in which these changes occur.

One of the reasons that the study of brain networks need more attention is that tinnitus potentially has a distributed representation. Tinnitus is a conscious percept that is based on conscious accessible activity in primary auditory cortex. This requires that: (1) there is increased spontaneous neural synchrony since only synchronous neural activity would become consciously accessible (Ward, 2011); (2) the primary areas would need to be activated by top-down signals (Meyer, 2011); (3) a combination of the two could be achieved if top-down signals synchronize activity in AI. The role that gamma-band activity plays in tinnitus needs to be further evaluated.

13.4 Psychophysical and functional consequences of tinnitus

Hamblen-Thomas (1937) described some (at that time) future evaluation methods (masking, recording emissions from the ear, and the noise spectrum recorded from the round window) to distinguish objective tinnitus from other ones and they basically cover what has been done so far (with the exception of brain imaging techniques). It is illustrative to read his thoughts on this, while we are contemplating directions for future improvements:

> There is much that may be done in the investigation of tinnitus. In some cases the noises may be heard by the observer through the otoscope, or, rarely, the movements of the contents of the middle ear have been reported in association with them. So far, a suitable microphone and valve-amplifier have not been produced so that the sounds can be recorded graphically, but it is hoped that a suitable microphone which is the main difficulty-will be available soon. One of the present difficulties with a sensitive microphone is the exclusion of extraneous sounds. In the case of many patients it is possible to match their particular tinnitus on the audiometer, piano, or tuning fork. It is interesting to note that, in testing patients with tinnitus, they show a hesitancy in responding to the audiometer at the particular frequency of their tinnitus. Again, the intensity of their tinnitus can be estimated by the amount of sound above normal at the particular frequency which is required to drown it. I am hoping that, in some cases of tinnitus, I may be able to pick up the action currents from the round window, as in the Wever and Bray phenomena, and record them: I am hopeful that such do exist in these cases. Until we can obtain more information about tinnitus we cannot hope to get far in its treatment. It is a very real discomfort to those who suffer from it, and whatever can be done towards relieving them is well worth while (Hamblen-Thomas, 1937).

One of the confounding aspects in tinnitus research is the interplay between the frequently co-occurring hyperacusis and tinnitus. Fortunately humans can indicate if they have hyperacusis and/or tinnitus, whereas in animals it has to be deduced from the startle-reflex test, which is sensitive to both hyperacusis and tinnitus (Sun et al., 2009). How to delineate brain changes due to tinnitus from those caused by hyperacusis and by hearing loss? In that respect the question arises if tinnitus without hearing loss is different from that accompanied by hearing loss. Does hyperacusis affect tinnitus loudness as well as annoyance? If hyperacusis can be modulated (Noreña and Chery-Croze, 2007), does the co-occurring tinnitus loudness decrease? Will tinnitus spectrum recording from, for example, the cochlear promontory become possible and can it be interpreted in terms of SFR, burst firing, and neural synchrony? Can it be modulated by attention or other cortical activity? Is the spectral power related to tinnitus loudness?

13.5 Targeted treatment requires a typology of tinnitus

Tinnitus induced in animals has a well-defined etiology, i.e. the tinnitus-inducing agent (mostly noise or salicylate) is known and well described. This is not the case in the majority of humans. My current impression is that drug treatments are more effective for certain well-described etiologies and in that case comparable between animals and humans. This is nicely illustrated by animal and human data obtained by the same authors, on the potential of gabapentin, a GABA agonist, to treat tinnitus. The animal group consisted of rats subjected to intense unilateral exposure to noise. Tinnitus was inferred from a frequency-specific shift in the discrimination of tones and noise from silence, which is expected to deteriorate after inducing tinnitus. This effect was reversibly attenuated by application of gabapentin (Bauer and Brozoski, 2001). In a prospective, placebo-controlled, single-blind clinical trial, of people with moderate or severe suffering from tinnitus resulting either from noise trauma or other conditions, gabapentin applied over 20 weeks at increasing dosage up to maximally 2.4 g/day, significantly reduced the tinnitus in the noise trauma sufferers, but not in the others (Bauer and Brozoski, 2006). This suggests that typology will be extremely important for designing effective clinical trials, and also that certain drugs may benefit one type of tinnitus but not others.

The etiology of their tinnitus is often unknown to the patient. This is illustrated by the fact that about half of tinnitus patients cannot name a cause for their tinnitus. For people complaining about tinnitus in general, the etiologies and underlying biological substrates of their tinnitus may be very different. In addition the way tinnitus manifests itself may or may not be related to the etiology. Tyler et al. (2008) used cluster analysis to identify four subgroups based on their symptoms: (1) constant distressing tinnitus, (2) varying tinnitus that is worse in noise, (3) tinnitus patients who can cope and whose tinnitus is not influenced by touch (somatic modulation), and (4) tinnitus patients who can cope but whose tinnitus is worse in quiet environments. This could be a pragmatic start for homogenization of patient groups for clinical trials. However, we don't know what differentiates the brains of people who suffer from tinnitus and people who experience it but do not care. Involvement of the limbic system is likely but more definitive statements cannot be made yet. Is there a relation of these clusters to the etiology? Can objective diagnostics such as the increasingly popular resting-state brain imaging be developed to assist the required typology? A related question is whether evaluating the outcomes of tinnitus treatment using objective means would be better or at least complementary to the current use of questionnaires?

13.6 **Outlook to new treatment options?**

Ultimately, regenerating cochlear hair cells (Brigande and Heller, 2009) and establishing a successful innervation with the remaining auditory nerve fibers is the long-awaited cure for hearing loss-related tinnitus. It is not unthinkable, however, that regenerated hair cells could produce spontaneous activity that in itself will be perceived as tinnitus, so we have to wait and see. But until then the paths laid out in the penultimate chapter in this book may provide short-term solutions. In case there is no hair cell loss, for example, in case of chronic low-dose salicylate application or pure somatic tinnitus, hair-cell regeneration may not be an option. Also in case of a restricted high-frequency hearing loss with unbearable tinnitus use of a short cochlear implant may potentially be a better solution.

It is not clear whether treatments that reverse some neural and behavioral correlates of tinnitus (e.g. enhanced acoustic environments (EAEs), vagus nerve stimulation with tone pairing) effective over time in case of a permanent hearing loss? We have seen that EAEs in the treatment of hyperacusis lose their effect after a few weeks post application. This is likely caused by the recurring imbalance resulting from the frequency-specific hearing loss, or by homeostatic mechanism in the brain that would revert the situation to what it was before the treatment.

Therefore an alternative could be to target cortical neural networks by either drug treatment, or local electrical (potentially deep brain) or magnetic stimulation. If there is a single neural network substrate for tinnitus as suggested (de Ridder et al., 2011) this would be the most promising solution especially if the network would be independent of tinnitus typology.

References

Abbott, S.D., Hughes, L.F., Bauer, C.A., Salvi, R., and Caspary, D.M. (1999) Detection of glutamate decarboxylase isoforms in rat inferior colliculus following acoustic exposure. *Neuroscience*, **93**, 1375–81.

Abel, M.D. and Levine, R.A. (2004) Muscle contractions and auditory perception in tinnitus patients and nonclinical subjects. *Cranio*, **22**, 181–91.

Abeles, M. (1982) *Local cortical circuits*. Berlin: Springer Verlag.

Abeles, M. (1991) Corticonics. *Neural circuits of the cerebral cortex*. Cambridge: Cambridge University Press.

Abeles, M., Bergmann, E., Margalit, E., and Vaadia, E. (1993) Spatio-temporal firing patterns in the frontal cortex of behaving monkeys. *Journal of Neurophysiology*, **70**, 1629–38.

Aid, T., Kazantseva, A., Piirsoo, M., Palm, K., and Timmusk, T. (2007) Mouse and rat BDNF gene structure and expression revisited. *Journal of Neuroscience Research*, **85**, 525–35.

Aigner, M., Lukas, J.R., Denk, M., and Mayr, R. (1997) Sensory innervation of the guinea pig extraocular muscles: a 1,10-dioctadecyl-3,3,3030-tetramethyl-indocarbocyanine perchlorate tracing and calcitonin generelated generelated peptide immunohistochemical study. *Journal of Comparative Neurology*, **380**, 16–22.

Aitkin, L.M., Dickhaus, H., Schult, W., and Zimmermann, M. (1978) External nucleus of inferior colliculus: auditory and spinal somatosensory afferents and their interactions. *Journal of Neurophysiology*, **41**, 837–47.

Aitkin, L.M., Kenyon, C.E., and Philpott, P. (1981) The representation of the auditory and somatosensory systems in the external nucleus of the cat inferior colliculus. *Journal of Comparative Neurology*, **196**, 25–40.

Al-Mana, D., Ceranic, B., Djahanbakhch, O., and Luxon, L.M. (2008) Hormones and the auditory system: a review of physiology and pathophysiology. *Neuroscience*, **153**, 881–900.

Alonso, J.M. and Martinez, L.M. (1989) Functional connectivity between simple cells and complex cells in cat striate cortex. *Nature Neuroscience*, **1**, 395–403.

Altschuler, R., Juiz, J., Shore, S., Bledsoe, S., Helfert, R., and Wenthold, R.J. (1991) Inhibitory amino acid synapses and pathways in the ventral cochlear nucleus. In: M.A. Merchan, J.M. Juiz, D.A. Godfrey, and E. Mugnaini (ed.), *The Mammalian Cochlear Nuclei: Organization and Function*, pp. 211–24. New York: Plenum Press.

Altschuler, R.A., Reeks, K.A., Fex, J., and Hoffman, D.W. (1988) Lateral olivocochlear neurons contain both enkephalin and dynorphin immunoreactivities: immunocytochemical co-localization studies. *Journal of Histochemistry and Cytochemistry*, **36**, 797–801.

Anari, M., Axelsson, A., Eliasson, A., and Magnusson, L. (1999) Hypersensitivity to sound—questionnaire data, audiometry and classification. *Scandinavian Audiology*, **28**, 219–30.

Anderson, J.E., Teitel, D., and Wu, Y.W. (2009) Venous hum causing tinnitus: case report and review of the literature. *Clinical Pediatrics*, **48**, 87–88.

Andersson, G., Carlbring, P., Kaldo, V., and Ström, L. (2004) Screening of psychiatric disorders via the Internet. A pilot study with tinnitus patients. *Nordic Journal of Psychiatry*, **58**, 287–91.

Andersson, G., Lyttkens, L., Hirvelè, C., Furmark, T., Tillfors, M., and Fredrikson, M. (2000) Regional cerebral blood flow during tinnitus: a PET case study with lidocaine and auditory stimulation. *Acta Otolaryngologica*, **120**, 967–72.

Argence, M., Saez, I., Sassu, R., Vassias, I., Vidal, P.P., and de Waele, C. (2006) Modulation of inhibitory and excitatory synaptic transmission in rat inferior colliculus after unilateral cochleectomy: an in situ and immunofluorescence study. *Neuroscience*, **141**, 1193–1207.

Arnold, W., Bartenstein, P., Oestreicher, E.W.R., and Schweiger, M. (1996) Focal metabolic activation in the predominant left auditory cortex in patients suffering from tinnitus: a PET study with [^{18}F] deoxyglucose. *Journal of Oto-Rhino-Laryngology and Related Specialties*, **58**, 195–99.

Ashton, H., Reid, K., Marsh, R., Johnson, I., Alter, K., and Griffiths, T. (2007) High frequency localised "hot spots" in temporal lobes of patients with intractable tinnitus: a quantitative electroencephalographic (QEEG) study. *Neuroscience Letters*, **426**, 23–28.

Atasoy, D., Ertunc, M., Moulder, K.L., Blackwell, J., Chung, C., Su, J., and Kavalali, E.T. (2008) Spontaneous and evoked glutamate release activates two populations of NMDA receptors with limited overlap. *Journal of Neuroscience*, **28**, 10151–66.

Atherley, G.R., Hempstock, T.I., and Noble, W.G. (1968) Study of tinnitus induced temporarily by noise. *Journal of the Acoustical Society of America*, **44**, 1503–1506.

Attias, J., Bresloff, I., and Furman, V. (1996) The influence of the efferent auditory system on otoacoustic emissions in noise induced tinnitus: clinical relevance. *Acta Otolaryngologica*, **116**, 534–39.

Attias, J., Urbach, D., Gold, S., and Shemesh, Z. (1993) Auditory event related potentials in chronic tinnitus patients with noise induced hearing loss. *Hearing Research*, **71**, 106–13.

Atzori, M., Lei, S., Evans, D.I., Kanold, P.O., Phillips-Tansey, E., McIntyre, O., and McBain, C.J. (2001) Differential synaptic processing separates stationary from transient inputs to the auditory cortex. *Nature Neuroscience*, **4**, 1230–37.

Axelsson, A. and Lindgren, F. (1981) Pop music and hearing. *Ear and Hearing*, **2**, 64–99.

Axelsson, A. and Ringdahl, A. (1989) Tinnitus—a study of its prevalence and characteristics. *British Journal of Audiology*, **23**, 53–62.

Azevedo, A.A. and Figueiredo, R.R. (2007) Treatment of tinnitus with acamprosate. *Progress in Brain Research*, **166**, 273–77.

Baguley, D.M. (2003) Hyperacusis. *Journal of the Royal Society of Medicine*, **96**, 582–85.

Baguley, D.M. and Atlas, M.D. (2007) Cochlear implants and tinnitus. *Progress in Brain Research*, **166**, 347–55.

Baguley, D.M., Humphriss, R.L., and Hodgson, C.A. (2000) Convergent validity of the tinnitus handicap inventory and the tinnitus questionnaire. *Journal of Laryngology and Otology*, **114**, 840–43.

Baguley, D.M., Humphriss, R.L., Axon, P.R., and Moffat, D.A. (2005a) Change in tinnitus handicap after translabyrinthine vestibular schwannoma excision. *Otology & Neurotology*, **26**, 1061–63.

Baguley, D.M., Humphriss, R.L., Axon, P.R., and Moffat, D.A. (2006a) The clinical characteristics of tinnitus in patients with vestibular schwannoma. *Skull Base*, **16**, 49–58.

Baguley, D.M., Jones, S., Wilkins, I., Axon, P.R., and Moffat, D.A. (2005b) The inhibitory effect of intravenous lidocaine infusion on tinnitus after translabyrinthine removal of vestibular schwannoma: a double-blind, placebo-controlled, crossover study. *Otology & Neurotology*, **26**, 169–76.

Baguley, D.M., Moffat, D.A., and Hardy. D.G. (1992) What is the effect of translabyrinthine acoustic schwannoma removal upon tinnitus? *Journal of Laryngology and Otology*, **106**, 329–31.

Baguley, D.M., Phillips, J., Humphriss, R.L., Jones, S., Axon, P.R., and Moffat, D.A. (2006b) The prevalence and onset of gaze modulation of tinnitus and increased sensitivity to noise after translabyrinthine vestibular schwannoma excision. *Otology & Neurotology*, **27**, 220–24.

Bai, U., Seidman, M.D., Hinojosa, R., and Quirk, W.S. (1997) Mitochondrial DNA deletions associated with aging and possibly presbycusis: a human archival temporal bone study. *American Journal of Otology*, **18**, 449–53.

Bair, W., Koch, C., Newsome, W., and Britten, K. (1994) Power spectrum analysis of bursting cells in area MT in the behaving monkey. *Journal of Neuroscience*, **14**, 2870–92.

Banay Schwartz, M., Lajtha, A., and Palkovits, M. (1989) Changes with aging in the levels of amino acids in rat CNS structural elements. I. Glutamate and related amino acids. *Neurochemistry Research*, **14**, 555–62.

Bao, S., Chan, V.T., and Merzenich, N.M. (2001) Cortical remodelling induced by activity of ventral tegmental dopamine neurons. *Nature*, **412**, 79–83.

Barcham, L.J. and Stephens, S.D. (1980) The use of an open-ended problems questionnaire in auditory rehabilitation. *British Journal of Audiology*, **14**, 49–54.

Barnea, G., Attias, J., Gold, S., and Shahar, A. (1990) Tinnitus with normal hearing sensitivity: extended high-frequency audiometry and auditory-nerve brain-stem-evoked responses. *Audiology*, **29**, 36–45.

Barto, A.G. (1985) Learning by statistical cooperation of self-interested neuron-like computing elements. *Human Neurobiology*, **4**, 229–56.

Bartos, M., Vida, I., and Jonas, P. (2007) Synaptic mechanisms of synchronized gamma oscillations in inhibitory interneuron networks. *Nature Reviews Neuroscience*, **8**, 45–56.

Basta, D. and Ernst, A. (2004) Effects of salicylate on spontaneous activity in inferior colliculus brain slices. *Neuroscience Research*, **50**, 237–43.

Basta, D., Goetze, R., and Ernst, A. (2008) Effects of salicylate application on the spontaneous activity in brain slices of the mouse cochlear nucleus, medial geniculate body and primary auditory cortex. *Hearing Research*, **240**, 42–51.

Basta, D., Tzschentke, B., and Ernst, A. (2005) Noise-induced cell death in the mouse medial geniculate body and primary auditory cortex. *Neuroscience Letters*, **381**, 199–204.

Bauer, C.A. and Brozoski, T.J. (2001) Assessing tinnitus and prospective tinnitus therapeutics using a psychophysical animal model. *Journal of the Association for Research in Otolaryngology*, **2**, 54–64.

Bauer, C.A. and Brozoski, T.J. (2006) Effect of gabapentin on the sensation and impact of tinnitus. *Laryngoscope*, **116**, 675–81.

Bauer, C.A. and Brozoski, T.J. (2007) Gabapentin. *Progress in Brain Research*, **166**, 287–301.

Bauer, C.A. and Brozoski, T.J. (2011) Effect of tinnitus retraining therapy on the loudness and annoyance of tinnitus: a controlled trial. *Ear & Hearing*, **32**, 145–55.

Bauer, C.A., Brozoski, T.J., Holder, T.M., and Caspary, D.M. (2000) Effects of chronic salicylate on GABAergic activity in rat inferior colliculus. *Hearing Research*, **147**, 175–82.

Bauer, C.A., Brozoski, T.J., and Myers, K, (2007) Primary afferent dendrite degeneration as a cause of tinnitus. *Journal of Neuroscience Research*, **85**, 1489–98.

Bauer, C.A., Brozoski, T.J., Rojas, R., Boley, J., and Wyder, M. (1999) Behavioral model of chronic tinnitus in rats. *Otolaryngology Head and Neck Surgery*, **121**, 457–62.

Bauer, C.A., Turner, J.G., Caspary, D.M., Myers, K.S., and Brozoski, T.J. (2008) Tinnitus and inferior colliculus activity in chinchillas related to three distinct patterns of cochlear trauma. *Journal of Neuroscience Research*, **86**, 2564–78.

Beaulieu, C. (2002) The basis of anisotropic water diffusion in the nervous system—a technical review. *NMR Biomedicine*, **15**, 435–55.

Beitz, E., Kumagami, H., Krippeit-Drew, P., Ruppersberg, J.P., and Schultz, J.E. (1999) Expression pattern of aquaporin water channels in the inner ear of the rat. The molecular basis for a water regulation system in the endolymphatic sac. *Hearing Research*, **132**, 76–84.

Beitz, E., Zenner, H.P., and Schultz, J.E. (2003) Aquaporin-mediated fluid regulation in the inner ear. *Cellular and Molecular Neurobiology*, **23**, 315–29.

Bektas, D. and Caylan, R. (2008) Non-pulsatile subjective tinnitus without hearing loss may be caused by undetectable sounds originating from venous system of the brain. *Medical Hypotheses*, **71**, 245–48.

Belelli, D., Harrison, N.L., Maguire, J., Macdonald, R.L., Walker, M.C. and Cope, D.W. (2009) Extrasynaptic GABAA receptors: form, pharmacology, and function. *Journal of Neuroscience*, **29**, 12757–63.

Bender, K.J. and Trussell, L.O. (2011) Synaptic plasticity in inhibitory neurons of the auditory brainstem. *Neuropharmacology*, **60**, 774–79.

Bendor, D. and Wang, X. (2005) The neuronal representation of pitch in primate auditory cortex. *Nature*, **436**, 1161–65.

Benedict, R.H., Lockwood, A.H., Shucard, J.L., Shucard, D.W., Wack, D., and Murphy, B.W. (1998) Functional neuroimaging of attention in the auditory modality. *Neuroreport*, **9**, 121–26.

Berens, P., Keliris, G.A., Ecker, A.S., Logothetis, N.K., and Tolias, A.S. (2008) Feature selectivity of the gamma-band of the local field potential in primate primary visual cortex. *Frontiers in Neuroscience*, **2**, 199–207.

Berliner, K.I., Shelton, C., Hitselberger, W.E., and Luxford, W.M. (1992) Acoustic tumors: effect of surgical removal on tinnitus. *American Journal of Otology*, **13**, 13–17.

Bernhardt, O., Gesch, D., Schwahn, C., Bitter, K., Mundt, T., Mack, F., Kocher, T., Meyer, G., Hensel, E., and John, U. (2004) Signs of temporomandibular disorders in tinnitus patients and in a population-based group of volunteers: results of the Study of Health in Pomerania. *Journal of Oral Rehabilitation*, **31**, 311–19.

Berridge, M.J. (1997) Elementary and global aspects of calcium signalling. *Journal of Physiology*, **499**, 291–306.

Bilak, M., Kim, J., Potashner, S.J., Bohne, B.A., and Morest, D.K. (1997) New growth of axons in the cochlear nucleus of adult chinchillas after acoustic trauma. *Experimental Neurology*, **147**, 256–68.

Blaustein, M.P. (1988) Calcium transport and buffering in neurons. *Trends in Neuroscience*, **11**, 438–43.

Bledsoe, S.C. Jr., Nagase, S., Miller, J.M., and Altschuler, R.A. (1995) Deafness-induced plasticity in the mature central auditory system. *Neuroreport*, 7, 225–29.

Bledsoe, S.C., Koehler, S., Tucci, D.L., Zhou, J., Le Prell, C., and Shore, S.E. (2009) Ventral cochlear nucleus responses to contralateral sound are mediated by commissural and olivocochlear pathways. *Journal of Neurophysiology*, **102**, 886–900.

Blood, A.J. and Zatorre, R.J. (2001) Intensely pleasurable responses to music correlate with activity in brain regions implicated in reward and emotion. *Proceedings of the National Acadamy of Science of the United states of America*, **98**, 11818–23.

Boettcher, F.A. and Salvi, R.J. (1991) Salicylate ototoxicity: review and synthesis. *American Journal of Otolaryngology*, **12**, 33–47.

Boettcher, F.A. and Salvi, R.J. (1993) Functional changes in the ventral cochlear nucleus following acute acoustic overstimulation. *Journal of the Acoustical Society of America*, **94**, 2123–34.

Bogoch, I.I., Ronald, A., House, R.A., and Kudla, I. (2005) Perceptions about hearing protection and noise-induced hearing loss of attendees of rock concerts. *Canadian Journal of Public Health*, **96**, 69–72.

Bohne, B.A., Gruner, M.M., and Harding, G.W. (1990) Morphological correlates of aging in the chinchilla cochlea. *Hearing Research*, **48**, 79–91.

Bowen, G.P., Lin, D., Taylor, M.K., and Ison, J.R. (2003) Auditory cortex lesions in the rat impair both temporal acuity and noise increment thresholds, revealing a common neural substrate. *Cerebral Cortex*, **13**, 815–22.

Bowery, N.G. (1993) $GABA_B$ receptor pharmacology. *Annual Reviews of Pharmacology and Toxicology*, 33, 109–47.

Bowman, D.M., Eggermont, J.J., and Smith, G.M. (1995) Effect of stimulation on burst firing in cat primary auditory cortex. *Journal of Neurophysiology*, **74**, 1841–55.

Braitenberg, V. (1977) *On the texture of brains*. Berlin: Springer Verlag.

Braitenberg, V. and Schüz, A. (1991) *Anatomy of the cortex. Statistics and Geometry*. Berlin: Springer Verlag.

Bramham, C.R. and Messaoudi, E. (2005) BDNF function in adult synaptic plasticity: the synaptic consolidation hypothesis. *Progress in Neurobiology*, **76**, 99–125.

Bray, A., Szymanski, M., and Mills, R. (2004) Noise induced hearing loss in dance music disc jockeys and an examination of sound levels in nightclubs. *Journal of Laryngology & Otology*, **118**, 123–28.

Brennan, J.F., Brown, C.A., and Jastreboff, P.J. (1996) Salicylate-induced changes in auditory thresholds of adolescent and adult rats. *Developmental Psychobiology*, **29**, 69–86.

Brennan, J.F. and Jastreboff, P.J. (1991) Generalization of conditioned suppression during salicylate-induced phantom auditory perception in rats. *Acta Neurobiologica Experientia*, **51**, 15–27.

Brigande, J.V. and Heller, S. (2009) Quo vadis, hair cell regeneration? *Nature Neurosciences*, **12**, 679–85.

Briner, W. and Willott, J.F. (1989) Ultrastructural features of neurons in the C57BL/6J mouse anteroventral cochlear nucleus: young mice versus old mice with chronic presbycusis. *Neurobiology of Aging*, **10**, 295–303.

Britvina, T. and Eggermont, J.J. (2008) Spectro-temporal receptive fields during spindling and non-spindling epochs in cat primary auditory cortex. *Neuroscience*, **154**, 1576–88.

Brosch, M., Budinger, E., and Scheich, H. (2002) Stimulus-related gamma oscillations in primate auditory cortex. *Journal of Neurophysiology*, **87**, 2715–25.

Brosch, M. and Schreiner, C.E. (1999) Correlations between neural discharges are related to receptive field properties in cat primary auditory cortex. *European Journal of Neuroscience*, **11**, 3517–30.

Brown, L.G. (1927) Arterial bruit causing tinnitus. *Proceedings of the Royal Society of Medicine*, **20**, 1448–49.

Brozoski, T.J. and Bauer, C.A. (2005) The effect of dorsal cochlear nucleus ablation on tinnitus in rats. *Hearing Research*, **206**, 227–36.

Brozoski, T.J., Bauer, C.A., and Caspary, D.M. (2002) Elevated fusiform cell activity in the dorsal cochlear nucleus of chinchillas with psychophysical evidence of tinnitus. *Journal of Neuroscience*, **22**, 2383–90.

Brozoski, T.J., Ciobanu, L., and Bauer, C.A. (2007) Central neural activity in rats with tinnitus evaluated with manganese-enhanced magnetic resonance imaging (MEMRI) *Hearing Research*, **228**, 168–79.

Brunnberg, E., Lindén-Boström, M., and Berglund, M. (2008) Tinnitus and hearing loss in 15–16-year-old students: mental health symptoms, substance use, and exposure in school. *International Journal of Audiology*, **47**, 688–94.

Budd, R.J. and Pugh, R. (1996) Tinnitus coping style and its relationship to tinnitus severity and emotional distress. *Journal of Psychosomatic Research*, **41**(4), 327–35.

Buechner, A., Brendel, M., Saalfeld, H., Litvak, L., Frohne-Buechner, C., and Lenarz, T. (2010) Results of a pilot study with a signal enhancement algorithm for HiRes 120 cochlear implant users. *Otolology & Neurotology*, **31**, 1386–90.

Bunzeck, N., Wuestenberg, T., Lutz, K., Heinze, H.J., and Jancke, L. (2005) Scanning silence: mental imagery of complex sounds. *Neuroimage*, **26**, 1119–27.

Burns, E.M. (1984) A comparison of variability among measurements of subjective tinnitus and objective stimuli. *Audiology*, **23**, 426–40.

Burns, E.M. and Keefe, D.H. (1992) Intermittent tinnitus resulting from unstable otoacoustic emissions. In: J.-M. Aran and R. Dauman (ed.), *Tinnitus 91*, pp. 89–94. Amsterdam: Kugler Publications.

Buzsáki, G. and Draguhn, A. (2004) Neuronal oscillations in cortical networks. *Science*, **304**, 1926–29.

Cacace, A.T. (2003) Expanding the biological basis of tinnitus: crossmodal origins and the role of neuroplasticity. *Hearing Research*, **175**, 112–32.

Cacace, A.T., Cousins, J.P., Parnes, S.M., Semenoff, D., Holmes, T., McFarland, D.J., Davenport, C., Stegbauer, K., and Lovely, T.J. (1999a) Cutaneous-evoked I. Phenomenology, psychophysics and functional imaging. *Audiology and Neurootology*, **4**, 247–57.

Cacace, A.T., Cousins, J.P., Parnes, S.M., Semenoff, D., Holmes, T., McFarland, D.J., Davenport, C., Stegbauer, K., and Lovely, T.J. (1999b) Cutaneous-evoked tinnitus. II. Review of neuroanatomical, physiological and functional imaging studies. *Audiology and Neurootology*, **4**, 258–68.

Cacace, A.T., Lovely, T.J., Winter, D.F., Parnes, S.M., and McFarland, D.J. (1994) Auditory perceptual and visual-spatial characteristics of gaze-evoked tinnitus. *Audiology*, **33**, 291–303.

Cai, S., Ma, W.L., and Young, E.D. (2009) Encoding intensity in ventral cochlear nucleus following acoustic trauma: implications for loudness recruitment. *Journal of the Association for Research in Otolaryngology*, **10**, 5–22.

Caperton, K.K. and Thompson, A.M. (2011) Activation of serotonergic neurons during salicylate-induced tinnitus. *Otology and Neurotology*, **32**, 301–307.

Carney, L.H. (1990) Sensitivities of cells in anteroventral cochlear nucleus of cat to spatiotemporal discharge patterns across primary afferents. *Journal of Neurophysiology*, **64**, 437–56.

Caspary, D.M., Holder, T.M., Hughes, L.F., Milbrandt, J.C., McKernan, R.M., and Naritoku, D.K. (1999) Age-related changes in GABAA receptor subunit composition and function in rat auditory system. *Neuroscience*, **93**, 307–312.

Caspary, D.M., Hughes, L.F., Schatteman, T.A., and Turner, J.G. (2006) Age-related changes in the response properties of cartwheel cells in rat dorsal cochlear nucleus. *Hearing Research*, **216–217**, 207–15.

Caspary, D.M., Ling, L., Turner, J.G., and Hughes, L.F. (2008) Inhibitory neurotransmission, plasticity and aging in the mammalian central auditory system. *Journal of Experimental Biology*, **211**, 1781–91.

Caspary, D.M., Pazara, K.E., Kössl, M., and Faingold, C.L. (1987) Strychnine alters the fusiform cell output from the dorsal cochlear nucleus. *Brain Research*, **417**, 273–82.

Caspary, D.M., Raza, A., Lawhorn Armour, B.A., Pippin, J., and Arneri, S.P. (1990) Immunocytochemical and neurochemical evidence for age-related loss of GABA in the inferior colliculus: implications for neural presbycusis. *Journal of Neuroscience*, **10**, 2363–72.

Caspary, D.M., Schatteman, T.A., and Hughes, L.F. (2005) Age-related changes in the inhibitory response properties of dorsal cochlear nucleus output neurons: role of inhibitory inputs. *Journal of Neuroscience*, **25**, 10952–59.

Cazals, Y. (2000) Auditory sensori-neural alterations induced by salicylate. *Progress in Neurobiology*, **62**, 583–631.

Cazals, Y., Li, X.Q., Aurousseau, C., and Didier, A. (1988) Acute effects of noradrenalin related vasoactive agents on the ototoxicity of aspirin: an experimental study in the guinea pig. *Hearing Research*, **36**, 89–96.

Cazals, Y., Horner, K.C., and Huang, Z.W. (1998) Alterations in average spectrum of cochleoneural activity by long-term salicylate treatment in the guinea pig: a plausible index of tinnitus. *Journal of Neurophysiology*, **80**, 2113–20.

Cazals, Y. and Huang, Z.W. (1996) Average spectrum of cochlear activity: a possible synchronized firing, its olivo-cochlear feedback and alterations under anesthesia. *Hearing Research*, **101**, 81–92.

Ceranic, B.J., Prasher, D.K., Raglan, E., and Luxon, L.M. (1998) Tinnitus after head injury: evidence from otoacoustic emissions. *Journal of Neurology Neurosurgery and Psychiatry*, **65**, 523–29.

Chang, H., Chen, K., Kaltenbach, J.A., Zhang, J., and Godfrey, D.A. (2002) Effects of acoustic trauma on dorsal cochlear nucleus neuron activity in slices. *Hearing Research*, **164**, 59–68.

Chapman, A.G. (1998) Glutamate receptors in epilepsy. *Progress in Brain Research*, **116**, 371–83.

Chapman, C.R. (1996) Limbic processes and the affective dimension of pain. *Progress in Brain Research*, **110**, 63–81.

Chen, G.D., Kermany, M.H., D'Elia, A., Ralli, M., Tanakam C., Bielefeld, E.C., Ding, D., Henderson, D., and Salvi, R. (2010) Too much of a good thing: long-term treatment with salicylate strengthens outer hair cell function but impairs auditory neural activity. *Hearing Research*, 265, 63–69.

Chen, G.D. and Jastreboff, P.J. (1995) Salicylate-induced abnormal activity in the inferior colliculus of rats. *Hearing Research*, **82**, 158–78.

Cheung, S.W. and Larson, P.S. (2010) Tinnitus modulation by deep brain stimulation in locus of caudate neurons (area LC) *Neuroscience*, **169**, 1768–78.

Chevrier, P., Vijayaragavan, K., and Chahine, M. (2004) Differential modulation of Nav1.7 and Nav1.8 peripheral nerve sodium channels by the local anesthetic lidocaine. *British Journal of Pharmacology*, **142**, 576–84.

Christensen, N., D'Souza, M., Zhu, X., and Frisina, R.D. (2009) Age-related hearing loss: aquaporin 4 gene expression changes in the mouse cochlea and auditory midbrain. *Brain Research*, **1253**, 27–34.

Clapham, J. (1972) *Smetana*. New York: Octagon Books.

Clarke, S., de Ribaupierre, F., Rouiller, E.M., and de Ribaupierre, Y. (1993) Several neuronal and axonal types form long intrinsic connections in the cat primary auditory cortical field (AI) *Anatomy and Embryology (Berlin)*, **188**, 117–38.

Coad, M.L., Lockwood, A., Salvi, R., and Burkard, R. (2001) Characteristics of patients with gaze-evoked tinnitus. *Otology & Neurotology*, **22**, 650–54.

Cohen, M.R. and Maunsell, J.H. (2009) Attention improves performance primarily by reducing interneuronal correlations. *Nature Neuroscience*, **12**, 1594–1600.

Colding-Jørgensen, E., Lauritzen, M., Johnsen, N.J., Mikkelsen, K.B., and Saermark, K. (1992) On the evidence of auditory evoked magnetic fields as an objective measure of tinnitus. *Electroencephalography and Clinical Neurophysiology*, **83**, 322–27.

Coles, A.R.R, Lutman, M.E., Axelsson, A., and Hazell, J.W.P. (1992) Tinnitus severity ratings: cross-sectional studies. In J.-M. Aran and R. Dauman (eds.) *Tinnitus 91*, pp. 453–55. Amsterdam/New York: Kugler Publications.

Colson, N.J., Fernandez, F., Lea, R.A., and Griffiths, L.R. (2007) The search for migraine genes: an overview of current knowledge. *Cellular and Molecular Life Sciences*, **64**, 331–44.

Condon, C.D. and Weinberger, N.M. (1991) Habituation produces frequency-specific plasticity of receptive fields in the auditory cortex. *Behavioral Neuroscience*, **105**, 416–30.

Coomes, D.L. and Schofield, B.R. (2004) Projections from the auditory cortex to the superior olivary complex in guinea pigs. *European Journal of Neuroscience*, **19**, 2188–2200.

Cope, T.E. and Baguley, D.M. (2009) Is musical hallucination an otological phenomenon? a review of the literature. *Clinical Otolaryngology*, **34**, 423–30.

Cope, T.E., Baguley, D.M., and Moore, B.C. (2011) Tinnitus loudness in quiet and noise after resection of vestibular schwannoma. *Otology & Neurotology*, **32**, 488–96.

Coupland, S.G., Ponton, C.W., Eggermont, J.J., Bowen, T., and Grant, R.M. (1991) Assessment of cisplatin-induced ototoxicity using derived-band ABRs. *International Journal of Pediatric Otorhinolaryngology*, **22**, 237–48.

Covell, W.P. and Rogers, J.B. (1957) Pathologic changes in the inner ear of senile guinea pigs. *Laryngoscope*, **67**, 118–29.

Cransac, H., Cottet-Emard, J.M., Hellström, S., and Peyrin, L. (1998) Specific sound-induced noradrenergic and serotonergic activation in central auditory structures. *Hearing Research*, **118**, 151–56.

Cransac, H., Peyrin, L., Cottet-Emard, J.M., Farhat, F., Pequignot, J.M., and Reber, A. (1996) Aging effects on monoamines in rat medial vestibular and cochlear nuclei. *Hearing Research*, **100**, 150–56.

Crippa, A., Lanting, C.P., van Dijk, P., and Roerdink, J.B.T.M. (2010) A diffusion tensor imaging study on the auditory system and tinnitus. *The Open Neuroimaging Journal*, **4**, 16–25.

Cruickshanks, K.J., Klein, R., Klein, B.E.K., Wiley, T.L., Nondahl, D.M., and Tweed, TS. (1998) Cigarette smoking and hearing loss. *Journal of the American Medical Association*, **279**, 1715–19.

Cruikshank, S.J., Lewis, T.J., and Connors, B.W. (2007) Synaptic basis for intense thalamocortical activation of feedforward inhibitory cells in neocortex. *Nature Neuroscience*, **10**, 462–68.

Dallos, P. and Harris, D. (1978) Properties of auditory nerve responses in absence of outer hair cells. *Journal of Neurophysiology*, **41**, 365–83.

Dauman, R. and Bouscau-Faure, F. (2005) Assessment and amelioration of hyperacusis in tinnitus patients. *Acta Otolaryngologica*, **125**, 503–509.

Davies, E. (2004) The pharmacological management of tinnitus. *Audiological Medicine*, **2**, 26–28.

Davis, A. (1995) The aetiology of tinnitus: risk factors for tinnitus in the UK population – A possible role for conductive pathologies. In: G.E. Reich, and J.A. Vernon (ed.), *Proceedings of the 5th International Tinnitus Seminar*, pp. 38–50. Portland, OR: American Tinnitus Association.

Davis, A. and El-Rafaie, A. (2000) Epidemiology of tinnitus. In: R.S. Tyler (ed.) *Tinnitus Handbook*, pp. 1–23. San Diego, CA: Singular Press.

Davis, A.C. (1989) The prevalence of hearing impairment and reported hearing disability among adults in Great Britain. *International Journal of Epidemiology*, **18**, 911–17.

Davis, H., Morgan, C.T., Hawkins, J.E. Jr., Galambos, R., and Smith, F.W. (1950) Temporary deafness following exposure to loud tones and noise. *Acta Otolaryngolica*, **Suppl. 88**, 1–56.

Davis, K.A. and Young, E.D. (1997) Granule cell activation of complex-spiking neurons in dorsal cochlear nucleus. *Journal of Neuroscience*, **17**, 6798–6806.

Davis, K.A. and Young, E.D. (2000) Pharmacological evidence of inhibitory and disinhibitory neuronal circuits in dorsal cochlear nucleus. *Journal of Neurophysiology*, **83**, 926–40.

Davis, P.B., Paki, B., and Hanley, P.J. (2007) Neuromonics tinnitus treatment: third clinical trial. *Ear & Hearing*, **28**, 242–59.

Davis, R.R. (2001) Noise-induced hearing loss. In: J.F. Willott (ed.), *Handbook of Mouse Auditory Research: from Behavior to Molecular Biology*, pp. 477–88. Boca Raton, FL: CRC Press.

Davis, R.R., Newlander, J.K., Ling, X., Cortopassi, G.A., Krieg, E.F., and Erway, L.C. (2001) Genetic basis for susceptibility to noise-induced hearing loss in mice. *Hearing Research*, **155**, 82–90.

Dawes, P.J. and Welch, D. (2010) Childhood hearing and its relationship with tinnitus at thirty-two years of age. *Annals of Otology, Rhinology and Laryngology*, **119**, 672–76.

Dawkins, R. (1976) *The Selfish Gene*. Oxford: Oxford University Press.

Day, R.O., Graham, G.G., Bieri, D., Brown, M., Cairns, D., Harris, G., Hounsell, J., Platt-Hepworth, S., Reeve, R., Sambrook, P.N., *et al.* (1989) Concentration-response relationships for salicylate-induced ototoxicity in normal volunteers. *British Journal of Clinical Pharmacology*, **28**, 695–702.

deCharms, R.C. and Merzenich, M.M. (1996) Primary cortical representation of sounds by the coordination of action-potential timing. *Nature*, **381**, 610–13.

De Ridder, D. (2010) Auditory nerve compression: a forgotten treatable cause for tinnitus. *Journal of Neurology Neurosurgery and Psychiatry*, **81**, 355.

De Ridder, D., De Mulder, G., Verstraeten, E., Van der Kelen, K., Sunaert, S., Smits, M., Kovacs, S., Verlooy, J., Van de Heyning, P., and Moller, A.R. (2006b) Primary and secondary auditory cortex stimulation for intractable tinnitus. *ORL Journal of Otorhinolaryngol and Related Specialties*, **68**, 48–54.

De Ridder, D., Elgoyhen, A.B., Romo, R., and Langguth, B. (2011) Phantom percepts: tinnitus and pain as persisting aversive memory networks. *Proceedings of the National Academy of Science of the United States of America*, **108**, 8075–80.

De Ridder, D., Fransen, H., Francois, O., Sunaertm, S., Kovacsm, S., and Van De Heyningm, P. (2006a) Amygdalo-hippocampal involvement in tinnitus and auditory memory. *Acta Otolaryngologica*, **Suppl. 556**, 50–53.

De Ridder, D. and Van de Heyning, P. (2007) The Darwinian plasticity hypothesis for tinnitus and pain. *Progress in Brain Research*, **166**, 55–60.

De Ridder, D., van der Loo, E., Van der Kelen, K., Menovsky, T., van de Heyning, P., and Møller, A. (2007a) Theta, alpha and beta burst transcranial magnetic stimulation: brain modulation in tinnitus. *International Journal of Medical Sciences*, **4**, 237–41.

De Ridder, D., van der Loo, E., Van der Kelen, K., Menovsky, T., van de Heyning, P., and Møller, A. (2007b) Do tonic and burst TMS modulate the lemniscal and extralemniscal system differentially? *International Journal of Medical Sciences*, **4**, 242–46.

Deer, B.C. and Hunter-Duvar, I. (1982) Salicylate ototoxicity in the chinchilla: a behavioral and electron microscope study. *Journal of Otolaryngology*, **11**, 260–64.

Dehaene, S. and Changeux, J.P. (2005) Ongoing spontaneous activity controls access to consciousness: a neuronal model for inattentional blindness. *PLoS Biology*, **3**(5): e141.

Dehaene, S., Changeux, J.P., Naccache, L., Sackur, J., and Sergent, C. (2006) Conscious, preconscious, and subliminal processing: a testable taxonomy. *Trends in Cognitive Science*, **10**, 204–211.

de Heer, A.M.R., Huygen, P.L.M., Collin, R.W.J., Oostrik, J., Kremer, H., and Cremers, C.W.R.J. (2009) Audiometric and vestibular features in a second Dutch DFNA20/26 family with a novel mutation in ACTG1. *Annals of Otology Rhinology & Laryngology*, **118**, 382–90.

Dehmel, S., Cui, Y.L., and Shore, S.E. (2008) Cross-modal interactions of auditory and somatic inputs in the brainstem and midbrain and their imbalance in tinnitus and deafness. *American Journal of Audiology*, **17**, S193–S209.

de Klaver, M.J., van Rijn, M.A., Marinus, J., Soede, W., de Laat, J.A., and van Hilten, J.J. (2007) Hyperacusis in patients with complex regional pain syndrome related dystonia. *Journal of Neurology Neurosurgery and Psychiatry*, **78**, 1310–1313.

de Lafuente, V. and Romo, R. (2005) Neuronal correlates of subjective sensory experience. *Nature Neuroscience*, **8**, 1698–1703.

de la Rocha, J., Doiron, B., Shea-Brown, E., Josic, K., and Reyes, A. (2007) Correlation between neural spike trains increases with firing rate. *Nature*, **448**, 802–806.

Del Bo, L. and Ambrosetti, U. (2007) Hearing aids for the treatment of tinnitus. *Progress in Brain Research*, **166**, 341–45.

Del Bo, L., Forti, S., Ambrosetti, U., Costanzo, S., Mauro, D., Ugazio, G., Langguth, B., and Mancuso, A. (2008) Tinnitus aurium in persons with normal hearing: 55 years later. *Otolaryngology–Head and Neck Surgery*, **139**, 391–94.

Delgutte, B. (1990) Physiological mechanisms of psychophysical masking: observations from auditory-nerve fibers. *Journal of the Acoustical Society of America*, **87**, 791–809.

Deniz, M., Bayazit, Y.A., Celenk, F., Karabulut, H., Yilmaz, A., Gunduz, B., Saridogan, C., Dagli, M., Erdal, E., and Menevse A. (2010) Significance of serotonin transporter gene polymorphism in tinnitus. *Otology & Neurotology*, **31**, 19–24.

Denk, D.M., Heinzl, H., Franz, P., and Ehrenberger, K. (1997) Caroverine in tinnitus treatment. A placebo-controlled blind study. *Acta Otolaryngologica*, **117**, 825–30.

Destexhe, A., Contreras, D., and Steriade, M. (1999) Spatiotemporal analysis of local field potentials and unit discharges in cat cerebral cortex during natural wake and sleep states. *Journal of Neuroscience*, **19**, 4595–4608.

Di Nardo, W., Cantore, I., Cianfrone, F., Melillo, P., Scorpecci, A., and Paludetti, G. (2007) Tinnitus modifications after cochlear implantation. *European Archive of Otorhinolaryngology*, **264**, 1145–49.

Dickenson, A.H. (1996) Balances between excitatory and inhibitory events in the spinal cord and chronic pain. *Progress in Brain Research*, **110**, 226–31.

Dickson, J.W. and Gerstein, G.L. (1974) Interactions between neurons in auditory cortex of the cat. *Journal of Neurophysiology*, **37**, 1239–61.

Didier, A., Miller, J.M., and Nuttall, A.L. (1993) The vascular component of sodium salicylate ototoxicity in the guinea pig. *Hearing Research*, **69**, 199–206.

Dieler, R., Shehata-Dieler, W.E., Richter, C.P., and Klinke, R. (1994) Effects of endolymphatic and perilymphatic application of salicylate in the pigeon. II: fine structure of auditory hair cells. *Hearing Research*, **74**, 85–98.

Dierks, T., Linden, D.E., Jandl, M., Formisano, E., Goebel, R., Lanfermann, H., and Singer, W. (1999) Activation of Heschl's gyrus during auditory hallucinations. *Neuron*, **22**(3), 615–21

Diesch, E., Andermann, M., Flor, H., and Rupp, A. (2010) Functional and structural aspects of tinnitus-related enhancement and suppression of auditory cortex activity. *Neuroimage*, **50**, 1545–59.

Dille, M.F., Konrad-Martin, D., Gallun, F., Helt, W.J., Gordon, J.S., Reavis, K.M., Bratt, G.W., and Fausti, S.A. (2010) Tinnitus onset rates from chemotherapeutic agents and ototoxic antibiotics: results of a large prospective study. *Journal of the American Academy of Audiology*, **21**, 409–17.

Dineen, R., Doyle, J., and Bench, J. (1997) Audiological and psychological characteristics of a group of tinnitus sufferers, prior to tinnitus management training. *British Journal of Audiology*, **31**, 27–38.

Dixon, G.J. and Roaf, R. (1946) Undulant fever complicated by acute transient attacks of aphasia and during convalescence, by deafness, tinnitus, and paraesthesia. *British Medical Journal*, **2**, 12.

Dobie, R.A. (2004) Clinical trials and drug therapy for tinnitus. In J.B. Snow Jr, (ed.), *Tinnitus: Theory and Management*, pp. 266–77. Hamilton: BC Dekker.

Dobie, R.A., Sakai, C.S., Sullivan, M.D., Katon, W.J., and Russo, J. (1993) Antidepressant treatment of tinnitus patients: report of a randomized clinical trial and clinical prediction of benefit. *American Journal of Oto*logy, **14**, 18–23.

Domeisen, H., Hotz, M.A., and Häusler, R. (1998) Caroverine in tinnitus treatment. *Acta Otolaryngologica*, **118**, 606–608.

Dong, S., Mulders, W.H., Rodger, J., and Robertson, D. (2009) Changes in neuronal activity and gene expression in guinea-pig auditory brainstem after unilateral partial hearing loss. *Neuroscience*, **159**, 1164–74.

Dong, S., Mulders, W.H., Rodger, J., Woo, S., and Robertson, D. (2010) Acoustic trauma evokes hyperactivity and changes in gene expression in guinea-pig auditory brainstem. *European Journal of Neuroscience*, **31**, 1616–28.

Douek, E.E., Dodson, H.C., and Bannister, L.H. (1983) The effects of sodium salicylate on the cochlea of guinea pigs. *Journal of Laryngology and Otology*, **97**, 793–99.

Douglas, R.J. and Martin, K.A., (1991) A functional microcircuit for cat visual cortex. *Journal of Physiology*, **440**, 735–69.

Duan, M., Agerman, K., Ernfors, P., and Canlon, B. (2000) Complementary roles of neurotrophin 3 and a N-methyl-D-aspartate antagonist in the protection of noise and aminoglycoside-induced ototoxicity. *Proceedings of the National Academy of Science of the United States of America*, **97**, 7597–7602.

Dunn, C.C., Perreau, A., Gantz, B., and Tyler, R.S. (2010) Benefits of localization and speech perception with multiple noise sources in listeners with a short-electrode cochlear implant. *Journal of the American Academy of Audiology*, **21**, 44–51.

Duong, T.Q., Silva, A.C., Lee, S.P., and Kim, S.G. (2000) Functional MRI of calcium-dependent synaptic activity: cross correlation with CBF and BOLD measurements. *Magnetic Resonance in Medicine*, **43**, 383–92.

Eggermont, J.J. (1977) Electrocochleography and recruitment. *Annals of Otology Rhinology and Laryngology*, **86**, 138–49.

Eggermont, J.J. (1984) Tinnitus: some thoughts about its origin. *Journal of Laryngol and Otolology*, **Suppl. 9**, 31–37.

Eggermont, J.J. (1985) Peripheral auditory adaptation and fatigue: a model oriented review. *Hearing Research*, **18**, 57–71.

Eggermont, J.J. (1989) Coding of free field intensity in the auditory midbrain of the leopard frog. I. Results for tonal stimuli. *Hearing Research*, **40**, 147–66.

Eggermont, J.J. (1990a) *The correlative brain; theory and experiment in neural interaction*. Berlin: Springer Verlag.

Eggermont, J.J. (1990b) On the pathophysiology of tinnitus; a review and a peripheral model. *Hearing Research*, **48**, 111–24.

Eggermont, J.J. (1992a) Neural interaction in cat primary auditory cortex. Dependence on recording depth, electrode separation and age. *Journal of Neurophysiology*, **68**, 1216–28.

Eggermont, J.J. (1992b) Stimulus induced and spontaneous rhythmic firing of single units in cat primary auditory cortex. *Hearing Research*, **61**, 1–11.

Eggermont, J.J. (1994) Neural interaction in cat primary auditory cortex II. Effects of sound stimulation. *Journal of Neurophysiology*, **71**, 246–70

Eggermont, J.J. (1997) Firing rate and firing synchrony distinguish dynamic from steady state sound. *NeuroReport*, **8**, 2709–13

Eggermont, J.J. (2000a) Physiological mechanisms and neural models. In: R.S. Tyler (ed.), *Tinnitus Handbook*, pp. 85–122. San Diego, CA Singular Press.

Eggermont, J.J. (2000b) Sound induced correlation of neural activity between and within three auditory cortical areas. *Journal of Neurophysiology*, **83**, 2708–22

Eggermont, J.J. (2001) Between sound and perception: reviewing the search for a neural code. *Hearing Research*, **157**, 1–42.

Eggermont, J.J. (2003) Central tinnitus. *Auris Nasus Larynx*, **30**, Suppl 1, 7–12.

Eggermont, J.J. (2004) The cortex and tinnitus. In: J.B. Snow (ed.), *Tinnitus: Theory and Management*, pp. 171–88. Hamilton: BC Decker.

Eggermont, J.J. (2005a) Tinnitus: neurobiological substrates. *Drug Discovery Today*, **10**, 1283–90.

Eggermont, J.J. (2005b) Correlated neural activity: epiphenomenon or part of the neural code? In: P. Heil, R. König, and E. Budinger (ed.), *Auditory cortex: towards a synthesis of human and animal research*, pp.255–73. Mahwah, NJ: Lawrence Erlbaum Associates.

Eggermont, J.J. (2006) Properties of correlated neural activity clusters in cat auditory cortex resemble those of neural assemblies. *Journal of Neurophysiology*, **96**, 746–64.

Eggermont, J.J. (2007a) Electric and magnetic fields of synchronous neural activity propagated to the surface of the head: peripheral and central origins of AEPs. In: R.R. Burkard, M. Don, and J.J. Eggermont (eds.), *Auditory Evoked Potentials*, pp. 2–21. Baltimore, MD: Lippincott Williams & Wilkins.

Eggermont, J.J. (2007b) Correlated neural activity as the driving force for functional changes in auditory cortex. *Hearing Research*, **229**, 69–80.

Eggermont, J.J. (2008) Role of auditory cortex in noise and drug-induced tinnitus. *American Journal of Audiology*, **27**, S162–67.

Eggermont, J.J. (2010) Tinnitus: processing of phantom sound. In: Koob G.F., Le Moal M., and Thompson R.F. (eds.)0, *Encyclopedia of Behavioural Neuroscience,* vol 3, pp. 405–411. Oxford: Academic Press.

Eggermont, J.J. (2012) Current issues in tinnitus. In R. Burkard and K. Tremblay (eds.), *Translational perspectives in auditory neuroscience*. San Diego, CA: Plural Publishing, Inc.

Eggermont, J.J., Brown, D.K., Ponton, C.W., and Kimberley, B.P. (1996) Comparison of distortion product otoacoustic emission (DPOAE) and auditory brain stem response (ABR) traveling wave delay measurements suggests frequency-specific synapse maturation. *Ear and Hearing*, **17**, 386–94.

Eggermont, J.J. and Kenmochi, M. (1998) Salicylate and quinine selectively increase spontaneous firing rates in secondary auditory cortex. *Hearing Research*, **117**, 149–60.

Eggermont, J.J. and Komiya, H. (2000) Moderate noise trauma in juvenile cats results in profound cortical topographic map changes in adulthood. *Hearing Research*, **142**, 89–101.

Eggermont, J.J., Munguia, R., Pienkowski, M., and Shaw, G. (2011) Comparison of LFP-based and spike-based spectro-temporal receptive fields and crosscorrelation in cat primary auditory cortex. *PLoS ONE*, **6**(5): e20046.

Eggermont, J.J. and Odenthal, D.W. (1974) Action potentials and summating potentials in the normal human cochlea. *Acta Otolaryngol*, **Suppl. 316**: 39–61.

Eggermont, J.J. and Roberts, L.E. (2004) The neuroscience of tinnitus. *Trends in Neuroscience*, **27**, 676–82.

Eggermont, J.J. and Sininger, Y. (1995) Correlated neural activity and tinnitus. In: J.A. Vernon and A.R. Møller (eds.), *Mechanisms of Tinnitus*, pp. 21–34. Boston, MA: Allyn and Bacon.

Eggermont, J.J. and Smith, G.M. (1995a) Synchrony between single-unit activity and local field potentials in relation to periodicity coding in primary auditory cortex. *Journal of Neurophysiology*, **73**, 227–45.

Eggermont, J.J. and Smith, G.M. (1995b) Rate covariance dominates spontaneous cortical unit-pair correlograms. *NeuroReport*, **6**, 2125–28.

Eggermont, J.J. and Smith, G.M. (1996) Neural connectivity only accounts for a small part of neural correlation in auditory cortex. *Experimental Brain Research*, **110**, 379– 92.

Eggermont, J.J., Smith, G.M., and Bowman, D. (1993) Spontaneous burst-firing in cat primary auditory cortex. Age and depth dependence and its effect on neural interaction measures. *Journal of Neurophysiology*, **69**, 1292–1313.

El-Kashlan, H.K. and Shore, S.E. (2004) Effects of trigeminal ganglion stimulation on the central auditory system. *Hearing Research*, **189**, 25–30.

Elgoyhen, A.B. and Langguth, B. (2010) Pharmacological approaches to the treatment of tinnitus. *Drug Discovery Today*, **15**, 300–305.

Engineer, N.D., Riley, J.R., Seale, J.D., Vrana, W.A., Shetake, J.A., Sudanagunta, S.P., Borland, M.S., and Kilgard, M.P. (2011) Reversing pathological neural activity using targeted plasticity. *Nature*, **470**, 101–104.

Enrico, P. and Goodey, R. (2011) Complications to Medical Treatment. In: A. Moller, B. Langguth, D. de Ridder, and T. Kleinjung (eds.), *Textbook of Tinnitus*, pp. 343–62. Berlin: Springer.

Erlandsson, S.I., (2000) Psychological profiles of tinnitus patients. In: R.S. Tyler (ed.) *Tinnitus Handbook*, pp. 25–57. San Diego, CA: Singular Press.

Erlandsson, S.I., Hallberg, L.R., and Axelsson, A. (1992) Psychological and audiological correlates of perceived tinnitus severity. *Audiology*, **31**, 168–79.

Ernst, M., Romeo, R.D., and Andersen, S.L. (2009) Neurobiology of the development of motivated behaviors in adolescence: a window into a neural systems model. *Pharmacology Biochemistry and Behavior*, **93**, 199–211.

Erway, L.C., Shiau, Y.W., Davis, R.R., and Krieg, E.F. (1996) Genetics of age-related hearing loss in mice. III. Susceptibility of inbred and F1 hybrid strains to noise-induced hearing loss. *Hearing Research*, **93**, 181–87.

Estes, W. K. and Skinner, B. F. (1941) Some quantitative properties of anxiety. *Journal of Experimental Psychology*, **29**, 390–400.

Evans, E.F. and Borerwe, T.A. (1982) Ototoxic effects of salicylates on the responses of single cochlear nerve fibres and on cochlear potentials. *British Journal of Audiology*, **16**, 101–108.

Evans, E.F., Wilson, J.P., and Borerwe, T.A. (1981) Animal models of tinnitus. *Ciba Foundation Symposium*, **85**, 108–38.

Everitt, B.S. (1998) *The Cambridge Dictionary of Statistics*. Cambridge: Cambridge University Press.

Fabijanska, A., Rogowski, M., Bartnik, G., and Skarzynski, H. (1999) Epidemiology of tinnitus and hyperacusis in Poland. In: J. Hazell (ed.) *Proceedings of the 6th International Tinnitus Seminar*, pp.1–23. London: The Tinnitus and Hyperacusis Centre.

Fahy, C., Nikolopoulos, T.P., and O'Donoghue, G.M. (2002) Acoustic neuroma surgery and tinnitus. *European Archive of Otorhinolaryngology*, **259**, 299–301.

Farb, C.R. and Ledoux, J.E. (1999) Afferents from rat temporal cortex synapse on lateral amygdala neurons that express NMDA and AMPA receptors. *Synapse*, **33**, 218–29.

Farhadi, M., Mahmoudian, S., Saddadi, F., Karimian, A.R., Mirzaee, M., Ahmadizadeh, M., Ghasemikian, K., Gholami, S., Ghoreyshi, E., Beyty, S., *et al.* (2010) Functional brain abnormalities localized in 55 chronic tinnitus patients: fusion of SPECT coincidence imaging and MRI. *Journal of Cerebral Blood Flow and Metabolism*, **30**, 864–70.

Farrant, M. and Nusser, Z. (2005) Variations on an inhibitory theme: phasic and tonic activation of GABA(A) receptors. *Nature Reviews Neuroscience*, **6**, 215–29.

Farri, A., Enrico, A., Lacilla, M., and Sartoris, A. (1999) Tinnitus during headache: clinical-instrumental evaluation. *Acta Otorhinolaryngolica Italia*, **19**, 70–75.

Fausti, S.A., Wilmington, D.J., Helt, P.V., Helt, W.J., and Konrad-Martin, D. (2005) Hearing health and care: the need for improved hearing loss prevention and hearing conservation practices. *Journal of Rehabilitation Research & Development*, **42**(Suppl 2), 45–62.

Fávero, M.L., Sanchez, T.G. Bento, R.F. and Nascimento, A.F. (2006) Contralateral suppression of otoacoustic emission in patients with tinnitus. *Brazilian Journal of Otorhinolaryngology*, **72**, 223–26.

Feldmann, H. (1971) Homolateral and contralateral masking of tinnitus by noise-bands and by pure tones. *Audiology*, **10**, 138–44.

Feng, H., Yin, S.H., Tang, A.Z., Cai, H.W., Chen, P., Tan, S.H., and Xie, L.H. (2010) Caspase-3 activation in the guinea pig cochlea exposed to salicylate. *Neuroscience Letters*, **479**, 34–39.

Fenoy, A.J., Severson, M.A., Volkov. I.O., Brugge, J.F., and Howard, M.A. 3rd. (2006) Hearing suppression induced by electrical stimulation of human auditory cortex. *Brain Research*, **1118**, 75–83.

Ferrucci, L., Guralnik, J.M., Penninx, B.W.J.H., Leveille, S. (1998) Cigarette smoke exposure and hearing loss. *Journal of the American Medical Association*, **280**, 963–64.

Fetz, E.E., Toyama, K., and Smith, W. (1991) Normal and altered states of function. In: A. Peters and E. G. Jones (eds.) *Cerebral Cortex*, vol. 9, pp. 1–47. New York: Plenum Press.

Figueiredo, R.R., Langguth, B., Mello de Oliveira, P., and Aparecida de Azevedo, A. (2008) Tinnitus treatment with memantine. *Otolaryngology Head and Neck Surgery*, **138**, 492–96.

Finlayson, P.G. and Kaltenbach, J.A. (2009) Alterations in the spontaneous discharge patterns of single units in the dorsal cochlear nucleus following intense sound exposure. *Hearing Research*, **256**, 104–17.

Fitzgerald, J.J., Robertson, D., and Johnstone, B.M. (1993) Effects of intra-cochlear perfusion of salicylates on cochlear microphonic and other auditory responses in the guinea pig. *Hearing Research*, **67**, 147–56.

Flor, H., Denke, C., Schaefer, M., and Grüsser, S. (2001) Effect of sensory discrimination training on cortical reorganisation and phantom limb pain. *Lancet*, **357**, 1763–64.

Flor, H., Elbert, T., Knecht, S., Wienbruch, C., Pantev, C., Birbaumer, N., Larbig, W., and Taub, E. (1995) Phantom limb pain as a perceptual correlate of cortical reorganization following arm amputation. *Nature*, **375**, 482–84.

Flor, H., Nikolajsen, L., and Jensen, T.H. (2006) Phantom limb pain: a case of maladaptive CNS plasticity? *Nature Reviews Neuroscience*, **7**, 873–81.

Folmer, R.L. and Carroll, J.R. (2006) Long-term effectiveness of ear-level devices for tinnitus. *Otolaryngology Head and Neck Surgery*, **134**, 132–37.

Folmer, R.L. and Griest, S.E. (2003) Chronic tinnitus resulting from head or neck injuries. *Laryngoscope*, **113**, 821–27.

Folmer, R.L. and Shi, Y.B. (2004) SSRI use by tinnitus patients: interactions between depression and tinnitus severity. *Ear Nose and Throat Journal*, **83**, 107–108.

Formby, C., Sherlock, L.P., and Gold, S.L. (2003) Adaptive plasticity of loudness induced by chronic attenuation and enhancement of the acoustic background. *Journal of the Acoustical Society of America*, **114**, 55–58.

Formisano, E., Kim, D.S., Di Salle, F., van de Moortele, P.F., Ugurbil, K., and Goebel, R. (2003) Mirror-symmetric tonotopic maps in human primary auditory cortex. *Neuron*, **40**, 859–69.

Förster, C.R. and Illing, R.B. (2000) Plasticity of the auditory brainstem: cochleotomy-induced changes of calbindin-D28k expression in the rat. *Journal of Comparative Neurology*, **416**, 173–87.

Fosbroke, J. (1831) Pathology and treatment of deafness. *Lancet*, **15**, 645–48.

Fowler, E.P. (1944) Head noises in normal and disordered ears. Significance, measurement, differentiation and treatment. *Archives of Otolaryngology*, **39**, 498–503.

Fox, M.D. and Raichle, M.E. (2007) Spontaneous fluctuations in brain activity observed with functional magnetic resonance imaging. *Nature Reviews Neuroscience*, **8**, 700–711.

Fox, M.D., Snyder, A.Z., Vincent, J.L., Corbetta, M., Van Essen, D.C., and Raichle, M.E. (2005) The human brain is intrinsically organized into dynamic, anticorrelated functional networks. *Proceedings of the National Academy of Science of the United States of America*, **102**, 9673–78.

Fransen, E. and Van Camp, G. (1999) The COCH gene: a frequent cause of hearing impairment and vestibular dysfunction? *British Journal of Audiology*, **33**, 297–98.

Fransen, E., Verstreken, M., Verhagen, W.I., Wuyts, F.L., Huygen, P.L., D'Haese, P., Robertson, N.G., Morton, C.C., McGuirt, W.T., Smith, R.J., *et al.* (1999) High prevalence of symptoms of Menière's disease in three families with a mutation in the COCH gene. *Human Molecular Genetics*, **8**, 1425–29.

Fredelius, L., Rask-Andersen, H., Johansson, B., Urquiza, R., Bagger-Sjöbäck, D., and Wersäll, J. (1988) Time sequence of degeneration pattern of the organ of Corti after acoustic overstimulation. A light microscopical and electrophysiological investigation in the guinea pig. *Acta Otolaryngologica*, **106**, 81–93.

Fredj, N.B. and Burrone, J. (2009) A resting pool of vesicles is responsible for spontaneous vesicle fusion at the synapse. *Nature Neuroscience*, **12**, 751–58.

Friederici, A.D. (2002) Towards a neural basis of auditory sentence processing. *Trends in Cognitive Sciences*, **6**, 78–84.

Friedland, D.R., Gaggl, W., Runge-Samuelson, C., Ulmer, J.L., and Kopell, B.H. (2007) Feasibility of auditory cortical stimulation for the treatment of tinnitus. *Otology & Neurotology*, **28**, 1005–12.

Fries, P., Nikoli, D., and Singer, W. (2007) The gamma cycle. *Trends in Neuroscience*, **30**, 309–16.

Frisina, R.D. and Walton, J. P. (2001) Aging of the mouse central auditory system. In: J.P. Willott (ed.), *Handbook of Mouse Auditory Research: from Behavior to Molecular Biology*, pp. 339–79. New York: CRC Press.

Frisina, R.D. and Walton, J.P. (2006) Age-related structural and functional changes in the cochlear nucleus. *Hearing Research*, **216–217**, 216–23.

Fritz, J., Shamma, S., Elhilali, M., and Klein, D. (2003) Rapid task-related plasticity of spectrotemporal receptive fields in primary auditory cortex. *Nature Neuroscience*, **6**, 1216–23.

Fujino, K. and Oertel, D. (2003) Bidirectional synaptic plasticity in the cerebellum-like mammalian dorsal cochlear nucleus. *Proceedings of the National Academy of Science of the United States of America*, **100**, 265–70.

Gao, M., Sossa, K., Song, L., Errington, L., Cummings, L., Hwang, H., Kuhl, D., Worley, P., and Lee, H.K. (2010) A specific requirement of Arc/Arg3.1 for visual experience-induced homeostatic synaptic plasticity in mouse primary visual cortex. *Journal of Neuroscience*, **30**, 7168–78.

Gates, G.A., Feeney, M.P., and Mills, D. (2008) Cross-sectional age-changes of hearing in the elderly. *Ear & Hearing*, **29**, 865–74.

Gates, G.A, Mills, D., Nam, B.H., D'Agostino, R., and Rubel, E.W. (2002) Effects of age on the distortion product otoacoustic emission growth functions. *Hearing Research*, **163**(1–2), 53–60.

Gates, G.A. and Mills, J.H. (2005) Presbycusis. *Lancet*, **366**, 1111–20.

Geisler, C.D. (1998) *From sound to Synapse*. Oxford: Oxford University Press.

Geisser, M.E., Glass, J.M., Rajcevska, L.D., Clauw, D.J., Williams, D.A., Kileny, P.R., and Gracely, R.H. (2008) A psychophysical study of auditory and pressure sensitivity in patients with fibromyalgia and healthy controls. *Journal of Pain*, **9**, 417–22.

Gil, Z., Connors, B.W., and Amitai, Y. (1999) Efficacy of thalamocortical and intracortical synaptic connections: quanta, innervation, and reliability. *Neuron*, **23**, 385–97.

Gil-Loyzaga, P., Bartolomé, V., Vicente-Torres, A., and Carricondo, F. (2000) Serotonergic innervation of the organ of Corti. *Acta Otolaryngologica*, **120**, 128–32.

Giraud, A.L., Chery-Croze, S., Fischer, G., Fischer, C., Vighetto, A., Gregoire, M.C., Lavenne, F., and Collet, L. (1999) A selective imaging of tinnitus. *Neuroreport*, **10**, 1–5.

Glasberg, B.R. and Moore, B.C. (2000) Frequency selectivity as a function of level and frequency measured with uniformly exciting notched noise. *Journal of the Acoustical Society of Amerca*, **108**, 2318–28.

Glitsch, M.D. (2008) Spontaneous neurotransmitter release and Ca^{2+} –how spontaneous is spontaneous neurotransmitter release? *Cell Calcium*, **43**, 9–15.

Goddard, G.V., McIntyre, D.C., and Leech, C.K. (1969) A permanent change in brain function resulting from daily electrical stimulation. *Experimental Neurology*, **25**, 295–330.

Godey, B., Schwartz, D., de Graaf, J.B., Chauvel, P., Liegeois-Chauvel, C. (2001) Neuromagnetic source localization of auditory evoked fields and intracerebral evoked potentials: a comparison of data in the same patients. *Clinical Neurophysiology*, **112**, 1850–59.

Gold, T. (1948) Hearing II. The physical basis of action of the cochlea. *Proceedings of the Royal Society of Edinburgh B*, **135**, 492–98.

Goodey, R.J. (1981) Drugs in the treatment of tinnitus. *Ciba Foundation Symposium*, **85**, 263–78.

Gopinath, B., Flood, V.M., McMahon, C.M., Burlutsky, G., Smith, W., and Mitchell, P. (2010b) The effects of smoking and alcohol consumption on age-related hearing loss: The Blue Mountains Hearing Study. *Ear & Hearing*, **31**, 277–82.

Gopinath, B., McMahon, C.M., Rochtchina, E., and Mitchell, P. (2009) Dizziness and vertigo in an older population: the Blue Mountains prospective cross-sectional study. *Clinical Otolaryngology*, **34**, 552–56.

Gopinath, B., McMahon, C.M., Rochtchina, E. Karpa, M.J., and Mitchell, P. (2010a) Incidence, persistence, and progression of tinnitus symptoms in older adults: The Blue Mountains Hearing Study. *Ear & Hearing*, **31**, 407–412.

Gourévitch, B. and Eggermont, J.J. (2007) Evaluating information transfer between auditory cortical neurons. *Journal of Neurophysiology*, **97**, 2533–43.

Gourévitch, B. and Eggermont, J.J. (2008) Spectrotemporal sound density dependent long-term adaptation in cat primary auditory cortex. *European Journal of Neuroscience*, **27**, 3310–21

Graham, J.M. (1979) Paediatric tinnitus. *Journal of Laryngology & Otology*, **4**, 114–120.

Graham, J.M. (1981) Tinnitus in children with hearing loss. *Ciba Foundation Symposium*, **85**, 172–92.

Graham, J. and Butler, J. (1984) Tinnitus in children. *Journal of Laryngology and Otology*, **Suppl 9**, 236–41.

Granjeiro, R.C., Kehrle, H.M., Bezerra, R.L., Almeida, V.F., Sampaio, A.L.L., and Oliveira, C.A. (2008) Transient and distortion product evoked oto-acoustic emissions in normal hearing patients with and without tinnitus. *Otolaryngology–Head and Neck Surgery*, **138**, 502–506.

Graybiel, A.M. (1973) The thalamo-cortical projection of the so-called posterior nuclear group: a study with anterograde degeneration methods in the cat. *Brain Research*, **49**, 229–44.

Greeson, J.N. and Raphael, R.M. (2009) Amphipath-induced nanoscale changes in outer hair cell plasma membrane curvature. *Biophysical Journal*, **96**, 510–20.

Griffiths, T.D. (2000) Musical hallucinosis in acquired deafness. Phenomenology and brain substrate. *Brain*, **123**, 2065–76.

Griffiths, T.D. (2001) The neural processing of complex sounds. *Annals of the New York Academy of Science*, **930**, 133–42.

Gristwood, R.E. and Venables, W.N. (2003) Otosclerosis and chronic tinnitus. *Annals of Otology, Rhinology and Laryngology*, **12**, 398–403.

Gross, J., Schnitzler, A., Timmermann, L., and Ploner, M. (2007) Gamma oscillations in human primary somatosensory cortex reflect pain perception. *PLoS Biology*, **5**, e133.

Gu, J.W., Halpin, C.F., Nam, E.C., Levine, R.A., and Melcher, J.R. (2010) Tinnitus, diminished sound-level tolerance, and elevated auditory activity in humans with clinically normal hearing sensitivity. *Journal of Neurophysiology*, **104**, 3361–70.

Guillery, R.W. and Sherman, S.M. (2002) Thalamic relay functions and their role in corticocortical communication: generalizations from the visual system. *Neuron*, **33**, 163–75.

Guitton, M.J., Caston, J., Ruel, J., Johnson, R.M., Pujol, R., and Puel, J.L. (2003) Salicylate induces tinnitus through activation of cochlear NMDA receptors. *Journal of Neuroscience*, **23**, 3944–52.

Guitton, M.J., Pujol, R., and Puel, J.L. (2005) m-Chlorophenylpiperazine exacerbates perception of salicylate-induced tinnitus in rats. *European Journal of Neuroscie*nce, **22**, 2675–78.

Guitton, M.J., Wang, J., and Puel, J.L. (2004) New pharmacological strategies to restore hearing and treat tinnitus. *Acta Otolaryngologica*, **124**, 411–415.

Gutschalk, A., Hamalainen, M.S., and Melcher, J.R. (2010) BOLD responses in human auditory cortex are more closely related to transient MEG responses than sustained ones. *Journal of Neurophysiology*, **103**, 2015–26.

Hackett, T.A, Takahata, T., and Balaram, P. (2011) VGLUT1 and VGLUT2 mRNA expression in the primate auditory pathway. *Hearing Research*, **274**, 129–41.

Haenggeli, C.A., Pongstaporn, T., Doucet, J.R., and Ryugo, D.K. (2005) Projections from the spinal trigeminal nucleus to the cochlear nucleus in the rat. *Journal of Comparative Neurology*, **484**, 191–205.

Halford, J.B. and Anderson, S.D. (1991) Tinnitus severity measured by a subjective scale, audiometry and clinical judgement. *Journal of Laryngology and Otology*, **105**, 89–93.

Hallam, R.S., Jakes, S.C., and Hinchcliffe, R. (1998) Cognitive variables in tinnitus annoyance. *British Journal of Clinical Psychology*, **27**, 213–22.

Hallam, R.S., Rachman, S., and Hinchcliffe, R. (1984) Psychological aspects of tinnitus. In: S. Rachman (ed.), *Contributions to Medical Psychology 3*, pp. 1–52. Oxford: Pergamon.

Haller, S., Wetzel, S.G., Radue, E.W., and Bilecen D. (2006) Mapping continuous neuronal activation without an ON-OFF paradigm: initial results of BOLD ceiling fMRI. *European Journal of Neuroscience*, **24**, 2672–78.

Hallett, M. (2007) Transcranial magnetic stimulation: a primer. *Neuron*, **55**, 187–99.

Halliwell, B., Gutteridge, J.M., and Cross, C.E. (1992) Free radicals, antioxidants, and human disease: where are we now? *Journal of Laboratory and Clinical Medicine*, **119**, 589–620.

Hämäläinen, J.A., Leppänen, P.H., Guttorm, T.K., and Lyytinen, H. (2007) N1 and P2 components of auditory event-related potentials in children with and without reading disabilities. *Clinical Neurophysiology*, **118**, 2263–75.

Hämäläinen, H., Kujala, T., Kekoni, J., Hurskainen, H., Pirilä, J., Wikström, H., and Huotilainen, M. (2007) Effects of unilateral hippocampus-amygdala-partial temporal lobe resection on auditory EEG/MEG responses: a case study. *Scandinavian Journal of Psychology*, **48**, 367–73.

Hamblen-Thomas, C. (1937) Physical aspects of tinnitus: (Sections of Otology and Laryngology) *Proceedings of the Royal Society of Medicine*, **30**, 1512–18.

Hari, R. and Salmelin, R. (1997) Human cortical oscillations: a neuromagnetic view through the skull. *Trends in Neuroscience*, **20**, 44–49.

Harrison, R.V. (1981) Rate-versus-intensity functions and related AP responses in normal and pathological guinea pig and human cochleas. *Journal of the Acoustical Society of America*, **70**, 1036–44.

Harrison, R.V. and Prijs, V.F. (1984) Single cochlear fibre responses in guinea pigs with long-term endolymphatic hydrops. *Hearing Research*, **14**, 79–84.

Hazell, J.W.P. and Sheldrake, J.B. (1992) Hyperacusis and tinnitus. In: J.-M. Aran and R. Dauman (eds.), *Tinnitus 91*, pp. 245–48. Amsterdam: Kugler Publications.

He, J. and Hu, B. (2002) Differential Distribution of burst and single-spike responses in auditory thalamus. *Journal of Neurophysiology*, **88**, 2152–56.

Hébert, S. and Carrier, J. (2007) Sleep complaints in elderly tinnitus patients: a controlled study. *Ear & Hearing*, **28**, 649–55.

Hébert, S. and Lupien, S.J. (2007) The sound of stress: blunted cortisol reactivity to psychosocial stress in tinnitus sufferers. *Neuroscience Letters*, **411**(2), 138–42.

Hébert, S. and Lupien, S.J. (2009) Salivary cortisol levels, subjective stress, and tinnitus intensity in tinnitus sufferers during noise exposure in the laboratory. *International Journal of Hygiene and Environmental Health*, **212**, 37–44.

Hébert, S., Paiement, P., and Lupien, S.J. (2004) A physiological correlate for the intolerance to both internal and external sounds. *Hearing Research*, **190**, 1–9.

Heeger, D.J. and Ress, D. (2002) What does fMRI tell us about neuronal activity? *Nature Reviews Neuroscience*, **3**, 142–51.

Heffner, H.E. and Harrington, I.A. (2002) Tinnitus in hamsters following exposure to intense sound. *Hearing Research*, **170**, 83–95.

Heffner, H.E., Heffner, R.S., Contos, C., and Ott, T. (1994) Audiogram of the hooded Norway rat. *Hearing Research*, **73**, 244–47.

Heffner, R.S., Koay, G., and Heffner. H.E. (2001) Audiograms of five species of rodents: implications for the evolution of hearing and the perception of pitch. *Hearing Research*, **157**, 138–52.

Hegerl, U. and Juckel, G. (1993) Intensity dependence of auditory evoked potentials as an indicator of central serotonergic neurotransmission: a new hypothesis. *Biological Psychiatry*, **33**, 173–87.

Heinz, M.G. and Young, E.D. (2004) Response growth with sound level in auditory-nerve fibers after noise-induced hearing loss. *Journal of Neurophysiology*, **91**, 784–95.

Heinz, M.G., Issa, J.B., and Young, E.D. (2005) Auditory-nerve rate responses are inconsistent with common hypotheses for the neural correlates of loudness recruitment. *Journal of the Association for Research in Otolaryngology*, **6**, 91–105.

Helfert, R.H., Krenning, J., Wilson, T.S., and Hughes, L.F. (2003) Age-related synaptic changes in the anteroventral cochlear nucleus of Fischer-344 rats. *Hearing Research*, **183**, 18–28.

Helfert, R.H., Sommer, T.J., Meeks, J., Hofstetter, P., and Hughes, L.F. (1999) Age-related synaptic changes in the central nucleus of the inferior colliculus of Fischer-344 rats. *Journal of Comparative Neurology*, **406**, 285–98.

Heller, M.F. and Bergman, M. (1953) Tinnitus aurium in normally hearing persons. *Annals of Otology Rhinology and Laryngology*, **62**, 73–83.

Henry, J.A., Dennis, K.C., and Schechter, M.A. (2005) General review of tinnitus: prevalence, mechanisms, effects, and management. *Journal of Speech Language and Hearing Research*, **48**, 1204–35.

Henry, J.A. and Meikle, M.B. (2000) Psychoacoustic measures of tinnitus. *Journal of the American Academy of Audiology*, **11**, 138–55.

Henry, J.A., Schechter, M.A., Nagler, S.M., and Fausti, S.A. (2002) Comparison of tinnitus masking and tinnitus retraining therapy. *Journal of the American Academy of Audiology*, **13**, 559–81.

Henry, J.L. and Wilson, P. H. (2001) *The psychological management of chronic tinnitus*. Needham Heights, MA: Allyn & Bacon.

Hequembourg, S. and Liberman, M.C. (2001) Spiral ligament pathology: a major aspect of age-related cochlear degeneration in C57BL/6 mice. *Journal of the Association for Research in Otolaryngology*, **2**, 118–29.

Hesse, G., Andres, R., Schaaf, H., and Laubert, A. (2008) DPOAE und laterale Inhibition bei chronischem Tinnitus. *HNO*, **56**, 694–700.

Hiller, W. and Goebel, G. (1992) A psychometric study of complaints in chronic tinnitus. *Journal of Psychosomatic Research*, **36**, 337–48.

Hiller, W., Goebel, G., and Rief, W. (1994) Reliability of self-rated tinnitus distress and association with psychological symptom patterns. *British Journal of Clinical Psychology*, **33**, 231–39.

Hinchcliffe, R. (1961) Prevalence of the commoner ear, nose and throat conditions in the adult rural population of Great Britain. *British Journal of Preventive and Social Medicine*, **15**, 128–40.

Hodgetts, W.E. and Liu, R. (2006) Can hockey playoffs harm your hearing? *Canadian Medical Association Journal*, **175**(12), 1541–42.

Hoffman, H.J. and Reed, G.W. (2004) Epidemiology of tinnitus. In J.B. Snow Jr (ed.), *Tinnitus: Theory and Management*, pp. 16–41. Hamilton: BC Dekker.

Hoke, M., Feldmann, H., Pantev, C., Lütkenhöner, B., and Lehnertz, K. (1989) Objective evidence of tinnitus in auditory evoked magnetic fields. *Hearing Research*, **37**, 281–316.

Holgers, K.-M. (2003) Tinnitus in 7-year-old children. *European Journal of Pediatrics*, **162**, 276–78.

Holgers, K.M., Erlandsson, S.I., and Barrenäs, M.L. (2000) Predictive factors for the severity of tinnitus. *Audiology*, **39**, 284–91.

Holgers, K.-M. and Juul, J. (2006) The suffering of tinnitus in childhood and adolescence. *International Journal of Audiology*, **45**, 267–72.

Holgers, K.M., Zöger, S., and Svedlund, K. (2003) Tinnitus suffering: a marker for a vulnerability in the serotonergic system? *Audiological Medicine*, **2**, 138–43.

Holgers, K.M., Zöger, S., and Svedlund, K. (2005) Predictive factors for development of severe tinnitus suffering—further characterisation. *International Journal of Audiology*, **44**, 584–92.

Holmans, P., Weissman, M.M., Zubenko, G.S., Scheftner, W.A., Crowe, R.R., Depaulo Jr., J.R., Knowles, J.A., Zubenko, W.N., Murphy-Eberenz, K., Marta, D.H., *et al.* (2007) Genetics of recurrent early-onset major depression (GenRED): final genome scan report. *American Journal of Psychiatry*, **164**, 248–58.

Holt, A.G., Asako, M., Lomax, C.A., MacDonaldm J.W., Tongm L., Lomax, M.I., and Altschuler, R.A. (2005) Deafness-related plasticity in the inferior colliculus: gene expression profiling following removal of peripheral activity. *Journal of Neurochemistry*, **93**, 1069–86.

Holt, A.G., Bissig, D., Mirza, N., Rajah, G., and Berkowitz, B. (2010) Evidence of key tinnitus-related brain regions documented by a unique combination of manganese-enhanced MRI and acoustic startle reflex testing. *PLoS One*, **5(12)**, e14260.

Horner, K.C. (2003) The emotional ear in stress. *Neuroscience and Biobehavioral Reviews*, **27**, 437–46.

Hoshiyama, M., Gunji, A., and Kakigi, R. (2001) Hearing the sound of silence: a magnetoencephalographic study. *Neuroreport*, **12**, 1097–1102.

House, J.W. and Brackman, D.E. (1981) Tinnitus: surgical treatment. In: D. Evered and G. Lawrenson (eds.), *CIBA Symposium 85: Tinnitus*, pp. 204–12. London: Pitman.

House, W.F. (1982) Letter to the editor. *American Journal of Otolaryngology*, **4**, 188.

Hua, Y., Sinha, R., Martineau, M., Kahms, M., and Klingauf, J. (2010) A common origin of synaptic vesicles undergoing evoked and spontaneous fusion. *Nature Neuroscience*, **13**, 1451–53.

Huang, D., Chen, P., Chen, S., Nagura, M., Lim, D.J., and Lin, X. (2002) Expression patterns of aquaporins in the inner ear: evidence for concerted actions of multiple types of aquaporins to facilitate water transport in the cochlea. *Hearing Research*, **165**, 85–95.

Huang, Y.Z., Edwards, M.J., Rounis, E., Bhatia, K.P., and Rothwell, J.C. (2005a) Theta burst stimulation of the human motor cortex. *Neuron*, **45**, 201–206.

Huang, Z.W., Luo, Y., Wu, Z., Tao, Z., Jones, R.O., and Zhao, H,B. (2005b) Paradoxical enhancement of active cochlear mechanics in long-term administration of salicylate. *Journal of Neurophysiology*, **93**, 2053–61.

Hughes, L.F., Turner, J.G., Parrish, J.L., and Caspary, D.M. (2010) Processing of broadband stimuli across A1 layers in young and aged rats. *Hearing Research*, **264**, 79–85.

Hultcrantz, M., Simonoska, R., and Stenberg, A.E. (2006) Estrogen and hearing: a summary of recent investigations. *Acta Otolaryngologica*, **126**, 10–11.

Hunter, K.P. and Willott, J.F. (1993) Effects of bilateral lesions of auditory cortex in mice on the acoustic startle response. *Physiology and Behavior*, **54**, 1133–39.

Husain, F.T., Medina, R.E., Davis, C.W., Szymko-Bennett, Y., Simonyan, K., Pajor, N.M., and Horwitz, B. (2011) Neuroanatomical changes due to hearing loss and chronic tinnitus: a combined VBM and DTI study. *Brain Research*, **1369**, 74–88.

Huss, M. and Moore, B.C.J. (2005) Dead regions and noisiness of pure tones. *International Journal of Audiology*, **44**, 599–611.

Idrizbegovic, E., Bogdanovic, N., Willott, J.F., and Canlon, B. (2004) Age-related increases in calcium-binding protein immunoreactivity in the cochlear nucleus of hearing impaired C57BL/6J mice. *Neurobiology of Aging*, **25**, 1085–93.

Idrizbegovic, E., Canlon, B., Bross, L.S., Willott, J.F., and Bogdanovic, N. (2001) The total number of neurons and calcium-binding protein-positive neurons during aging in the cochlear nucleus of CBA/CaJ mice: a quantitative study. *Hearing Research*, **158**, 102–115.

Im, G.J., Jung, H.H., Chae, S.W., Cho, W.S., and Kim, S.J. (2007) Differential gene expression profiles in salicylate ototoxicity of the mouse. *Acta Otolaryngologica*, **127**, 459–69.

Ingham, N.J., Comis, S.D., and Withington, D.J. (1999) Hair cell loss in the aged guinea pig cochlea. *Acta Otolaryngologica*, **119**, 42–47.

Irvine, D.R., Rajan, R., and McDermott, H.J. (2000) Injury-induced reorganization in adult auditory cortex and its perceptual consequences. *Hearing Research*, **147**, 188–99.

Irvine, D.R., Rajan, R., and Smith, S. (2003) Effects of restricted cochlear lesions in adult cats on the frequency organization of the inferior colliculus. *Journal of Comparative Neurology*, **467**, 354–74.

Ison, J.R., Allen, P.D., and O'Neill, W.E. (2007) Age-related hearing loss in C57BL/6J mice has both frequency-specific and non-frequency-specific components that produce a hyperacusis-like exaggeration of the acoustic startle reflex. *Journal of the Association for Research in Otolaryngology*, **8**, 539–50.

Jacobson, G.P., Ahmad, B.K., Moran, J., Newman, C.W., Tepley, N., and Wharton, J. (1991) Auditory evoked cortical magnetic field (M100-M200) measurements in tinnitus and normal groups. *Hearing Research*, **56**, 44–52.

Jain, R. and Shore, S. (2006) External inferior colliculus integrates trigeminal and acoustic information: unit responses to trigeminal nucleus and acoustic stimulation in the guinea pig. *Neuroscience Letters*, **395**, 71–75.

Jäncke, L. and Shah, N.J. (2004) Hearing syllables by seeing visual stimuli. *European Journal of Neuroscience*, **19**, 2603–2608.

Jansen, E.J., Helleman, H.W., Dreschler, W.A., and de Laat JA. (2009) Noise induced hearing loss and other hearing complaints among musicians of symphony orchestras. *International Archives of Occupational and Environmental Health*, **82**, 153–64.

Janssen, T., Boege, P., Oestreicher, E., and Arnold, W. (2000) Tinnitus and 2f1-f2 distortion product otoacoustic emissions following salicylate overdose. *Journal of the Acoustical Society of America*, **107**, 1790–92.

Jaskelioff, M., Muller, F.L., Paik, J.H., Thomas, E., Jiang, S., Adams, A.C., Sahin, E., Kost-Alimova, M., Protopopov, A., Cadiñanos, J., *et al.* (2010) Telomerase reactivation reverses tissue degeneration in aged telomerase-deficient mice. *Nature*, **469**, 102–106.

Jastreboff, P.J. (1990) Phantom auditory perception (tinnitus): mechanisms of generation and perception. *Neuroscience Research*, **8**, 221–54.

Jastreboff, P.J. and Brennan, J.F. (1988) Specific effects of nimodipine on the auditory system. *Annals of the New York Academy of Science*, **522**, 716–18.

Jastreboff, P.J. and Sasaki, C.T. (1986) Salicylate-induced changes in spontaneous activity of single units in the inferior colliculus of the guinea pig. *Journal of the Acoustical Society of America*, **80**, 1384–91.

Jastreboff, P.J., Brennan, J.F., Coleman, J.K., and Sasaki, C.T. (1988a) Phantom auditory sensation in rats: an animal model for tinnitus. *Behavioral Neuroscience*, **102**, 811–22.

Jastreboff, P.J., Brennan, J.F., and Sasaki, C.T. (1988b) An animal model for tinnitus. *Laryngoscope*, **98**, 280–86.

Jastreboff, P.J., Brennan, J.F., and Sasaki, C.T. (1991) Quinine-induced tinnitus in rats. *Archive of Otolaryngology Head and Neck Surgery*, **117**, 1162–66.

Jastreboff, P.J. and Brennan, J.F. (1994) Evaluating the loudness of phantom auditory perception (tinnitus) in rats. *Audiology*, 33, 202–17.

Jastreboff, P.J., Gray W.C., and Gold S.L. (1996) Neurophysiological approach to tinnitus patients. *American Journal of Otolology*, **17**, 236–40.

Jastreboff, P.J., Hazell, J.W., and Graham, R.L. (1994) Neurophysiological model of tinnitus: dependence of the minimal masking level on treatment outcome. *Hearing Research*, **80**, 216–32.

Jastreboff, P.J. and Hazell, J.W.P. (1993) A neurophysiological approach to tinnitus: clinical implications. *British Journal of Audiology*, **27**, 7–17.

Javel, E., McGee, J.A., Horst, J.W., and Farley, G.R. (1988) Temporal mechanismsin auditory stimulus coding. In: G.M. Edelman, W.E. Gall, and W.M. Cowan (eds.), *Auditory function. Neurobiological basis of hearing*, pp. 515–58. New York: J. Wiley.

Jin, Y.M., Godfrey, D.A., Wang, J., and Kaltenbach, J.A. (2006) Effects of intense tone exposure on choline acetyltransferase activity in the hamster cochlear nucleus. *Hearing Research*, **216–217**, 168–75.

Johnson, D.H. and Kiang, N.Y. (1976) Analysis of discharges recorded simultaneously from pairs of auditory nerve fibers. *Biophysical Journal*, **16**, 719–34.

Johnson, K.R., Zheng, Q.Y., and Erway, L.C. (2000) A major gene affecting age-related hearing loss is common to at least 10 inbred strains of mice. *Genomics*, **70**, 171–80.

Johnson, R.M., Brummett, R., and Schleuning, A. (1993) Use of alprazolam for relief of tinnitus. A double-blind study. *Archive of Otolaryngology Head and Neck Surgery*, **119**, 842–45.

Johnsrude, I.S., Giraud, A.-L., and Frackowiak, R.S.J. (2002) Functional imaging of the auditory system: the use of positron emission tomography. *Audiology & Neurootology*, 7, 251–76.

Joris, P.X., Carney, L.H., Smith, P.H., and Yin, T.C. (1994) Enhancement of neural synchronization in the anteroventral cochlear nucleus. I. Responses to tones at the characteristic frequency. *Journal of Neurophysiology*, **71**, 1022–36.

Juliano, L.M. and Griffiths, R.R. (2004) A critical review of caffeine withdrawal: empirical validation of symptoms and signs, incidence, severity, and associated features. *Psychopharmacology (Berlin)*, **176**, 1–29.

Jung, T.T., Miller, S.K., Rozehnal S., Woo, H.Y., Park, Y.M., and Baer, W. (1992) Effect of round window membrane application of salicylate and indomethacin on hearing and levels of arachidonic acid metabolites in perilymph. *Acta Otolaryngologica*, **Suppl 493**, 81–87.

Jung, T.T., Rhee, C.K., Lee, C.S.L., Park, Y.S, and Choi, D.C. (1993) Ototoxicity of salicylate, nonsteroidal anti-inflammatory drugs and quinine. *Otolaryngology Clinics of North America*, **26**, 791–810.

Kahlbrock, N. and Weisz, N. (2008) Transient reduction of tinnitus intensity is marked by concomitant reductions of delta band power. *BMC Biology*, **6**, 4.

Kaltenbach, J.A. (2000) Neurophysiologic mechanisms of tinnitus. *Journal of the American Academy of Audiology*, **11**, 125–37.

Kaltenbach, J.A. and Afman, C.E. (2000) Hyperactivity in the dorsal cochlear nucleus after intense sound exposure and its resemblance to tone-evoked activity: a physiological model for tinnitus. *Hearing Research*, **140**, 165–72.

Kaltenbach, J.A., Godfrey, D.A., Neumann, J.B/, McCaslin, D.L., Afman, C.E., and Zhang, J. (1998) Changes in spontaneous neural activity in the dorsal cochlear nucleus following exposure to intense sound: relation to threshold shift. *Hearing Research*, **124**, 78–84.

Kaltenbach, J.A., Zacharek, M.A., Zhang, J., and Frederick, S. (2004) Activity in the dorsal cochlear nucleus of hamsters previously tested for tinnitus following intense tone exposure. *Neuroscience Letters*, **355**, 121–25.

Kaltenbach, J.A., Zhang, J., and Afman, C.E. (2000) Plasticity of spontaneous neural activity in the dorsal cochlear nucleus after intense sound exposure. *Hearing Research*, **147**, 282–92.

Kaltenbach, J.A., Zhang, J., and Finlayson, P. (2005) Tinnitus as a plastic phenomenon and its possible neural underpinnings in the dorsal cochlear nucleus. *Hearing Research*, **206**, 200–26.

Kameda, K., Shono, T., Hashiguchi, K., Yoshida, F., and Sasaki, T. (2010) Effect of tumor removal on tinnitus in patients with vestibular schwannoma. *Journal of Neurosurgery*, **112**, 152–57.

Kamke, M.R., Brown, M., and Irvine, D.R. (2003) Plasticity in the tonotopic organization of the medial geniculate body in adult cats following restricted unilateral cochlear lesions. *Journal of Comparative Neurology*, **459**, 355–67.

Kanold, P.O. and Young, E.D. (2001) Proprioceptive information from the pinna provides somatosensory input to cat dorsal cochlear nucleus. *Journal of Neuroscience*, **21**, 7848–58.

Karlsson, K., Lundquist, P.G., and Olaussen, T. (1983) The hearing of symphony orchestra musicians. *Scandinavian Audiology*, **12**, 257–64.

Kasai, K., Asada, T., Yumoto, M., Takeya, J., and Matsuda, H. (1999) Evidence for functional abnormality in the right auditory cortex during musical hallucinations. *Lancet*, **354**, 1703–1704.

Katon, W., Sullivan, M., Russo, J., Dobie, R., and Sakai, C. (1993) Depressive symptoms and measures of disability: a prospective study. *Journal of Affective Disorders*, **27**, 245–54.

Katzenell, U. and Segal, S. (2001) Hyperacusis: review and clinical guidelines. *Otology & Neurotology*, 22, 321–26.

Kay, F., Searchfield, G.D., Coad, G., and Kobayashi, K. (2008) Towards improving the assessment of tinnitus pitch. *The New Zealand Audiological Society Bulletin*, **18**, 27–41.

Kay, N.J. (1981) Oral chemotherapy in tinnitus. *British Journal of Audiology*, **15**, 123–24.

Kazee, A.M., Han, L.Y., Spongr, V.P., Walton, J.P., Salvi, R.J., and Flood, D.G. (1995) Synaptic loss in the central nucleus of the inferior colliculus correlates with sensorineural hearing loss in the C57BL/6 mouse model of presbycusis. *Hearing Research*, **89**, 109–20.

Keithley, E.M. and Feldman, M.L. (1979) Spiral ganglion cell counts in an age-graded series of rat cochleas. *Journal of Comparative Neurology*, **188**, 429–42.

Keithley, E.M., Ryan, A.F., and Feldman, M.L. (1992) Cochlear degeneration in aged rats of four strains. *Hearing Research*, **59**, 171–78.

Keithley, E.M., Ryan, A.F., and Woolf, N.K. (1989) Spiral ganglion cell density in young and old gerbils. *Hearing Research*, **38**, 125–33.

Kemp, D.T. (1978) Stimulated acoustic emissions from within the human auditory system. *Journal of the Acoustical Society of America*, **64**, 1386–91.

Kemp, E.H. (1935) A critical review of experiments on the problem of stimulation deafness. *Psychological Bulletin*, **32**, 325–432.

Kemp, S. and George, R.N. (1992) Diaries of tinnitus sufferers. *British Journal of Audiology*, **26**, 381–86.

Kennedy, C., Sakurada, O., Shinohara, M., J. Jehle, J., and Sokoloff, L. (1978) Local cerebral glucose utilization in the normal conscious macaque monkey. *Annals of Neurology*, **4**, 293–301.

Kennedy, V., Wilson, C., and Stephens, D. (2004) Quality of life and tinnitus. *Audiological Medicine*, **2**, 29–40.

Kennedy, V., Chéry-croze, S., Stephens, D., Kramer, S., Thai-van, H.g and Collet, L. (2005) Development of the International Tinnitus Inventory (ITI): a patient-directed problem questionnaire. *Audiological Medicine*, **3**, 228–37.

Khalfa, S., Bruneau, N., Rogé, B., Georgieff, N., Veuillet, E., Adrien, J.L., Barthélémy, C., and Collet, L. (2004) Increased perception of loudness in autism. *Hearing Research*, **198**, 87–92.

Khedr, E.M., Rothwell, J.C., and El-Atar, A. (2009) One-year follow up of patients with chronic tinnitus treated with left temporoparietal rTMS. *European Journal of Neurology*, **16**, 404–408.

Kiang, N.Y. (1975) Stimulus representation in the discharge patterns of auditory neurons. In: D.B. Towers (ed.) *The Nervous System, Vol 3. Human communication and its disorders*, pp. 81–96. New York: Raven Press.

Kiang, N.Y., Liberman, M.C., and Levine, R.A. (1976) Auditory-nerve activity in cats exposed to ototoxic drugs and high-intensity sounds. *Annals of Otology Rhinology and Laryngology*, **85**, 752–68.

Kiang, N.Y., Moxon, E.C., and Levine, R.A. (1970) Auditory-nerve activity in cats with normal and abnormal cochleas. *Ciba Foundation Symposium*, 241–73.

Kiang, N. Y. S., Watanabe, T., Thomas, E.C., and Clark, L.F. (1965) *Discharge Patterns of Single Fibers in the Cat's Auditory Nerve*. Research Monograph No. 35. Cambridge, MA: MIT Press.

Kilgard, M. P., and Merzenich, M. M. (1998) Cortical map reorganization enabled by nucleus basalis activity. *Science*, **279**, 1714–1718.

Kim, J.J., Gross, J., Morest, D.K., and Potashner, S.J. (2004a) Quantitative study of degeneration and new growth of axons and synaptic endings in the chinchilla cochlear nucleus after acoustic overstimulation. *Journal of Neuroscience Research*, **77**, 829–42.

Kim, J.J., Gross, J., Potashner, S.J., and Morest, D.K. (2004b) Fine structure of degeneration in the cochlear nucleus of thechinchilla after acoustic overstimulation. *Journal of Neuroscience Research*, 77, 798–816.

Kim, J.J., Gross, J., Potashner, S.J., and Morest, D.K. (2004c) Fine structure of long-term changes in the cochlear nucleus after acoustic overstimulation: chronic degeneration and new growth of synaptic endings. *Journal of Neuroscience Research*, **77**, 817–28.

Kimpo, R.R., Theunissen, F.E., and Doupe, A.J. (2003) Propagation of correlated activity through multiple stages of a neural circuit. *Journal of Neuroscience*, **23**, 5750–61.

Kimura, M. and Eggermont, J.J. (1999) Effects of acute pure tone induced hearing loss on response properties in three auditory cortical fields in cat. *Hearing Research*, **135**, 146–62.

Kirkegaard, M., Murai, N., Risling, M., Suneson, A., Järlebark, L., Ulfendahl, M. (2006) Differential gene expression in the rat cochlea after exposure to impulse noise. *Neuroscience*, **142**, 425–35.

Kistler, W.M. and Gerstner, W. (2002) Stable propagation of activity pulses in populations of spiking neurons. *Neural Computation*, **14**, 987–97.

Kizawa, K., Kitahara, T., Horii, A., Maekawa, C., Kuramasu, T., Kawashima, T., Nishiike, S., Doi, K., and Inohara, H. (2010) Behavioral assessment and identification of a molecular marker in a salicylate-induced tinnitus in rats. *Neuroscience*, **165**, 1323–32.

Kleinjung, T., Eichhammer, P., Landgrebe, M., Sand, P., Hajak, G., Steffens, T., Strutz, J., and Langguth, B. (2008) Combined temporal and prefrontal transcranial magnetic stimulation for tinnitus treatment: a pilot study. *Otolaryngology Head and Neck Surgery*, **138**, 497–501.

Kleinjung, T., Fischer, B., Langguth, B., Eichhammer, P., Hajak, G., and Sand, P. (2006) BDNF gene variants in subjects with chronic tinnitus. *Otolaryngology Head and Neck Surgery*, **135**(S1), 111.

Kleinjung, T., Frank, E., Vielsmeier, V., Landgrebe, M., Langguth, B., and Sand, P. (2009) BDNF and GDNF variants predict tinnitus Severity. *Otolaryngology Head and Neck Surgery*, SP317.

Klopf, A.H. (1982) *The hedonistic neuron*. Washington, DC: Hemisphere publishing Corporation.

Knipper, M., Zimmermann, U., and Müller, M. (2010) Molecular aspects of tinnitus. *Hearing Research*, **266**, 60–69.

Knobel, K.A., and Sanchez, T.G. (2008) Influence of silence and attention on tinnitus perception. *Otolaryngology Head and Neck Surgery*, **138**, 18–22.

Koch, M. (1999) The neurobiology of startle. *Progress in Neurobiology*, **59**, 107–28.

Koehler, S.D., Pradhan, S., Manis, P.B., and Shore, S.E. (2011) Somatosensory inputs modify auditory spike timing in dorsal cochlear nucleus principal cells. *European Journal of Neuroscience*, **33**(3), 409–20.

Koerber, K.C., Pfeiffer, R.R., Warr, W.B., and Kiang, N.Y. (1966) Spontaneous spike discharges from single units in the cochlear nucleus after destruction of the cochlea. *Experimental Neurology*, **16**, 119–30.

Kotak, V.C., Fujisawa, S., Lee, F.A., Karthikeyan, O., Aoki, C., and Sanes, D.H. (2005) Hearing loss raises excitability in the auditory cortex. *Journal of Neuroscience*, **25**, 3908–18.

Krahe, R. and Gabbiani, F. (2004) Burst firing in sensory systems. *Nature Reviews Neuroscience*, **5**, 13–23.

Kral, A. and Eggermont, J.J. (2007) What's to lose and what's to learn: development under auditory deprivation, cochlear implants and limits of cortical plasticity. *Brain Research Reviews*, **56**, 259–69.

Kraus, K.S., Ding, D., Zhou, Y., and Salvi, R.J. (2009) Central auditory plasticity after carboplatin-induced unilateral inner ear damage in the chinchilla: up-regulation of GAP-43 in the ventral cochlear nucleus. *Hearing Research*, **255**, 33–43.

Kroener-Herwig, B., Biesinger, E., Gerhards, F., Goebel, G., Verena Greimel, K., and Hiller, W. (2000) Retraining therapy for chronic tinnitus. A critical analysis of its status. *Scandinavian Audiology*, **29**, 67–78.

Kujawa, S.G., Fallon, M., and Bobbin, R.P. (1992) Intracochlear salicylate reduces low-intensity acoustic and cochlear microphonic distortion products. *Hearing Research*, **64**, 73–80.

Kujawa, S.G. and Liberman, M.C. (2006) Acceleration of age-related hearing loss by early noise exposure: evidence of a misspent youth. *Journal of Neuroscience*, **26**, 2115–23.

Kujawa, S.G. and Liberman, M.C. (2009) Adding insult to injury: cochlear nerve degeneration after "temporary" noise-induced hearing loss. *Journal of Neuroscien*ce, **29**, 14077–85.

Kuk, F.K., Tyler, R.S., Russell, D., and Jordan, H. (1990) The psychometric properties of a tinnitus handicap questionnaire. *Ear & Hearing*, **11**, 434–45.

Kumagai, M. (1992) Effect of intravenous injection of aspirin on the cochlea. *Hokkaido Igaku Zasshi*, **67**, 216–33.

Kvestad, E., Czajkowski, N., Engdahl, B., Hoffman, H.J., and Tambs, K. (2010) Low heritability of tinnitus results from the Second Nord-Trøndelag Health Study. *Archives of Otolaryngology, Head Neck Surgery*, **136**, 178–82.

Landgrebe, M., Langguth, B., Rosengarth, K., Braun, S., Amelie Koch, A., Kleinjung, T., May, A., de Ridder, D., and Hajak, G. (2009) Structural brain changes in tinnitus: grey matter decrease in auditory and non-auditory brain areas. *NeuroImage*, **46**, 213–18.

Lang, H., Schulte, B.A., Zhou, D., Smythe, N., Spicer, S.S., and Schmiedt, R.A. (2006) Nuclear factor kappaB deficiency is associated with auditory nerve degeneration and increased noise-induced hearing loss. *Journal of Neuroscience*, **26**, 3541–50.

Langers, D.R., van Dijk, P., Schoenmaker, E.S., and Backes, W.H. (2007) fMRI activation in relation to sound intensity and loudness. *Neuroimage*, **35**, 709–18.

Langner, G. (1992) Periodicity coding in the auditory system. *Hearing Research*, **60**, 115–42.

Langner, G. (1997) Neural processing and representation of periodicity pitch. *Acta Otolaryngologica*, **Suppl. 532**, 68–76.

Langner, G. and Schreiner, C.E. (1988) Periodicity coding in the inferior colliculus of the cat. I. Neuronal mechanisms. *Journal of Neurophysiology*, **60**, 1799–822.

Lanting, C.P., De Kleine, E., Bartels, H., and Van Dijk, P. (2008) Functional imaging of unilateral tinnitus using fMRI. *Acta Otolaryngologica*, **128**, 415–21.

Lanting, C.P., De Kleine, E., and Van Dijk, P. (2009) Neural activity underlying tinnitus generation: results from PET and fMRI. *Hearing Research*, **255**, 1–13.

Lanting, C.P., De Kleine, E., Eppinga, R.N., and van Dijk, P. (2010) Neural correlates of human somatosensory integration in tinnitus. *Hearing Research*, **267**, 78–88.

Latremoliere, A. and Woolf, C.J. (2009) Central sensitization: a generator of pain hypersensitivity by central neural plasticity. *The Journal of Pain*, **10**(9), 895–926.

LeDoux, J.E. (1991) Emotion and the limbic system concept. *Concepts in Neuroscience*, **2**, 169–99.

Leaver, A.M., Renier, L., Chevillet, M.A., Morgan, S., Kim, H.J., and Rauschecker, J.P. (2011) Dysregulation of limbic and auditory networks in tinnitus. *Neuron*, **69**, 33–43.

Lee, C.C. and Sherman, S.M. (2011) On the classification of pathways in the auditory midbrain, thalamus, and cortex. *Hearing Research*, **276**, 79–87.

Lee, C.C., Schreiner, C.E., Imaizumi, K., and Winer, J.A. (2004) Tonotopic and heterotopic projection systems in physiologically defined auditory cortex. *Neuroscience*, **128**, 871–87.

Lee, S.L., Abraham, M., Cacace, A.T., and Silver, S.M. (2008) Repetitive transcranial magnetic stimulation in veterans with debilitating tinnitus: a pilot study. *Otolaryngology Head and Neck Surgery*, **138**, 398–99.

Lee, Y.J., Bae, S.J., Lee, S.H., Lee. J.J., Lee, K.Y., Kim, M.N., Kim, Y.S., Baik, S.K., Woo, S., and Chang, Y. (2007) Evaluation of white matter structures in patients with tinnitus using diffusion tensor imaging. *Journal of Clinical Neurosci*ence, **14**, 515–519.

Lehtelä, L., Salmelin, R., and Hari, R. (1997) Evidence for reactive magnetic 10-Hz rhythm in the human auditory cortex. *Neuroscience Lett*ers, **222**, 111–114.

Lenarz, T. (1996) Treatment of tinnitus with lidocain and tocainide. *Scandinavian Audiology*, **Suppl 26**, 49–51.

Lenarz, T., Schreiner, C.E., Snyder, R.L., and Ernst, A. (1993) Neural mechanisms of tinnitus. *European Archive of Otorhinolaryngology*, **249**, 441–46.

Lenarz, T., Schreiner, C., Snyder, R.L., and Ernst, A. (1995) Neural mechanisms of tinnitus: the pathological ensemble of spontaneous activity of the auditory system. In: J.A. Vernon and A.R. Møller (eds.), *Mechanisms of Tinnitus*, pp. 101–113. Boston, MA: Allyn and Bacon.

Lennox, B.R., Park, S.B., Jones, P.B., and Morris, P.G. (1999) Spatial and temporal mapping of neural activity associated with auditory hallucinations. *Lancet*, **353**, 644.

Lenz, F.A., Garonzik, I.M., Zirh, T.A., and Dougherty, P.M. (1998) Neuronal activity in the region of the thalamic principal sensory nucleus (ventralis caudalis) in patients with pain following amputations. *Neuroscience*, **86**, 1065–81.

Leske, M.C. (1981) Prevalence estimates of communicative disorders in the US: language learning and vestibular disorders. *The American Speech-Language and Hearing Association*, **23**, 229–37.

Lessmann, V. (1998) Neurotrophin-dependent modulation of glutamatergic synaptic transmission in the mammalian CNS. *General Pharmacology*, **31**, 667–74.

Levine, R.A. (1999) Somatic (craniocervical) tinnitus and the dorsal cochlear nucleus hypothesis. *American Journal of Otolaryngol*ogy, **20**, 351–62.

Levine, R.A., Abel, M., and Cheng, H. (2003) CNS somatosensory-auditory interactions elicit or modulate tinnitus. *Experimental Brain Research*, **153**, 643–48.

Levo, H., Blomstedt, G., and Pyykkö, I. (2000) Tinnitus and vestibular schwannoma surgery. *Acta Otolaryngologica*, **Suppl 543**, 28–29.

Lew, H.L., Jerger, J.F., Guillory, S.B., and Henry, J.A. (2007) Auditory dysfunction in traumatic brain injury. *Journal of Rehabilitation Research & Development*, **44**, 921–28.

Li, H.S. and Borg, E. (1991) Age-related loss auditory sensitivity in two mouse genotypes. *Acta Otolaryngologica*, **111**, 827–34.

Li, H. and Mizuno, N. (1997) Single neurons in the spinal trigeminal and dorsal column nuclei project to both the cochlear nucleus and the inferior colliculus by way of axon collaterals: a fluorescent retrograde double-labeling study in the rat. *Neuroscience Research*, **29**(2), 135–42.

Li, H. and Steyger, P.S. (2009) Synergistic ototoxicity due to noise exposure and aminoglycoside antibiotics. *Noise & Health*, **11**, 26–32.

Li, J. and Verkman, A.S. (2001) Impaired hearing in mice lacking aquaporin-4 water channels. *Journal of Biological Chemistry*, **276**, 31233–37.

Li, L., Du, Y., Li, N., Wu, X., and Wu, Y. (2009) Top–down modulation of prepulse inhibition of the startle reflex in humans and rats. *Neuroscience and Biobehavioral Reviews*, **33**, 1157–67.

Liberman, M.C. (1978) Auditory-nerve response from cats raised in a low-noise chamber. *Journal of the Acoustical Society of America*, **63**, 442–55.

Liberman, M.C. (1987) Chronic ultrastructural changes in acoustic trauma: serial-section reconstruction of stereocilia and cuticular plates. *Hearing Research*, **26**, 65–88.

Liberman, M.C. and Beil, D.G. (1979) Hair cell condition and auditory nerve response in normal and noise-damaged cochleas. *Acta Otolaryngologica*, **88**, 161–76.

Liberman, M.C. and Dodds, L.W. (1984) Single-neuron labeling and chronic cochlear pathology. II. Stereocilia damage and alterations of spontaneous discharge rates. *Hearing Research*, **16**, 43–53.

Liberman, M.C. and Dodds, L.W. (1987) Acute ultrastructural changes in acoustic trauma: serial-section reconstruction of stereocilia and cuticular plates. *Hearing Research*, **26**, 45–64.

Liberman, M.C. and Kiang, N.Y. (1978) Acoustic trauma in cats. Cochlear pathology and auditory-nerve activity. *Acta Otolaryngol*, **Suppl 358**, 1–63.

Liberman, M.C. and Kiang, N.Y. (1984) Single-neuron labeling and chronic cochlear pathology. IV. Stereocilia damage and alterations in rate- and phase-level functions. *Hearing Research*, **16**, 75–90.

Lim, D.J. (1986) Functional structure of the organ of Corti: a review. *Hearing Research*, **22**, 117–46.

Lin, X., Chen, S., and Tee, D. (1998) Effects of quinine on the excitability and voltage-dependent currents of isolated spiral ganglion neurons in culture. *Journal of Neurophysiology*, **79**, 2503–12.

Ling, L.L., Hughes, L.F., and Caspary, D.M. (2005) Age-related loss of the GABA synthetic enzyme glutamic acid decarboxylase in rat primary auditory cortex. *Neuroscience*, **132**, 1103–13.

Liu, J., Li, X., Wang, L., Dong, Y., Han, H., and Liu, G. (2003) Effects of salicylate on serotoninergic activities in rat inferior colliculus and auditory cortex. *Hearing Research*, **175**, 45–53.

Liu, X.Z. and Yan, D. (2007) Ageing and hearing loss. *Journal of Pathology*, **211**, 188–97.

Liu, Y. and Li, X. (2004a) Effects of salicylate on transient outward and delayed rectifier potassium channels in rat inferior colliculus neurons. *Neuroscience Letters*, **369**, 115–20.

Liu, Y. and Li, X. (2004b) Effects of salicylate on voltage-gated sodium channels in rat inferior colliculus neurons. *Hearing Research*, **193**, 68–74.

Liu, Y., Li, X., Ma, C., Liu, J., and Lu, H. (2005) Salicylate blocks L-type calcium channels in rat inferior colliculus neurons. *Hearing Research*, **205**, 271–76.

Liyanage, S.H., Singh, A., Savundra, P., and Kalan, A. (2006) Pulsatile tinnitus. *Journal of Laryngology and Otology*, **120**(2), 93–97.

Llinás, R. R. (1988) The intrinsic electrophysiological properties of mammalian neurons: insights into central nervous system function. *Science*, **242**, 1654–64.

Llinás, R. and Jahnsen, H. (1982) Electrophysiology of mammalian thalamic neurones in vitro. *Nature*, **297**, 406–408.

Llinás, R.R., Ribary, U., Jeanmonod, D., Kronberg, E., and Mitra, P.P. (1999) Thalamocortical dysrhythmia: a neurological and neuropsychiatric syndrome characterized by magnetoencephalography. *Proceedings of the National Academy of Science of the United States of America*, **96**, 15222–27.

Llinás, R.R. and Steriade, M. (2006) Bursting of thalamic neurons and states of vigilance. *Journal of Neurophysiology*, **95**, 3297–3308.

Llinás, R., Urbano, F.J., Leznik, E., Ramírez, R.R., and van Marle, H.J. (2005) Rhythmic and dysrhythmic thalamocortical dynamics: GABA systems and the edge effect. *Trends in Neuroscience*, **28**, 325–33.

Lobarinas, E. and Falk, J.L. (1998) Schedule-induced polydipsic consumption of hypertonic NaCl solutions: effects of chlordiazepoxide. *Physiology and Behaviour*, **63**, 419–23.

Lobarinas, E., Sun, W., Cushing, R., Salvi, R. (2004) A novel behavioral paradigm for assessing tinnitus using schedule-induced polydipsia avoidance conditioning (SIP-AC) *Hearing Research*, **190**, 109–114.

Lockwood, A.H., Salvi, R.J., Coad, M.L., Towsley, M.L., Wack, D.S., and Murphy, B.W. (1998) The functional neuroanatomy of tinnitus. Evidence for limbic system links and neural plasticity. *Neurology*, **50**, 114–20.

Lockwood, A. H., Wack, D. S., Burkard, R. F., Coad, M. L., Reyes, S. A., Arnold, S. A., and Salvi, R. J. (2001) The functional anatomy of gaze-evoked tinnitus and sustained lateral gaze. *Neurology*, **56**, 472–80.

Loeb, M. and Smith, R.P. (1967) Relation of induced tinnitus to physical characteristics of the inducing stimuli. *Journal of the Acoustical Society of America*, **42**, 453–55.

Logothetis, N.K. (2002) The neural basis of the blood-oxygen-level-dependent functional magnetic resonance imaging signal. *Philosophical Transactions of the Royal Society of London B Biological Sciences*, **357**, 1003–37.

Logothetis, N.K., Pauls, J., Augath, M., Trinath, T., and Oeltermann, A. (2001) Neurophysiological investigation of the basis of the fMRI signal. *Nature*, **412**, 150–57.

Long, G.R. and Tubis, A. (1988) Modification of spontaneous and evoked otoacoustic emissions and associated psychoacoustic microstructure by aspirin consumption. *Journal of the Acoustical Society of America*, **84**, 1343–53.

Lonsbury-Martin, B., and Martin, G.K. (2004) Otoacoustic emissions and tinnitus. In: J.B. Snow (ed.), *Tinnitus: theory and management*, pp. 69–78. Hamilton: BC Decker.

Lopes da Silva, F.H., Kamphuis, W., Titulaer, M., Vreugdenhil, M., and Wadman, W.J. (1995) An experimental model of progressive epilepsy: the development of kindling of the hippocampus of the rat. *Italian Journal of Neurological Sciences*, **16**, 45–57.

Lorenz, I., Müller, N., Schlee, W., Hartmann, T., and Weisz, N. (2009) Loss of alpha power is related to increased gamma synchronization-A marker of reduced inhibition in tinnitus? *Neuroscience Letters*, **453**(3), 225–28.

Lowry, L.D., Eisenman, L.M., and Saunders, J.C. (2004) An absence of tinnitus. *Otology & Neurotology*, **25**, 474–78.

Lu, J., Lobarinas, E., Deng, A., Goodey, R., Stolzberg, D., Salvi, R.J., and Sun, W. (2011) Gabaergic neural activity involved in salicylate-induced auditory cortex gain enhancement. *Neuroscience*, **189**, 187–98.

Lu, Y.-G, Tang, Z.-Q., Ye, Z.-Y., Wang, H.-T., Huang, Y.-N., K.-Q Zhou, K.-Q., Zhang, M., Xu, T.-L., and Chen, L. (2009) Salicylate, an aspirin metabolite, specifically inhibits the current mediated by glycine receptors containing a1-subunits. *British Journal of Pharmacology*, **157**, 1514–22.

Luczak, A., Barthó, P., and Harris, K.D. (2009) Spontaneous events outline the realm of possible sensory responses in neocortical populations. *Neuron*, **62**, 413–25.

Luczak, A., Barthó, P., Marguet, S.L., Buzsáki, G., and Harris, K.D. (2007) Sequential structure of neocortical spontaneous activity in vivo. *Proceedings of the National Academy of Science of the United States of America*, **104**, 347–52.

Luo, F., Wang, Q., Kashani, A., and Yan, J. (2008) Corticofugal modulation of initial sound processing in the brain. *Journal of Neuroscience*, **28**, 11615–21.

Lurie, M.H., Davis, H., and Hawkins, J.E. jr. (1944) Acoustic trauma of the organ of Corti in the guinea pig. *Laryngoscope*, **54**, 375–86.

Lustig, L.R. (2006) Nicotinic acetylcholine receptor structure and function in the efferent auditory system. *Anatomical Record A*, **288**, 424–34.

Lütkenhöner, B. and Steinsträter, O. (1998) High-precision neuromagnetic study of the functional organization of the human auditory cortex. *Audiology and Neurootology*, **3**, 191–213.

Lütkenhöner, B., Krumbholz, K., and Seither-Preisler, A. (2003a) Studies of tonotopy based on wave N100 of the auditory evoked field are problematic. *Neuroimage*, **19**, 935–49.

Lütkenhöner, B., Krumbholz, K., Lammertmann, C., Seither-Preisler, A., Steinstrater, O., and Patterson, R.D. (2003b) Localization of primary auditory cortex in humans by magnetoencephalography. *Neuroimage*, **18**, 58–66.

Ma, W.L., Hidaka, H., and May, B.J. (2006) Spontaneous activity in the inferior colliculus of CBA/J mice after manipulations that induce tinnitus. *Hearing Research*, **212**, 9–21.

Ma, W.L. and Young, E.D. (2006) Dorsal cochlear nucleus response properties following acoustic trauma: response maps and spontaneous activity. *Hearing Research*, **216–217**, 176–88.

Ma, X., Suga, N. (2001) Plasticity of bat's central auditory system evoked by focal electric stimulation of auditory and/or somatosensory cortices. *Journal of Neurophysiology*, **85**, 1078–87.

MacDonald, K.D. and Barth, D.S. (1995) High frequency (gamma-band) oscillating potentials in rat somatosensory and auditory cortex. *Brain Research*, **694**, 1–12.

MacNaughton Jones, H. (1981) The differential diagnosis and prognosis of tinnitus aurium (noises in the head and ears) *Lancet*, **16**, 1084–86.

MacNaughton-Jones, H. (1890) The etiology of tinnitus aurium. *The British Medical Journal*, **20**, 667–72.

Maffei, G., Miani, P. (1962) Experimental tobacco poisoning. Resultant structural modifications of the cochlea and tuba acustica. *Arch Otolaryngol*, **75**, 386–96.

Mahlke, C. and Wallhäusser-Franke, E. (2004) Evidence for tinnitus-related plasticity in the auditory and limbic system, demonstrated by arg3.1 and c-fos immunocytochemistry. *Hearing Research*, **195**, 17–34.

Mammano, F. and Ashmore, J.F. (1993) Reverse transduction measured in the isolated cochlea by laser Michelson interferometry. *Nature*, **365**, 838–41.

Manabe, Y., Yoshida, S., Saito, H., and Oka, H. (1997) Effects of lidocaine on saliylate-induced discharge of neurons in the inferior colliculus of the guinea pig. *Hearing Research*, **103**, 192–98.

Marcondes, R.A., Sanchez, T.G., Kii, M.A., Ono, C.R., Buchpiguel. C.A., Langguth, B., and Marcolin, M.A. (2010) Repetitive transcranial magnetic stimulation improve tinnitus in normal hearing patients: a double-blind controlled, clinical and neuroimaging outcome study. *European Journal of Neurology*, **17**, 38–44.

Markaryan, A., Nelson, E.G., and Hinojosa, R. (2009) Quantification of the mitochondrial DNA common deletion in presbycusis. *Laryngoscope*, **119**, 1184–89.

Marriage, J. and Barnes, N.M. (1995) Is central hyperacusis a symptom of 5-hydroxytryptamine (5-HT) dysfunction? *Journal of Laryngology and Otology*, **109**, 915–21.

Martin, G.K., Lonsbury-Martin, B.L., Probst, R., and Coats, A.C. (1988) Spontaneous otoacoustic emissions in a nonhuman primate. I. Basic features and relations to other emissions. *Hearing Research*, **33**, 49–68.

Martin, W.H. (1994) Spectral analysis of brain activity in the study of tinnitus. In: J.A. Vernon and A.R. Moller (eds.), *Mechanisms of Tinnitus*, pp. 163 80. Boston, MA: Allyn and Bacon.

Martin, W.H., Schwegler, J.W., Scheibelhoffer, J., and Ronis, M.L. (1993) Salicylate-induced changes in cat auditory nerve activity. *Laryngoscope*, **103**, 600–604.

Martin, W.H., Schwegler, J.W., Shi, Y.-B., Pratt, H., and Adler, S. (1995) Developing an objective measurement for evaluating tinnitus: spectral averaging. In: G.E. Reich and J.A. Vernon (eds.), *Proceedings of the Fifth International Tinnitus Seminar*, pp.127–34. Portland, OR: American Tinnitus Association.

Martinez-Devesa, P., Perera, R., Theodoulou, M., and Waddell, A. (2010) Cognitive behavioural therapy for tinnitus. *Cochrane Database of Systematic Reviews*, **9**, CD005233.

Martinez Devesa, P., Waddell, A., Perera, R., and Theodoulou, M. (2007) Cognitive behavioural therapy for tinnitus. *Cochrane Databaseof Systematic Reviews*, **1**, CD005233.

Mazurek, B., Haupt, H., Joachim, R., Klapp, B.F., Stöver, T., and Szczepek, A.J. (2010) Stress induces transient auditory hypersensitivity in rats. *Hearing Research*, **259**, 55–63.

McCormick, D.A. and Wang, Z. (1991) Serotonin and noradrenaline excite GABAergic neurones of the guinea-pig and cat nucleus reticularis thalami. *Journal of Physiology*, **442**, 235–55.

McFadden, D. (1982) *Tinntus; Facts, theories and treatments. Report of the working group 89, Committee on hearing bioacoustics and biomechanics*. Washington, DC: National Research Council. National Academy Press

McFadden, S.L., Quaranta, N., and Henderson, D. (1997) Suprathreshold measures of auditory function in the aging chinchilla. *Hearing Research*, **111**, 127–35.

McKenna, L. (1998) Psychological treatments for Tinnitus, in J.A. Vernon (ed.), *Tinnitus, treatment and relief*, pp.140–55. Needham Heights, MA: Allyn & Bacon.

McKenna, L. (2004) Models of tinnitus suffering and treatment compared and contrasted. *Audiological Medicine*, **2**, 41–53.

McMahon, C.M., Kifley, A., Rochtchina, E., Newall, P., and Mitchell, P. (2008) The contribution of family history to hearing loss in an older population. *Ear & Hearing*, **29**, 578–84.

McMahon, C.M. and Patuzzi, R.B. (2002) The origin of the 900 Hz spectral peak in spontaneous and sound-evoked round-window electrical activity. *Hearing Research*, **173**, 134–52.

Meikle, M.B. (1995) The interaction of central and peripheral mechanisms in tinnitus. In J.A. Vernon and A.R. Møller (eds.), *Mechanisms of Tinnitus*, pp. 181–206. Needham Heights, MA: Allyn & Bacon.

Meikle, M. B. and Griest, S. E. (1989) Gender-based differences in characteristics of tinnitus. *Hearing Journal*, **42**, 68–76.

Meikle, M and Taylor-Walsh, E. (1984) Characteristics of tinnitus and related observations in over 1800 tinnitus clinic patients. *Journal of Laryngology and Otology*, **Suppl 9**, 17–21.

Meikle, M.B., Vernon, J., and Johnson, R.M. (1984) The perceived severity of tinnitus. Some observations concerning a large population of tinnitus clinic patients. *Otolaryngology Head and Neck Surgery*, **92**, 689–96.

Melcher, J.R., Levine, R.A., Bergevin. C., and Norris, B. (2009) The auditory midbrain of people with tinnitus: abnormal sound-evoked activity revisited. *Hearing Research*, **257**, 63–74.

Melcher, J.R., Sigalovsky, I.S., Guinan, J.J., Jr., and Levine, R.A. (2000) Lateralized tinnitus studied with functional magnetic resonance imaging: abnormal inferior colliculus activation. *Journal of Neurophysiology*, **83**, 1058–72.

Melloni, L., Molina, C., Pena, M., Torres, D., Singer, W., and Rodriguez, E. (2007) Synchronization of neural activity across cortical areas correlates with conscious perception. *Journal of Neuroscience*, **27**, 2858–65.

Meltser, I. and Canlon, B. (2010) The expression of mitogen-activated protein kinases and brain-derived neurotrophic factor in inferior colliculi after acoustic trauma. *Neurobiology of Disease*, **40**, 325–30.

Meltser, I., Tahera, Y., and Canlon, B. (2010) Differential activation of mitogen-activated protein kinases and brain-derived neurotrophic factor after temporary or permanent damage to a sensory system. *Neuroscience*, **165**, 1439–46.

Meyer, K. (2011) Primary sensory cortices, top-down projections and conscious experience. *Progress in Neurobiology*, **94**, 408–17.

Mhatre, A.N., Stern, R.E., Li, J., and Lalwani, A.K. (2002) Aquaporin 4 expression in the mammalian inner ear and its role in hearing. *Biochemical and Biophysical Research Communications*, **297**, 987–96.

Middleton, J.W., Kiritani, T., Pedersen, C., Turner, J.G., Shepherd, G.M., and Tzounopoulos T. (2011) Mice with behavioral evidence of tinnitus exhibit dorsal cochlear nucleus hyperactivity because of decreased GABAergic inhibition. *Proceedings of the National Academy of Science of the United States of America*, **108**, 7601–7606.

Milbrandt, J. C., Albin, R. L., and Caspary, D. M. (1994) Age-related decrease in GABAB receptor binding in the Fischer 344 rat inferior colliculus. *Neurobiology of Aging*, **15**, 699–703.

Milbrandt, J. C., Albin, R. L., Turgeon, S. M., and Caspary, D. M. (1996) GABAA receptor binding in the aging rat inferior colliculus. *Neuroscience*, **73**, 449–58.

Milbrandt, J. C. and Caspary, D. M. (1995) Age-related reduction of [3H]strychnine binding sites in the cochlear nucleus of the Fischer 344 rat. *Neuroscience*, **67**, 713–719.

Milbrandt, J.C., Holder, T.M., Wilson, M.C., Salvi, R.J., and Caspary, D.M. (2000) GAD levels and muscimol binding in rat inferior colliculus following acoustic trauma. *Hearing Research*, **147**, 251–60.

Milbrandt, J. C., Hunter, C., and Caspary, D. M. (1997) Alterations of GABAA receptor subunit mRNA levels in the aging Fischer 344 rat inferior colliculus. *Journal of Comparative Neurology*, **379**, 455–65.

Miller, L.M., Escabí, M.A., Read, H.L., and Schreiner, C.E. (2001) Functional convergence of response properties in the auditory thalamocortical system. *Neuron*, **32**, 151–60.

Milligan, E.D. and Watkins, L.R. (2009) Pathological and protective roles of glia in chronic pain. *Nature Reviews Neuroscience*, **10**, 23–36.

Mills, J.H., Boettcher, F.A., and Dubno, J.R. (1997) Interaction of noise-induced permanent threshold shift and age-related threshold shift. *Journal of the Acoustical Society of America*, **101**, 1681–86.

Mills, R.P., Albert, D.M., and Brain, C.E. (1986) Tinnitus in childhood. *Clinical Otolaryngology*, **11**, 431–34.

Mills, R.P. and Cherry, J.R. (1984) Subjective tinnitus in children with otological disorders. *Internationl Journal of Pediatric Otorhinolaryngology*, **7**, 21–27.

Mirz, F., Pedersen, B., Ishizu, K., Johannsen, P., Ovesen, T., Stodkilde-Jorgensen, H., and Gjedde, A. (1999) Positron emission tomography of cortical centers of tinnitus. *Hearing Research*, **134**, 133–44.

Mitchell, C.R. (1983) The masking of tinnitus with pure tones. *Audiology*, **22**, 73–87.

Mitzdorf, U. (1985) Current source-density method and application in cat cerebral cortex: investigation of evoked potentials and EEG phenomena. *Physiological Reviews*, **65**, 37–100.

Moazami-Goudarzi, M., Michels, L., Weisz, N., and Jeanmonod, D. (2010) Temporo-insular enhancement of EEG low and high frequencies in patients with chronic tinnitus. QEEG study of chronic tinnitus patients. *BMC Neuroscience*, **11**, 40.

Moelker, A., Vogel, M.W., and Pattynama, P.M. (2003) Efficacy of passive acoustic screening: implications for the design of imager and MR-suite. *Journal of Magnetic Resonance Imaging*, **17**, 270–75.

Moffat, G., Adjout, K., Gallego, S., Thai-Van, H., Collet, L., and Noreña, A.J. (2009) Effects of hearing aid fitting on the perceptual characteristics of tinnitus. *Hearing Research*, **254**, 82–91.

Moller, A.R. and Rollins, P.R. (2002) The non-classical auditory pathways are involved in hearing in children but not in adults. *Neuroscience Letters*, **319**, 41–44.

Mollison, W.M. (1916) Audible tinnitus. *Proceedings of the Royal Society of Medicine*, **9**, 7.

Moore, B.C. (2000) Use of a loudness model for hearing aid fitting. IV. Fitting hearing aids with multi-channel compression so as to restore "normal" loudness for speech at different levels. *British Journal of Audiology*, **34**, 165–77.

Moore, B.C.J. and Vinay, S. (2010) The relationship between tinnitus pitch and the edge frequency of the audiogram in individuals with hearing impairment and tonal tinnitus. *Hearing Research*, **261**, 51–56.

Moore, B.C. and Glasberg, B.R. (2004) A revised model of loudness perception applied to cochlear hearing loss. *Hearing Research*, **188**, 70–88.

Moore, B.C., Huss, M., Vickers, D.A., Glasberg, B.R., and Alcántara, J.I. (2000) A test for the diagnosis of dead regions in the cochlea. *British Journal of Audiology*, **34**, 205–24.

Morest, D.K., Kim, J., Potashner, S.J., and Bohne, B.A. (1998) Long-term degeneration in the cochlear nerve and cochlear nucleus of the adult chinchilla following acoustic overstimulation. *Microscopic Research Techniques*, **41**, 205–216.

Morgenstern, L. (2005) The bells are ringing. Tinnitus in their own words. *Perspectives in Biology and Medicine*, **48**, 396–407.

Morley, B.J. and Happe, H.K. (2000) Cholinergic receptors: dual roles in transduction and plasticity. *Hearing Research*, **147**, 104–12.

Morton, N.E. (1991) Genetic epidemiology of hearing impairment. *Annals of the New York Academy of Science*, **630**, 16–31.

Mrena, R., Savolainen, S., Kuokkanen, J.T., and Ylikoski, J. (2002) Characteristics of tinnitus induced by acute acoustic trauma: A Long-term follow-up. *Audiology & Neurotoology*, **7**, 122–30.

Mugnaini, E. (1985) GABA neurons in the superficial layers of the rat dorsal cochlear nucleus: light and electron microscopic immunocytochemistry. *Journal of Comparative Neurology*, **235**, 61–81.

Muhlau, M., Rauschecker, J.P., Oestreicher, E., Gaser, C., Rottinger, M., Wohlschlager, A.M., Simon, F., Etgen, T., Conrad, B., and Sander, D. (2006) Structural brain changes in tinnitus. *Cerebral Cortex*, **16**, 1283–88.

Mühlnickel, W., Elbert, T., Taub, E., and Flor, H. (1988) Reorganization of auditory cortex in tinnitus. *Proceedings of the National Academy of Science of the United States of America*, **95**, 10340–43.

Mulders, W.H. and Robertson, D. (2009) Hyperactivity in the auditory midbrain after acoustic trauma: dependence on cochlear activity. *Neuroscience*, **164**, 733–46.

Mulders, W.H. and Robertson, D. (2011) Progressive centralization of midbrain hyperactivity after acoustic trauma. *Neuroscience*, **192**, 753–60.

Mulders, W.H.A.M., Seluakumaran, K., and Robertson, D. (2010) Efferent pathways modulate hyperactivity in inferior colliculus. *Journal of Neuroscience*, **30**, 9578–87.

Mulert, C., Jager, L., Proppm S., Karch, S., Stormann, S., Pogarell, O., Möller, H.J., Juckel, G., and Hegerl, U. (2005) Sound level dependence of the primary auditory cortex: simultaneous measurement with 61-channel EEG and fMRI. *Neuroimage*, **28**, 49–58.

Mullen, S. and Penfield, W. (1959) Illusions of comparative interpretation and emotion; production by epileptic discharge and by electrical stimulation in the temporal cortex. *AMA Archives of Neurology and Psychiatry*, **81**, 269–84.

Muly, S.M., Gross, J.S., Morest, D.K., and Potashner, S.J., (2002) Synaptophysin in the cochlear nucleus following acoustic trauma. *Experimental Neurology*, **177**, 202–221.

Muly, S.M., Gross, J.S., and Potashner, S.J. (2004) Noise trauma alters D-[3H]aspartate release and AMPA binding in chinchilla cochlear nucleus. *Journal of Neuroscience Research*, **75**, 585–96.

Murakami, N., Ishibashi, H., Katsurabayashi, S., and Akaike, N. (2002) Calcium channel subtypes on single GABAergic presynaptic terminal projecting to rat hippocampal neurons. *Brain Research*, **951**, 121–29.

Murugasu, E. and Russell, I.J. (1995) Salicylate ototoxicity: the effects on basilar membrane displacement, cochlear microphonics and neural responses in the basal turn of the guinea pig cochlea. *Auditory Neuroscience*, **1**, 139–50.

Myers, E.N. and Bernstein, J.M. (1965) Salicylate ototoxicity; a clinical and experimental study. *Archives of Otolaryngology*, **82**, 483–93.

Møller, A.R. (1984) Pathophysiology of tinnitus. *Annals of Otology Rhinology and Laryngology*, **93**, 39–44.

Møller, A.R. (1997) Similarities between chronic pain and tinnitus. *American Journal of Otology*, **18**, 577–85.

Møller, A.R. (2007) Tinnitus and pain. *Progress in Brain Research*, **166**, 47–53.

Møller, A.R., Møller, M.B., and Yokota, M. (1992) Some forms of tinnitus may involve the extralemniscal auditory pathway. *Laryngoscope*, **102**, 1165–71.

Mühlnickel, W., Elbert, T., Taub, E., and Flor, H. (1998) Reorganization of auditory cortex in tinnitus. *Proceedings of the National Academy of Science of the United States of America*, **95**, 10340–43.

Müller, M., Klinke, R., Arnold, W., and Oestreicher, E. (2003) Auditory nerve fibre responses to salicylate revisited. *Hearing Research*, **183**, 37–43.

Näätänen, R. and Picton, T. (1987) The N1 wave of the human electric and magnetic response to sound: a review and an analysis of the component structure. *Psychophysiology*, **24**, 375–425.

Nageris, B.I., Attias, J., and Raveh, E. (2010) Test-retest tinnitus characteristics in patients with noise-induced hearing loss. *American Journal of Otolaryngology–Head and Neck Medicine and Surgery*, **31**, 181–84.

Nelson, J.J. and Chen, K. (2004) The relationship of tinnitus, hyperacusis, and hearing loss. *Ear Nose and Throat Journal*, **83**, 472–76.

Newman, C.W., Jacobson, G.P., and Spitzer, J.B. (1996) Development of the Tinnitus Handicap Inventory. *Archives of Otolaryngology Head and Neck Surgery*, **122**, 143–48.

Newman, C.W., Sandridge, S.A., and Jacobson, G.P. (1998) Psychometric adequacy of the Tinnitus Handicap Inventory (THI) for evaluating treatment outcome. *Journal of the American Academy of Audiology*, **9**, 153–60.

Newman, C.W. and Sandridge, S.A. (2004) Tinnitus Questionnaires. In: J.B. Snow Jr. (ed.), *Tinnitus: Theory and Management*, pp. 237–54. Hamilton: BC Dekker.

Nie, L. (2008) KCNQ4 mutations associated with nonsyndromic progressive sensorineural hearing loss. *Current Opinion in Otolaryngology & Head and Neck Surgery*, **16**, 441–44.

Niedermeyer, E. (1997) Alpha rhythms as physiological and abnormal phenomena. *International Journal of Psychophysiology*, **26**, 31–49.

Niedzielski, A.S. and Wenthold, R.J. (1995) Expression of AMPA, kainate, and NMDA receptor subunits in cochlear and vestibular ganglia. *Journal of Neuroscience*, **15**, 2338–53.

Nigam, A. and Samuel, P.R. (1984) Hyperacusis and Williams syndrome. *Journal of Laryngology and Otology*, **108**, 494–96.

Nitsche, M.A., Cohen, L.G., Wassermann, E.M., Priori, A., Lang, N., Antal, A., Paulus, W., Hummel, F., Boggio, P.S., Fregni, F., and Pascual-Leone, A. (2008) Transcranial direct current stimulation: state of the art 2008. *Brain Stimulation*, **1**, 206–23.

Noben-Trauth, K., Zheng, Q.Y., and Johnson, K.R. (2003) Association of cadherin 23 with polygenic inheritance and genetic modification of sensorineural hearing loss. *Nature Genetics*, **35**, 21–23.

Nobili, R., Mammano, F., and Ashmore, J. (1998) How well do we understand the cochlea? *Trends in Neuroscience*, **21**, 159–67.

Noble, W. (1998) *Self-assessment of hearing and related function*. London: Whurr Publishers Ltd.

Noble, W. (2000) Self-reports about tinnitus and about cochlear implants. *Ear & Hearing*, **21**, 50S–9S.

Nodar, R.H. (1972) Tinnitus aurium in schoolage children, a survey. *Journal of Auditory Research*, **12**, 133–35.

Nodar, V.R. and LeZak, M. (1984) Pediatric tinnitus (a thesis revised) *Journal of Laryngology & Otology*, **9(Suppl)**, 234–35.

Nondahl, D.M., Cruickshanks, K.J., Wiley, T.L., Klein, R., Klein, B.E., and Tweed, T.S. (2002) Prevalence and 5-year incidence of tinnitus among older adults: the epidemiology of hearing loss study. *Journal of the American Academy of Audiology*, **13**, 323–31.

Nondahl, D.M., Cruickshanks, K.J., Wiley, T.L, Klein, B.E.K., Klein, R., Chappell, R., and Tweed, T.S. (2010) The ten-year incidence of tinnitus among older adults. *International Journal of Audiology*, **49**, 580–85.

Noreña, A.J. (2011) An integrative model of tinnitus based on a central gain controlling neural sensitivity. *Neuroscience and Biobehavioral Reviews*, **35**, 1089–1109.

Noreña, A.J. and Chery-Croze, S. (2007) Enriched acoustic environment rescales auditory sensitivity. *Neuroreport*, **18**, 1251–55.

Noreña, A., Cransac, H., and Chéry-Croze, S. (1999) Towards an objectification by classification of tinnitus. *Clinical Neurophysiology*, **110**, 666–75.

Noreña, A.J. and Eggermont, J.J. (2003a) Changes in spontaneous neural activity immediately after an acoustic trauma: implications for neural correlates of tinnitus. *Hearing Research*, **183**, 137–53.

Noreña, A. and Eggermont, J.J. (2003b) Neural correlates of an auditory after image in primary auditory cortex. *Journal of the Association for Research in Otolaryngology*, **4**, 312–28.

Noreña, A.J. and Eggermont, J.J. (2005) Enriched acoustic environment after noise trauma reduces hearing loss and prevents cortical map reorganization. *Journal of Neuroscience*, **25**, 699–705.

Noreña, A.J. and Eggermont, J.J. (2006) Enriched acoustic environment after noise trauma abolishes neural signs of tinnitus. *Neuroreport*, **17**, 559–63.

Noreña, A.J., Gourévitch, B., Aizawa, N., and Eggermont, J.J. (2006) Spectrally enhanced acoustic environment disrupts frequency representation in cat auditory cortex. *Nature Neuroscience*, **9**, 932–39.

Noreña, A., Micheyl, C., Chery-Croze, S., and Collet, L. (2002) Psychoacoustic characterization of the tinnitus spectrum: implications for the underlying mechanisms of tinnitus. *Audiology & Neurootology*, **7**, 358–69.

Noreña, A.J., Moffat, G., Blanc, J.L., Pezard, L., and Cazals, Y. (2010) Neural cjhanges in the auditory cortex of awake guina pigs after two tinnitus inducers: salicylate and acoustic trauma. *Neuroscience*, **166**, 1194–1209.

Noreña, A.J., Tomita, M., and Eggermont, J.J. (2003) Neural changes in cat auditory cortex after a transient pure-tone trauma. *Journal of Neurophysiology*, **90**, 2387–2401.

Nottet, J.B., Moulin, A., Brossard, N., Suc, B., and Job, A. (2006) Otoacoustic emissions and persistent tinnitus after acute acoustic trauma. *Laryngoscope*, **116**, 970–75.

Nystrom, C., Matousek, M., and Hallstrom, T. (1986) Relationships between EEG and clinical characteristics in major depressive disorder. *Acta Psychiatrica Scandinavica*, **73**, 390–94.

Ochi, K. and Eggermont, J.J. (1996) Effects of salicylate on neural activity in cat primary auditory cortex. *Hearing Research*, **95**, 63–76.

Ochi, K. and Eggermont, J.J. (1997) Effects of quinine on neural activity in cat primary auditory cortex. *Hearing Research*, **97**, 105–118.

Oertel, D. and Young, E.D. (2004) What's a cerebellar circuit doing in the auditory system? *Trends in Neuroscience*, **27**, 104–110.

Oertel, D., Wright, S., Cao, X.J., Ferragamo, M., and Bal, R. (2010) The multiple functions of T stellate / multipolar/chopper cells in the ventral cochlear nucleus. *Hearing Research*, **276**, 61–69.

Oertel, D., Wu, S.H., and Hirsch, J.A. (1988) Electrical characteristics of cells and neuronal circuitry in the cochlear nuclei studied with interacellular recordings from brain slices. In: G.M. Edelman, W.E. Gall, and W.M. Cowan (eds.), *Auditory function. Neurobiological bases of hearing*, pp. 313–36. New York: John Wiley & Sons.

Ohlemiller, K.K. (2004) Age-related hearing loss: the status of Schuknecht's typology. *Current Opinion in Otolaryngology, Head and Neck Surgery*, **12**, 439–43.

Ohlemiller, K.K. (2006) Contributions of mouse models to understanding of age- and noise-related hearing loss. *Brain Research*, **1091**, 89–102.

Ohlemiller, K.K., McFadden, S.L., Ding, D.-L., Lear, P.M., and Ho, Y.-S. (2000) Targeted mutation of the gene for cellular glutathione peroxidase (Gpx1) increases noiseinduced hearing loss in mice. *Journal of the Association for Research in Otolaryngology*, **1**, 243–54.

Okamoto, H., Stracke, H., Stoll, W., and Pantev, C. (2010) Listening to tailor-made notched music reduces tinnitus loudness and tinnitus-related auditory cortex activity. *Proceedings of the National Academy of Science (of the United States of America)*, **107**, 1207–1210.

Orekhova, E.V., Stroganova, T.A., Prokofyev, A.O., Nygren, G., Gillberg, C., and Elam, M. (2008) Sensory gating in young children with autism: relation to age, IQ, and EEG gamma oscillations. *Neuroscience Letters*, **434**, 218–23.

Osaki, Y., Nishimura, H., Takasawa, M., Imaizumi, M., Kawashima, T., Iwaki, T., Oku, N., Hashikawa, K., Doi, K., Nishimura, T., Hatazawa, J., and Kubo, T. (2005) Neural mechanism of residual inhibition of tinnitus in cochlear implant users. *Neuroreport*, **16**, 1625–28.

Oxenham, A.J. and Bacon, S.P. (2003) Cochlear compression: perceptual measures and implications for normal and impaired hearing. *Ear & Hearing*, **24**, 352–66.

Palmer, K.T., Griffin, M.J., Syddall, H.E., Davis, A., Pannett, B., and Coggon, D. (2002) Occupational exposure to noise and the attributable burden of hearing difficulties in Great Britain. *Occupational and Environmental Medicine*, **59**, 634–39.

Palombi, P.S. and Caspary, D.M. (1996) Physiology of the aged Fischer 344 rat inferior colliculus: responses to contralateral monaural stimuli. *Journal of Neurophysiology*, **76**, 3114–25.

Pan, T., Tyler, R.S., Ji, H., Coelho, C., Gehringer, A.K., and Gogel, S.A. (2009) The relationship between tinnitus pitch and the audiogram. *International Journal of Audiology*, **48**, 277–94.

Panford-Walsh, R., Singer, W., Rüttiger, L., Hadjab, S., Tan, J., Geisler, H.S., Zimmermann, U., Köpschall, I., Rohbock, K., Vieljans, A., *et al.* (2008) Midazolam reverses salicylate-induced changes in brain-derived neurotrophic factor and arg3.1 expression: implications for tinnitus perception and auditory plasticity. *Molecular Pharmacology*, **74**, 595–604.

Pang, L.Q. (1971) The otological aspects of whiplash injuries. *Laryngoscope*, **81**, 1381–87.

Pantev, C., Lütkenhöner, B., Hoke, M., and Lehnertz, K. (1986) Comparison between simultaneously recorded auditory-evoked magnetic fields and potentials elicited by ipsilateral, contralateral and binaural tone burst stimulation. *Audiology*, **25**, 54–61.

Pantev, C., Roberts, L.E., Elbertm T., Ross, B., and Wienbruch, C. (1996) Tonotopic organization of the sources of human auditory steady-state responses. *Hearing Research*, **101**, 62–74.

Parra, L.C. and Pearlmutter, B.A. (2007) Illusory percepts from auditory adaptation. *Journal of the Acoustical Society of America*, **121**, 1632–41.

Parving, A., Hein, H.O., Suadicani, P., Ostri, B., and Gyntelberg, F. (1993) Epidemiology of hearing disorders: some factors affecting hearing. The Copenhagen Male Study. *Scandinavian Audiology*, **22**, 101–107.

Patuzzi, R.B., Brown, D.J., McMahon, C.M., and Halliday, A.F. (2004) Determinants of the spectrum of the neural electrical activity at the round window: transmitter release and neural depolarisation. *Hearing Research*, **190**, 87–108.

Paul, A.K., Lobarinas, E., Simmons, R., Wack, D., Luisi, J.C., Spernyak, J., Mazurchuk, R., Abdel-Nabi, H. and Salvi, R. (2009) Metabolic imaging of rat brain during pharmacologically-induced tinnitus. *Neuroimage*, **44**, 312–318.

Pemble, S., Schroeder, K.R., Spencer, S.R., Meyer, D.J., Hallier, E., Bolt, H.M., Ketterer, B., and Taylor, J.B. (1994) Human glutathione S-transferase theta (GSTT1): cDNA cloning and the characterization of a genetic polymorphism. *Biochemical Journal*, **300**, 271–76.

Penfield, W. and Rasmussen, T. (1950) *The Cerebral Cortex of Man—A Clinical Study of Localization of Brain Function*. New York: Macmillan.

Peng, B.G., Chen, S., and Lin, X. (2003) Aspirin selectively augmented N-methyl-D-aspartate types of glutamate responses in cultured spiral ganglion neurons of mice. *Neuroscience Letters*, **343**, 21–24.

Penner, M.J. (1980) Two-tone forward masking patterns and tinnitus. *Journal of Speech and Hearing Research*, **23**, 779–86.

Penner, M.J. (1992) Linking spontaneous otoacoustic emissions and tinnitus. *Audiology*, **26**, 115–23.

Penner, M.J. (1993) Synthesizing tinnitus from sine waves. *Journal of Speech and Hearing Research*, **36**, 1300–1305.

Penner, M.J. and Burns, E.M. (1987) The dissociation of SOAEs and Tinnitus. *Journal of Speech and Hearing Research*, **30**, 396–403.

Penner, M.J. and Coles, R.R. (1992) Indications for aspirin as a palliative for tinnitus caused by SOAEs: a case study. *British Journal of Audiology*, **26**, 91–96.

Penner, M.J. and Jastreboff, P.J. (1996) Tinnitus: psychophysical observations in humans and an animal model. In: T.R. Van De Water, A.N. Popper and R.R. Fay (eds.), *Clinical aspects of hearing*, pp. 258–304. New York: Springer Verlag.

Pettersson, K. and Gustafsson, J.A. (2001) Role of estrogen receptor beta in estrogen action. *Annual Reviews of Physiology*, **63**, 165–92.

Phillips, D.P. (1987) Stimulus intensity and loudness recruitment: neural correlates. *Journal of the Acoustical Society of America*, **82**, 1–12.

Phillips, H.C, and Hunter, M. (1982) A laboratory technique for the assessment of pain behaviour. *Journal of Behavioural Medicine*, **5**, 283–94.

Picton, T.W., Alain, C., Otten, L., Ritter, W., and Achim, A. (2000) Mismatch negativity: different water in the same river. *Audiology & Neurootology*, **5**, 111–39.

Pienkowski, M. and Eggermont, J.J. (2009) Recovery from reorganization induced in adult cat primary auditory cortex by a band-limited spectrally enhanced acoustic environment. *Hearing Research*, **257**, 24–40.

Pilgramm, M., Rychlick, R., Lebisch, H., Siedentop, H., Goebel, G., and Kirchhoff, D. (1999) Tinnitus in the Federal Republic of Germany: a representative epidemiological study. In: J. Hazell (ed.), *Proceedings of the 6th International Tinnitus Seminar, Cambridge*, pp. 64–7. London: The Tinnitus and Hyperacusis Centre.

Pinchoff, R.J., Burkard, R.F., Salvi, R.J., Coad, M.L., and Lockwood, A.H. (1998) Modulation of tinnitus by voluntary jaw movements. *American Journal of Otology*, **19**, 785–89.

Plath, N., Ohana, O., Dammermann, B., Errington, M.L., Schmitz, D., Gross, C., Mao, X., Engelsberg, A., Mahlke, C., Welzl, H., *et al.* (2006) Arc/Arg3.1 is essential for the consolidation of synaptic plasticity and memories. *Neuron*, **52**, 437–44.

Plewnia, C., Bartels, M., Gerloff, C. (2003) Transient suppression of tinnitus by transcranial magnetic stimulation. *Annals of Neurology*, **53**, 263–66.

Plewnia, C., Reimold, M., Najib, A., Brehm, B., Reischl, G., Plontke, S.K., and Gerloff, C. (2007) Dose-dependent attenuation of auditory phantom perception (tinnitus) by PET-guided repetitive transcranial magnetic stimulation. *Human Brain Mapping*, **28**, 238–46.

Polley, D.B., Steinberg, E.E., and Merzenich, M.M. (2006) Perceptual learning directs auditory cortical map reorganization through top-down influences. *Journal of Neuroscience*, **26**, 4970–82.

Ponton, C.W., Don M., and Eggermont, J.J. (1992) Place-specific derived cochlear microphonics. *Scandinavian Audiology*, **21**, 131–42.

Ponton, C.W., Moore, J.K., and Eggermont, J.J. (1999) Prolonged deafness limits auditory system developmental plasticity: evidence from an evoked potentials study in children with cochlear implants. *Scandinavian Audiology*, **28**, Suppl 51, 13–22.

Poreisz, C., Paulus, W., Moser, T., and Lang, N. (2009) Does a single session of theta-burst transcranial magnetic stimulation of inferior temporal cortex affect tinnitus perception? *BMC Neuroscience*, **10**: 54.

Portmann, M., Cazals, Y., Negrevergne, M., and Aran, J.M. (1979) Temporary tinnitus suppression in man through electrical stimulation of the cochlea. *Acta Otolaryngologica*, **87**, 294–99.

Portmann, M., Negrevergne, M. Aran, J-M., and Cazals, Y. (1983) Electrical stimulation of the ear: clinical applications. *Annals of Otology Rhinology and Laryngology*, **92**, 621–22.

Potashner, S.J., Suneja, S.K., and Benson, C.G. (1997) Regulation of D-aspartate release and uptake in adult brain stem auditory nuclei after unilateral middle ear ossicle removal and cochlear ablation. *Experimental Neurology*, **148**, 222–35.

Potier, M., Hoquet, C., Lloyd, R., Nicolas-Puel, C., Uziel, A., and Puel, J.-L. (2009) The Risks of Amplified Music for Disc-Jockeys Working in Nightclubs. *Ear & Hearing*, **30**, 291–93.

Prijs, V.F. (1986) Single-unit response at the round window of the guinea pig. *Hearing Research*, **21**, 127–33.

Puel, J.L., Bledsoe Jr., S.C., Bobbin, R.P., Ceasar, G., and Fallon, M. (1989) Comparative actions of salicylate on the amphibian lateral line and guinea pig cochlea. *Comparative Biochemistry and Physiology C*, **93**, 73–80.

Puel, J.L., Bobbin, R.P., and Fallon, M. (1990) Salicylate, mefenamate, meclofenamate, and quinine on cochlear potentials. *Otolaryngology Head and Neck Surgery*, **102**, 66–73.

Puel, J.L., d'Aldin, C., Ruel, J., Ladrech, S., and Pujol, R. (1997) Synaptic repair mechanisms responsible for functional recovery in various cochlear pathologies. *Acta Otolaryngologica*, **117**, 214–218.

Puel, J.L., Ruel, J., Gervais d'Aldin, C., and Pujol, R. (1998) Excitotoxicity and repair of cochlear synapses after noise-trauma induced hearing loss. *Neuroreport*, **9**, 2109–14.

Pugh, J.E. Jr, Moody, D.B., and Anderson, D.J. (1979) Electrocochleography and experimentally induced loudness recruitment. *Archives of Otorhinolaryngology*, **224**, 241–55.

Pugh, R., Budd, R.J., and Stephens, S.D.G. (1995) Patients reports of the effect of alcohol on tinnitus. *British Journal of Audiology*, **29**, 279–83.

Pujol, R., Gervais d'Aldin, C., Saffiedine, S., Eybalin, M., and Puel, J-L. (2005) Repair of inner hair cell-auditory nerve synapses and recovery of function after an excitotoxuc injury. In: R.J. Salvi, D. Henderson, F. Fiorino, V. Colletti (eds.), *Auditory plasticity and regeneration*, pp. 100–107. New York: Thieme Medical Publishers Inc.

Qiu, C., Salvi, R., Ding, D., and Burkard, R. (2000) Inner hair cell loss leads to enhanced response amplitudes in auditory cortex of unanesthetized chinchillas: evidence for increased system gain. *Hearing Research*, **139**, 153–71.

Quaranta, A., Assennato, G., and Sallustio, V. (1996) Epidemiology of hearing problems among adults in Italy. *Scandivavian Audiology*, **25**(Suppl 42), 7–11

Quaranta, N., Wagstaff, S., and Baguley, D.M. (2004) Tinnitus and cochlear implantation. *International Journal of Audiology*, **43**, 245–51.

Rabinowitz, P.M., Pierce Wise, J. Sr., Hur Mobo, B., Antonucci, P.G., Powell, C., and Slade, M. (2002) Antioxidant status and hearing function in noise-exposed workers. *Hearing Research*, **173**, 164–71.

Rabinowitz, P.M., Slade, M.D., Galusha, D., Dixon-Ernst, C., and Cullen, M.R. (2006) Trends in the prevalence of hearing loss among young adults entering an industrial workforce 1985 to 2004. *Ear & Hearing*, **27**, 369–75.

Racine, R.J., Chapman, C.A., Teskey, G.C., and Milgram, N.W. (1995) Postactivation potentiation in the neocortex. III. Kindling-induced potentiation in the chronic preparation. *Brain Research*, **702**, 77–86.

Rajan, R. and Irvine, D.R. (1998) Absence of plasticity of the frequency map in dorsal cochlear nucleus of adult cats after unilateral partial cochlear lesions. *Journal of Comparative Neurology*, **399**, 35–46.

Rajan, R., Irvine, D.R., Wise, L.Z., and Heil, P. (1993) Effect of unilateral partial cochlear lesions in adult cats on the representation of lesioned and unlesioned cochleas in primary auditory cortex. *Journal of Comparative Neurology*, **338**, 17–49

Rauschecker, J.P. (1999) Auditory cortical plasticity: a comparison with other sensory systems. *Trends in Neuroscience*, **22**, 74–80.

Rauschecker, J.P., Leaver, A.M, and Mühlau, M. (2010) Tuning out the noise: limbic-auditory interactions in tinnitus. *Neuron*, **66**, 819–26.

Rausell, E., Bickford, L., Manger, P.R., Woods, T.M., and Jones, E.G. (1998) Extensive divergence and convergence in the thalamocortical projection to monkey somatosensory cortex. *Journal of Neuroscience*, **18**, 4216–32.

Ravicz, M.E., Melcher, J.R., and Kiang, N.Y. (2000) Acoustic noise during functional magnetic resonance imaging. *Journal of the Acoustical Society of America*, **108**, 1683–96.

Raza, A., Milbrandt, J. C., Arneric, S. P., and Caspary, D. M. (1994) Age-related changes in brainstem auditory neurotransmitters: measures of GABA and acetylcholine function. *Hearing Research*, **77**, 221–30.

Read, H.L., Winer, J.A., and Schreiner, C.E. (2001) Modular organization of intrinsic connections associated with spectral tuning in cat auditory cortex. *Proceedings of the National Academy of Science of the United States of America*, **98**, 8042–47.

Relkin, E.M. and Doucet, J.R. (1997) Is loudness simply proportional to the auditory nerve spike count? *Journal of the Acoustical Society of America*, **101**, 2735–40.

Rescorla, R. and Wagner, A. (1972) A theory of Pavlovian conditioning: variations in the effectiveness of reinforcement and nonreinforcement. In A. Black and W. Prokasy (eds.), *Classical conditioning II: Current research and theory*, pp. 64–99. New York: AppletonCentury-Crofts.

Reser, D.H. and Van de Water, T.R. (1997) Implications of neurotrophin supported auditory neuron survival for maintenance of the tonotopic organization of the central auditory pathway. *Acta Otolaryngologica*, **117**, 239–43.

Reyes, A.D. (2003) Synchrony-dependent propagation of firing rate in iteratively constructed networks in vitro. *Nature Neuroscience*, **6**, 593–99.

Reyes, S.A., Salvi, R.J., Burkard, R.F., Coad, M.L., Wack, D.S., Galantowicz, P.J., and Lockwood, A.H. (2002) Brain imaging of the effects of lidocaine on tinnitus. *Hearing Research*, **171**, 43–50.

Ridding, M.C. and Rothwell, J.C. (2007) Therapeutic use of rTMS. *Nature Reviews Neuroscience*, **8**, 559–67.

Riga, M., Papadas, T., Werner, J.A., and Dalchow, C.V. (2007) A clinical study of the efferent auditory system in patients with normal hearing who have acute tinnitus. *Otology & Neurotology*, **28**, 185–90.

Roberts, L.E. (2007) Residual inhibition. *Progress in Brain Research*, **166**, 487–95.

Roberts, L.E. (2011) Neural synchrony and neural plasticity in tinnitus. In A.R. Møller, T. Kleinjung, B. Langguth, and D. De Ridder (eds.), *Textbook of Tinnitus*, pp. 103–112. New York: Springer Science + Business Media.

Roberts, L.E., Moffat, G., and Bosnyak, D.J. (2006) Residual inhibition functions in relation to tinnitus spectra and auditory threshold shift. *Acta Otolaryngology Supplement*, **556**, 27–33.

Roberts, L.E., Moffat, G., Baumann, M., Ward, L.M., and Bosnyak, D.J. (2008) Residual inhibition functions overlap tinnitus spectra and the region of auditory threshold shift. *Journal of the Association for Research in Otolaryngology*, **9**, 417–35.

Robertson, D. and Irvine, D.R.F. (1989) Plasticity of frequency organization in auditory cortex of guinea pigs with partial unilateral deafness. *Journal of Comparative Neurology*, **282**, 456–71.

Robertson, D. and Johnstone, B.M. (1980) Acoustic trauma in the guinea pig cochlea: early changes in ultrastructure and neural threshold. *Hearing Research*, **3**, 167–79.

Robinson, S. (2007) Antidepressants for treatment of tinnitus. *Progress in Brain Research*, **166**, 263–71.

Robles, L. and Ruggero, M.A. (2001) Mechanics of the mammalian cochlea. *Physiological Review*, **81**, 1305–52

Rodieck, R.W., Kiang, N.Y.-S., and Gerstein, G.L. (1962) Some quantitative methods for the study of spontaneous activity of single cells. *Biophysical Journal*, **2**, 351–68.

Roeser, R.J., Ballachanda, B.B. (1997) Physiology, pathophysiology, and anthropology/epidemiology of human ear canal secretions. *Journal of the American Academy of Audiology*, **8**, 391–400.

Roeser, R.J. and Price, D.R. (1980) Clinical experience with tinnitus maskers. *Ear & Hearing*, **1**, 63–68.

Rosanowski, F., Eysholdt, U., and Hoppe, U. (2006) Influence of leisure-time noise on outer hair cell activity in medical students. *International Archive of Occupational and Environmental Health*, **80**, 25–31.

Rossi, S., Ferro, M., Cincotta, M., Ulivelli, M., Bartalini, S., Miniussi, C., Giovannelli, F., and Passero, S. (2007) A real electro-magnetic placebo (REMP) device for sham transcranial magnetic stimulation (TMS) *Clinical Neurophysiology*, **118**, 709–716.

Roy, S.A. and Alloway, K.D. (2001) Coincidence detection or temporal integration? What the neurons in somatosensory cortex are doing. *Journal of Neuroscience*, **21**, 2462–73.

Rubak, T., SKock, S., Koefoed-Nielsen, B., Lund, S.P., Bonde, J.P., and Kolstad, H.A. (2008) The risk of tinnitus following occupational noise exposure in workers with hearing loss or normal hearing. *International Journal of Audiology*, **47**, 109–114.

Ruckenstein, M.J., Hedgepeth, C., Rafter, K.O., Montes, M.L., and Bigelow, D.C. (2001) Tinnitus suppression in patients with cochlear implants. *Otology & Neurotology*, **22**, 200–204.

Ruel, J., Chabbert, C., Nouvian, R., Bendris, R., Eybalin, M., Leger, C.L., Bourien, J., Mersel, M., and Puel, J.L. (2008) Salicylate enables cochlear arachidonic-acid-sensitive NMDA receptor responses. *Journal of Neuroscience*, **28**, 7313–23.

Ruggero, M.A. and Rich, N.C. (1991) Furosemide alters organ of Corti mechanics: evidence for feedback of outer hair cells upon the basilar membrane. *Journal of Neuroscience*, **11**, 1057–67.

Rutherford, L.C., Nelson, S.B., and Turrigiano, G.G. (1998) BDNF has opposite effects on the quantal amplitude of pyramidal neuron and interneuron excitatory synapses. *Neuron*, **2**, 521–30.

Ruytjens, L., Willemsen, A.T., Van Dijk, P., Wit, H.P., and Albers, F.W. (2006) Functional imaging of the central auditory system using PET. *Acta Otolaryngologica*, **126**, 1236–44.

Rüttiger, L., Ciuffani, J., Zenner, H.P., and Knipper, M. (2003) A behavioral paradigm to judge acute sodium salicylate-induced sound experience in rats: a new approach for an animal model on tinnitus. *Hearing Research*, **180**, 39–50.

Sachs, M.B. and Abbas, P.J. (1974) Rate versus level functions for auditory-nerve fibers in cats: tone-burst stimuli. *Journal of the Acoustical Society of America*, **56**(6), 1835–47.

Sahley, T.L. and Nodar, R.H. (2001) A biochemical model of peripheral tinnitus. *Hearing Research*, **152**, 43–54.

Sakai, M., (2007) Habituation enhances auditory perceptual capacity in adult rats. *Behavioral Brain Research*, **181**, 1–11.

Saltzmann, M. and Ersner, M.S. (1947) A hearing aid for the relief of tinnitus aurum. *Laryngoscope*, **48**, 358–66.

Salvi, R.J., Hamernik, R.P., Henderson, D., and Ahroon, W.A. (1983) Neural correlates of sensorineural hearing loss. *Ear & Hearing*, **4**, 115–29.

Salvi, R.J., Saunders, S.S., Gratton, M.A., Arehole, S., and Powers, N. (1990) Enhanced evoked response amplitudes in the inferior colliculus of the chinchilla following acoustic trauma. *Hearing Research*, **50**, 245–57.

Salvi, R.J., Wang, J., and Ding, D. (2000) Auditory plasticity and hyperactivity following cochlear damage. *Hearing Research*, **147**, 261–74.

Sanchez, L. (2004) The epidemiology of tinnitus. *Audiological Medicine*, **2**, 8–17.

Sanchez, L., Boyd, C., and Davis, A. (1999) Prevalence and problems of tinnitus in the elderly. In: J. Hazell (ed.), *Proceedings of the 6th International Tinnitus Seminar, Cambridge*, pp. 58–63. London: The Tinnitus and Hyperacusis Centre.

Sand, P.G., Langguth, B., Kleinjung, T., and Eichhammer, P. (2007) Genetics of chronic tinnitus. *Progress in Brain Research*, **166**, 159–68.

Sand, P.G., Luettich, A., Kleinjung, T., Hajak, G., and Langguth, B. (2010) *An Examination of KCNE1 Mutations and Common Variants in Chronic Tinnitus Genes*, **1**, 23–37.

Saunders, G.H. and Griest, S.S. (2009) Hearing loss in veterans and the need for hearing loss prevention programs. *Noise & Health*, **11**, 14–21.

Saunders, J.C., Bock, G.R., James, R., and Chen, C.S. (1972) Effects of priming for audiogenic seizure on auditory evoked responses in the cochlear nucleus and inferior colliculus of BALB-c mice. *Experimental Neurology*, **37**(2), 388–94.

Saunders, J.C., Dear, S.P., and Schneider, M.E. (1985) The anatomical consequences of acoustic injury: a review and tutorial. *Journal of the Acoustical Society of America*, **78**, 833–60.

Savastano, M. (2007) Characteristics of tinnitus in childhood. *European Journal of Pediatrics*, **166**, 797–801.

Savastano, M. (2008) Tinnitus with or without hearing loss: are its characteristics different? *European Archive of Otorhinolaryngology*, **265**, 1295–1300.

Savastano, M., Marioni, G., and de Filippis, C. (2009) Tinnitus in children without hearing impairment. *International Journal of Pediatric Otorhinolaryngology*, **73** (Suppl 1), S13–15.

Scarff, C.J., Reynolds, A., Goodyear, B.G., Ponton, C.W., Dort, J.C., and Eggermont, J.J. (2004) Simultaneous 3-T fMRI and high-density recording of human auditory evoked potentials. *Neuroimage*, 23, 1129–42.

Schaette, R. and Kempter, R. (2006) Development of tinnitus-related neuronal hyperactivity through homeostatic plasticity after hearing loss: a computational model. *European Journal of Neuroscience*, 23, 3124–38.

Schaette, R., Kempter, R. (2008) Development of hyperactivity after hearing loss in a computational model of the dorsal cochlear nucleus depends on neuron response type. *Hearing Research*, **240**, 57–72.

Schaette, R., König, O., Hornig, D., Gross, M., and Kempter, R. (2010) Acoustic stimulation treatments against tinnitus could be most effective when tinnitus pitch is within the stimulated frequency range. *Hearing Research*, **269**, 95–101.

Schlee, W., Hartmann, T., Langguth, B., and Weisz, N. (2009a) Abnormal resting-state cortical coupling in chronic tinnitus. *BMC Neuroscience*, **10**: 11.

Schlee, W., Mueller, N., Hartmann, T., Keil, J., Lorenz, I., and Weisz, N. (2009b) Mapping cortical hubs in tinnitus. *BMC Biology*, **23**; 7:80.

Schlee, W., Weisz, N., Bertrand, O., Hartmann, T., and Elbert, T. (2008) Using auditory steady state responses to outline the functional connectivity in the tinnitus brain. *PLoS One*, **3**(11), e3720.

Schmiedt, R.A., Mills, J.H., and Boettcher, F.A. (1996) Age-related loss of activity of auditory-nerve fibers. *Journal of Neurophysiology*, **76**, 2799–2803.

Schmuziger, N., Patscheke, J., and Probst, R. (2006) Hearing in nonprofessional pop/rock musicians. *Ear & Hearing*, **27**, 321–30.

Schneider, P., Andermann, M., Wengenroth, M., Goebel, R., Flor, H., Rupp, A., and Diesch, E. (2009) Reduced volume of Heschl's gyrus in tinnitus. *Neuroimage*, **45**, 927–39.

Schofield, B.R. (2010) Structural organization of the descending auditory pathway. In: A. Rees and A. R. Palmer (eds.), *The Oxford Handbook of Auditory Science: The Auditory Brain*, pp. 43–64. Oxford: Oxford University Press.

Schofield, B.R. and Coomes, D.L. (2005) Projections from auditory cortex contact cells in the cochlear nucleus that project to the inferior colliculus. *Hearing Research*, **206**, 3–11.

Schreiner, C.E. and Langner, G. (1988) Periodicity coding in the inferior colliculus of the cat. II. Topographical organization. *Journal of Neurophysiology*, **60**, 1823–40.

Schreiner, C.E. and Snyder, R.L. (1987) A physiological animal model of peripheral tinnitus, in: H. Feldman (ed.), *Proceedings III International Tinnitus Seminar, Munster*, pp. 100–106. Karlsruhe: Harsch Verlag.

Searchfield, G.D., Morrison-Low, J., and Wise, K. (2007) Object identification and attention training for treating tinnitus. *Progress in Brain Research*, **166**, 441–60.

Searchfield, G.D., Muñoz, D.J., and Thorne, P.R. (2004) Ensemble spontaneous activity in the guinea-pig cochlear nerve. *Hear Res*, **192**(1–2), 23–35.

Searchfield, G.D., Kaur, M., and Martin, W.H. (2010) Hearing aids as an adjunct to counseling: tinnitus patients who choose amplification do better than those that don't. *International Journal of Audiology*, **49**, 574–79.

Sée, G. (1877) Etudes sur l'acide salicylique et les salicylates; traitement du rhumatisme aigu et chronique de la goutte, et de diverses affections du systeme nerveux sensitif par les salicylates. *Bulletin de l'Académie Nationale de Médecine (Paris)*, **26**, 689–706.

Seidman, M.D., Ahmad, N., and Bai, U. (2002) Molecular mechanisms of age-related hearing loss. *Ageing Research Reviews*, **1**, 331–43.

Seidman, M.D., Bai, U., Khan, M.J., and Quirk, W.S. (1997) Mitochondrial DNA deletions associated with ageing and presbyacusis. *Archives of Otolaryngology Head and Neck Surgery*, **123**, 1039–45.

Seidman, M.D., de Ridder, D., Elisevich, K., Bowyer, S.M., Darrat, I., Dria, J., Stach, B., Jiang, Q., Tepley, N., Ewing, J., Seidman, M., and Zhang, J. (2008) Direct electrical stimulation of Heschl's gyrus for tinnitus treatment. *Laryngoscope*, **118**, 491–500.

Seki, S. and Eggermont, J.J. (2003) Changes in spontaneous firing rate and neural synchrony in cat primary auditory cortex after localized tone-induced hearing loss. *Hearing Research*, **180**, 28–38.

Sellick, P.M., Patuzzi, R., and Johnstone, B.M. (1982) Measurement of basilar membrane motion in the guinea pig using the Mössbauer technique. *Journal of the Acoustical Society of America*, **72**, 131–41.

Sendowski, I., Raffin, F., and Clarençon, D. (2006) Spectrum of neural electrical activity in guinea pig cochlea: effects of anaesthesia regimen, body temperature and ambient noise. *Hearing Research*, **211**, 63–73.

Sens, P.M. and de Almeida C.I. (2007) Participation of the cerebellum in auditory processing. *Brazilian Journal of Otorhinolaryngology*, **73**, 266–70.

Seung, H.S. (2003) Learning in spiking neural networks by reinforcement of stochastic synaptic transmission. *Neuron*, **40**, 1063–73.

Sexton, S. (1880) Note on tinnitus aurium. *British Medical Journal*, **1**(1017), 963–65.

Shadlen, M.N. and Movshon, J.A. (1999) Synchrony unbound: a critical evaluation of the temporal binding hypothesis. *Neuron*, **24**, 67–77.

Shargorodsky, J., Curhan, G.C., Wildon, R., and Farwell, W.R. (2010) Prevalence and characteristics of tinnitus among US adults. *The American Journal of Medicine*, **123**, 711–18.

Shargorodsky, J., Curhan, S.G., Curhan, G.C., Eavey, R. (2010) Change in prevalence of hearing loss in US adolescents. *Journal of the American Medical Association*, **304**, 772–78.

Sharp A. (1910) Audible tinnitus. *Proceedings of the Royal Society of Medicine*, **3**, 54–55.

Shehata-Dieler, W.E., Richter, C.P., Dieler, R., and Klinke, R. (1994) Effects of endolymphatic and perilymphatic application of salicylate in the pigeon. I: single fiber activity and cochlear potentials. *Hearing Research*, **74**, 77–84.

Shepherd, J.D., Rumbaugh, G., Wu, J., Chowdhury, S., Plath, N., Kuhl, D., Huganir, R.L., and Worley, P.F. (2006) Arc/Arg3.1 mediates homeostatic synaptic scaling of AMPA receptors. *Neuron*, **52**, 475–84.

Shergill, S.S., Bullmore, E., Simmons, A., Murray, R., and McGuire, P. (2000) Functional anatomy of auditory verbal imagery in schizophrenic patients with auditory hallucinations. *American Journal of Psychiatry*, **157**, 1691–93.

Sherlock, L.P. and Formby, C. (2005) Estimates of loudness, loudness discomfort, and the auditory dynamic range: normative estimates, comparison of procedures, and test-retest reliability. *Journal of the American Academy of Audiology*, **16**, 85–100.

Sherman, S.M. (2001) Tonic and burst firing: dual modes of thalamocortical relay. *Trends in Neuroscience*, **24**, 122–26.

Shi, Y., Burchiel, K.J., Anderson, V.C. and Martin, W.H. (2009) Deep brain stimulation effects in patients with tinnitus. *Otolaryngology – Head and Neck Surgery*, **141**, 285–87.

Shore, S.E. (2005) Multisensory integration in the dorsal cochlear nucleus: unit responses to acoustic and trigeminal ganglion stimulation. *European Journal of Neuroscience*, **21**, 3334–48.

Shore, S.E., Vass, Z., Wys, N.L., and Altschuler, R.A. (2000) Trigeminal ganglion innervates the auditory brainstem. *Journal of Comparative Neurology*, **419**, 271–85.

Shore, S.E. and Zhou. J. (2006) Somatosensory influence on the cochlear nucleus and beyond. *Hearing Research*, **216–217**, 90–99.

Shore, S., El-Kashlan, H.K., and Lu, J. (2003) Effects of trigeminal ganglion stimulation on unit activity of ventral cochlear nucleus neurons. *Neuroscience*, **119**, 1085–1101.

Shore, S., Zhou, J., and Koehler, S. (2007) Neural mechanisms underlying somatic tinnitus. *Progress in Brain Research*, **166**, 107–23.

Shore, S.E., Koehler, S., Oldakowski, M., Hughes, L.F., and Syed, S. (2008) Dorsal cochlear nucleus responses to somatosensory stimulation are enhanced after noise-induced hearing loss. *European Journal of Neuroscience*, **27**, 155–68.

Shulman, A. and Strashun, A. (1999) Descending auditory system / cerebellum / tinnitus. *International Tinnitus Journal*, **5**, 92–106.

Siebert, W.M. (1965) Some implications of the stochastic behavior of primary auditory neurons. *Kybernetik*, **2**, 206–15.

Silbersweig, D.A., Stern, E., Frith, C., Cahill, C., Holmes, A., Grootoonk, S., Seaward, J., McKenna, P., Chua, S.E., Schnorr, L., *et al.* (1995) A functional neuroanatomy of hallucinations in schizophrenia. *Nature*, **378**, 176–79.

Simpson, J.J. and Davies, W.E. (1999) Recent advances in the pharmacological treatment of tinnitus. *Trends in Pharmacological Science*, **20**, 12–18.

Simpson, J.J. and Davies, W.E. (2000) A review of evidence in support of a role for 5-HT in the perception of tinnitus. *Hearing Research*, **145**, 1–7.

Simpson, J.J., Gilbert, A.M., Weiner, G.M., and Davies, W.E. (1999) The assessment of lamotrigine, an antiepileptic drug, in the treatment of tinnitus. *American Journal of Otology*, **20**, 627–31.

Sindhusake, D., Golding, M., Newall, P., Rubin, G., Jakobsen, K., and Mitchell, P. (2003) Risk factors for tinnitus in a population of older adults: the Blue Mountains Hearing Study. *Ear & Hearing*, **24**, 501–507.

Singer, W. (1999) Neuronal synchrony: a versatile code for the definition of relations? *Neuron*, **24**, 49–65.

Singer, W. and Gray, C.M. (1995) Visual feature integration and the temporal correlation hypothesis. *Annual Reviews of Neuroscience*, **18**, 555–86.

Singer, W., Panford-Walsh, R., Watermann, D., Hendrich, O., Zimmermann, U., Kopschall, I., Rohbock, K., and Knipper, M. (2008) Salicylate alters the expression of calcium response transcription factor 1 in the cochlea: implications for brain-derived neurotrophic factor transcriptional regulation. *Molecular Pharmacology*, **73**, 1085–91.

Sininger, Y.S., Eggermont, J.J., and King, A.J. (1992) Spontaneous activity from the peripheral auditory system in tinnitus. In: J.-M. Aran and R. Dauman (ed.) *Tinnitus 91*, pp. 341–45. Amsterdam: Kugler Publishers.

Slepecky, N. (1986) Overview of mechanical damage to the inner ear: noise as a tool to probe cochlear function. *Hearing Research*, **22**, 307–21.

Smiley, J.F., and Goldman-Rakic, P.S. (1996) Serotonergic axons in monkey prefrontal cerebral cortex synapse predominantly on interneurons as demonstrated by serial section electron microscopy. *Journal of Comparative Neurology*, **367**, 431–43.

Smith, L., Gross, J., and Morest, D.K. (2002) Fibroblast growth factors (FGFs) in the cochlear nucleus of the adult mouse following acoustic overstimulation. *Hearing Research*, **169**, 1–12.

Smith, P.H. and Rhode, W.S. (1989) Structural and functional properties distinguish two types of multipolar cells in the ventral cochlear nucleus. *Journal of Comparative Neurology*, **282**, 595–616.

Smits, M., Kovacs, S., de Ridder, D., Peeters, R.R., van Hecke, P., and Sunaert, S. (2007) Lateralization of functional magnetic resonance imaging (fMRI) activation in the auditory pathway of patients with lateralized tinnitus. *Neuroradiology*, **49**, 669–79.

Sokoloff, L. (1979) Mapping of local cerebral functional activity by measurement of local cerebral glucose utilization with [14C]deoxyglucose. *Brain*, **102**, 653–68.

Spangler, K.M. and Warr, W.B. (1991) The descending auditory system. In: R.A. Altschuler, R.P. Bobbin, B.M. Clopton, and D.W. Hoffman (eds.) *Neurobiology of Hearing: The Central Auditory System*, pp.27–45. New York: Raven Press.

Spoendlin, H. (1976) Organisation of the auditory receptor. *Revue de Laryngologie Otologie et Rhinologie (Bord)*, **97** Suppl, 453–62.

Spongr, V.P., Boettcher, F.A., Saunders, S.S., and Salvi, R.J. (1992) Effects of noise and salicylate on hair cell loss in the chinchilla cochlea. *Archives of Otolaryngology Head and Neck Surgery*, **118**, 157–64.

Spongr, V.P., Flood, D.G., Frisina, R.D., and Salvi, R.J. (1997) Quantitative measures of hair cell loss in CBA and C57BL/6 mice throughout their life spans. *Journal of the Acoustical Society of America*, **101**, 3546–53.

Spoor, A. and Passchier-Vermeer, W. (1969) Spread in hearing levels of nonnoise exposed people at various ages. *Audiology*, **8**(2), 328 –36.

Stamataki, S., Francis, H.W., Lehar, M., May, B.J., and Ryugo, D.K. (2006) Synaptic alterations at inner hair cells precede spiral ganglion cell loss in aging C57BL/6J mice. *Hearing Research*, **221**, 104–118.

Steriade, M. (2006) Grouping of brain rhythms in corticothalamic systems. *Neuroscience*, **137**, 1087–1106.

Steriade, M. and Amzica, F. (1996) Intracortical and corticothalamic coherency of fast spontaneous oscillations. *Proceedings of the National Academy of Science (of the United States of America)*, **93**, 2533–38.

Steriade, M., Contreras, D., Curró Dossi, R., and Nuñez, A. (1993) The slow (<1 Hz) oscillation in reticular thalamic and thalamocortical neurons: scenario of sleep rhythm generation in interacting thalamic and neocortical networks. *Journal of Neuroscience*, **13**, 3284–99.

Steriade, M., Dossi, R.C., Paré, D., and Oakson, G. (1991) Fast oscillations (20–40 Hz) in thalamocortical systems and their potentiation by mesopontine cholinergic nuclei in the cat. *Proceedings of the National Academy of Science of the United States of America*, **88**, 4396–4400.

Stevens, S.S. (1955) The measurement of loudness. *Journal of the Acoustical Society of America*, **27**, 815–29.

Stewart, M., Borer, S.E., and Lehman, M. (2009) Shooting habits of U.S. waterfowl hunters. *Noise Health*, **11**, 8–13.

Stillman, J.A., Zwislocki, J.J., Zhang, M., and Cefaratti, L.K. (1993) Intensity just-noticeable differences at equal-loudness levels in normal and pathological ears. *Journal of the Acoustical Society of America*, **93**, 425–34.

Stolzberg, D., Chen, G.-D., Allman, B.L., and Salvi, R.J. (2011) Salicylate-induced peripheral auditory changes and tonotopic reorganization of auditory cortex. *Neuroscience*, **180**, 157–64.

Stouffer, J.L. and Tyler, R.S. (1990a) Characterization of tinnitus by tinnitus patients. *Journal of Speech and Hearing Disorders*, **55**, 439–53.

Stouffer, J.L. and Tyler, R.S. (1990b) Subjective tinnitus loudness. *Hearing Instruments*, **41**, 17–19.

Strange, R.C., Spiteri, M.A., Ramachandran, S., and Fryer, A.A. (2001) Glutathione-S-transferase family of enzymes. *Mutation Research*, **482**, 21–26

Stypulkowski, P.H. (1990) Mechanisms of salicylate ototoxicity. *Hearing Research*, **46**, 113–46.

Su, Y.Y., Luo, B., Wang, H.T., and Chen, L. (2009) Differential effects of sodium salicylate on current-evoked firing of pyramidal neurons and fast-spiking interneurons in slices of rat auditory cortex. *Hearing Research*, **253**, 60–66.

Suga, N., Gao, E., Zhang, Y., Ma, X., and Olsen, J.F. (2000) The corticofugal system for hearing: recent progress. *Proceedings of the National Academy of Science of the United States of America*, **97**, 11807–11814.

Sullivan, M., Katon, W., Russo, J., Dobie, R., and Sakai, C. (1993) A randomized trial of nortriptyline for severe chronic tinnitus. Effects on depression, disability, and tinnitus symptoms. *Archives of Internal Medicine*, **153**, 2251–59.

Sullivan, M.J., Rarey, K.E., and Conolly, R.B. (1987) Comparative ototoxicity of gentamicin in the guinea pig and two strains of rats. *Hearing Research*, **31**, 161–67.

Sumner, C.J., Tucci, D.L., and Shore, S.E. (2005) Responses of ventral cochlear nucleus neurons to contralateral sound after conductive hearing loss. *Journal of Neurophysiology*, **94**, 4234–43.

Sun, W., Lu, J., Stolzberg, D., Gray, L., Deng, A., Lobarinas, E., and Salvi, R.J. (2009) Salicylate increases the gain of the central auditory system. *Neuroscience*, **159**, 325–34.

Suneja, S.K., Benson, C.G., and Potashner, S.J. (1998a) Glycine receptors in adult guinea pig brain stem auditory nuclei: regulation after unilateral cochlear ablation. *Experimental Neurology*, **154**, 473–88.

Suneja, S.K. and Potashner, S. J. (2003) ERK and SAPK signaling in auditory brainstem neurons after unilateral cochlear ablation. *Journal of Neuroscience Research*, **73**, 235–45.

Suneja, S.K., Potashner, S.J., and Benson, C.G. (1998b) Plastic changes in glycine and GABA release and uptake in adult brain stem auditory nuclei after unilateral middle ear ossicle removal and cochlear ablation. *Experimental Neurology*, **151**, 273–88.

Suneja, S.K., Potashner, S.J., and Benson, C.G. (2000) AMPA receptor binding in adult guinea pig brain stem auditory nuclei after unilateral cochlear ablation. *Experimental Neurology*, **165**, 355–69.

Suneja, S.K., Yan, L., and Potashner, S.J. (2005) Regulation of NT-3 and BDNF levels in guinea pig auditory brain stem nuclei after unilateral cochlear ablation. *Journal of Neuroscience Research*, **80**(3), 381–90.

Surr, R.K., Montgomery, A.A., and Mueller, H.G. (1985) Effect of amplification on tinnitus among new hearing aid users. *Ear & Hearing*, **6**, 71–75.

Sweatt, J.D. (2001) The neuronal MAP kinase cascade: a biochemical signal integration system subserving synaptic plasticity and memory. *Journal of Neurochemistry*, **76**, 1–10.

Sweetow, R.W. and Levy, M.C. (1990) Tinnitus severity scaling for diagnostic / therapeutic usage. *Hearing Instruments*, **41**, 20–46.

Swerdlow, N.R., Geyer, M.A., and Braff, D.L. (2001) Neural circuit regulation of prepulse inhibition of startle in the rat: current knowledge and future challenges. *Psychopharmacology (Berlin)*, **156**, 194–215.

Syka, J. (2002) Plastic changes in the central auditory system after hearing loss, restoration of function, and during learning. *Physiological Review*, **82**, 601–36.

Syka, J. (2010) The Fischer 344 rat as a model of presbycusis. *Hearing Research*, **264**, 70–78.

Szczepaniak, W.S. and Møller, A.R. (1995) Evidence of decreased GABAergic influence on temporal integration in the inferior colliculus following acute noise exposure: a study of evoked potentials in the rat. *Neuroscience Letters*, **196**, 77–80.

Sztuka, A., Pospiech, L., Gawron, W., and Dudek, K. (2010) DPOAE in estimation of the function of the cochlea in tinnitus patients with normal hearing. *Auris Nasus Larynx*, **37**, 55–60.

Tabuchi, A., Nakaoka, R., Amano, K., Yukimine, M., Andoh, T., Kuraishi, Y., and Tsuda, M. (2000) Differential activation of brain-derived neurotrophic factor gene promoters I and III by Ca^{2+} signals evoked via L-type voltage-dependent and N-methyl-D-aspartate receptor Ca^{2+} channels. *Journal of Biological Chemistry*, **275**, 17269–75.

Tadros, S. F., D'Souza, M., Zettel, M. L., Zhu, X., Lynch-Erhardt, M., and Frisina, R.D. (2007) Serotonin 2B receptor: upregulated with age and hearing loss in mouse auditory system. *Neurobiology of Aging*, **28**, 1112–23.

Takeno, S., Harrison, R.V., Mount, R.J., Wake, M., Harada, Y. (1994) Induction of selective inner hair cell damage by carboplatin. *Scanning Microscopy*, **8**, 97–106.

Takumida, M., Ishibashi, T., Hamamotom T., Hirakawam K., and Anniko, M. (2009) Age-dependent changes in the expression of klotho protein, TRPV5 and TRPV6 in mouse inner ear. *Acta Otolaryngologica*, **129**, 1340–50.

Tan, J., Rüttiger, L., Panford-Walsh, R., Singer, W., Schulze, H., Kilian, S.B., Hadjab, S., Zimmermann, U., Köpschall, I., Rohbock, K., and Knipper, M. (2007) Tinnitus behavior and hearing function correlate with the reciprocal expression patterns of BDNF and Arg3.1/arc in auditory neurons following acoustic trauma. *Neuroscience*, **145**, 715–26.

Tanriverdi, N., Sayilgan, M.A., and Ozçürümez, G. (2001) Musical hallucinations associated with abruptly developed bilateral loss of hearing. *Acta Psychiatrica Scandinavica*, **103**, 153–55.

Terry, A.M., Jones, D.M., Davis, B.R., and Slater, R. (1983) Parametric studies of tinnitus masking and residual inhibition. *British Journal of Audiology*, **17**, 245–56.

Thai-Van, H., Micheyl, C., Moore, B.C., and Collet, L. (2003) Enhanced frequency discrimination near the hearing loss cut-off: a consequence of central auditory plasticity induced by cochlear damage? *Brain*, **126**, 2235–45.

Thomas, D.R. (2006) 5-HT5A receptors as a therapeutic target. *Pharmacology & Therapeutics*, **111**, 707–14.

Tilley, H. (1910) Audible tinnitus. *Proceedings of the Royal Society of Medicine*, **3**, 1–2.

Timmann, D., Musso, C., Kolb, F.P., Rijntjes, M., Jüptner, M., Müller, S.P., Diener, H.C., and Weiller, C. (1998) Involvement of the human cerebellum during habituation of the acoustic startle response: a PET study. *Journal of Neurology Neurosurgery and Psychiatry*, **65**, 771–73.

Tomita, M. and Eggermont, J.J. (2005) Cross-correlation and joint spectro-temporal receptive field properties in auditory cortex. *Journal of Neurophysiology*, **93**, 378–92.

Tonndorf, J. (1987) The analogy between tinnitus and pain: a suggestion for a physiological basis of chronic tinnitus. *Hearing Research*, **28**, 271–75.

Toyama, K., Kimura, M., and Tanaka, K. (1981) Cross-correlation analysis of interneuronal connectivity in cat visual cortex. *Journal of Neurophysiology*, **46**, 191–201.

Tran Ba Huy, P., Ferrary, E., Escoubet, B., and Sterkers, O. (1987) Strial prostaglandins and leukotrienes. Biochemical characteristics and interrelationship with furosemide. *Acta Otolaryngologica*, **103**, 558–66.

Trenado, C., Haab, L., Reith, W., and Strauss, D.J. (2009) Biocybernetics of attention in the tinnitus decompensation: an integrative multiscale modeling approach. *Journal of Neuroscience Methods*, **178**, 237–47.

Tsodyks, M., Kenet, T., Grinvald, A., and Arieli, A. (1999) Linking spontaneous activity of single cortical neurons and the underlying functional architecture. *Science*, **286**, 1943–46.

Tucker, D.A., Phillips, S.L., Ruth, R.A., Clayton, W.A., Royster, E., and Todd, A.D. (2005) The effect of silence on tinnitus perception. *Otolaryngology Head and Neck Surgery*, **132**, 20–24.

Turner, J.G. and Parrish, J. (2008) Gap detection methods for assessing salicylate-induced tinnitus and hyperacusis in rats. *American Journal of Audiology*, **17**, S185–192.

Turner, J.G., Brozoski, T.J., Bauer, C.A., Parrish, J.L., Myers, K., Hughes, L.F., and Caspary, D.M. (2006) Gap detection deficits in rats with tinnitus: a potential novel screening tool. *Behavioral Neuroscience*, **120**, 188–95.

Turner, J.G., Hughes, L.F., and Caspary, D.M. (2005) Divergent response properties of layer-V neurons in rat primary auditory cortex. *Hearing Research*, **202**, 129–40.

Turrigiano, G. (1999) Homeostatic plasticity in neuronal networks: the more things change, the more they stay the same. *Trends in Neuroscience*, **22**, 221–27.

Turrigiano, G. (2007) Homeostatic signaling: the positive side of negative feedback. *Current Opinion in Neurobiology*, **17**, 318–24.

Turrigiano, G.G., Leslie, K.R., Desai, N.S., Rutherfordm L.C., and Nelson S.B. (1998) Activity-dependent scaling of quantal amplitude in neocortical neurons. *Nature*, **391**, 892–96.

Tuz, H.H., Onder, E.M., and Kisnisci, R.S. (2003) Prevalence of otologic complaints in patients with temporomandibular disorder. *American Journal of Orthodontics and Dentofacial Orthopedics*, **123**(6), 620–23.

Tyler, R.S. (1993) Tinnitus disability and handicap questionnaires. *Seminars in Hearing*, **14**, 377–84.

Tyler, R.S. and Baker, L.J. (1983) Difficulties experienced by tinnitus sufferers. *Journal of Speech and Hearing Disorders*, **48**, 150–54.

Tyler, R.S. and Conrad-Armes, D. (1983) The determination of tinnitus loudness considering the effects of recruitment. *Journal of Speech and Hearing Research*, **26**, 59–72.

Tyler, R.S. and Conrad-Armes, D. (1984) Masking of tinnitus compared to masking of pure tones. *Journal of Speech and Hearing Research*, **27**, 106–111.

Tyler, R., Coelho, C., Tao, P., Ji, H., Noble, W., Gehringer, A., and Gogel, S. (2008) Identifying tinnitus subgroups with cluster analysis. *American Journal of Audiology*, **17**, S176–184.

Tzounopoulos, T., Kim, Y., Oertel, D., and Trussell, L.O. (2004) Cell-specific, spike timing-dependent plasticities in the dorsal cochlear nucleus. *Nature Neuroscience*, **7**, 719–25.

Uhlhaas, P.J. and Singer, W. (2006) Neural synchrony in brain disorders: relevance for cognitive dysfunctions and pathophysiology. *Neuron*, **52**, 155–68.

Utz, K.S., Dimova, V., Oppenländer, K., and Kerkhoff, G. (2010) Electrified minds: transcranial direct current stimulation (tDCS) and galvanic vestibular stimulation (GVS) as methods of non-invasive brain stimulation in neuropsychology—a review of current data and future implications. *Neuropsychologia*, **48**, 2789–2810.

Valentine, P.A. and Eggermont, J.J. (2001) Spontaneous burst firing in three auditory cortical fields and its relation to local field potentials. *Hearing Research*, **154**, 146–57.

Valentine, P.A., Teskey, G.C., and Eggermont, J.J. (2004) Kindling changes burst firing, neural synchrony and tonotopic organization of cat primary auditory cortex. *Cerebral Cortex*, **14**, 827–39.

Van de Heyning, P., Vermeire, K., Diebl, M., Nopp, P., Anderson, I., and De Ridder, D. (2008) Incapacitating unilateral tinnitus in single-sided deafness treated by cochlear implantation. *Annals of Otology Rhinology and Laryngology*, **117**, 645–52.

Van der Loo, E., Gais, S., Congedo, M., Vanneste, S., Plazier, M., Menovsky, T., Van de Heyning, P., and De Ridder, D. (2009) Tinnitus intensity dependent gamma oscillations of the contralateral auditory cortex. *PLoS One*, **4**(10), e7396.

Vane, J.R. (1971) Inhibition of prostaglandin synthesis as a mechanism of action for aspirin-like drugs. *Nature New Biology*, **231**, 232–35.

Vane, J.R., Bakhle, Y.S., and Botting, R.M. (1998) Cyclooxygenases 1 and 2. *Annual Reviews of Pharmacology and Toxicology*, **38**, 97–120.

van Heusden, E. and Smoorenburg, G.F. (1983) Responses from AVCN units in the cat before and after inducement of an acute noise trauma. *Hearing Research*, **11**, 295–326.

Vanneste, S., Plazier, M., Ost, J., van der Loo, E., Van de Heyning, P., and De Ridder, D. (2010a) Bilateral dorsolateral prefrontal cortex modulation for tinnitus by transcranial direct current stimulation: a preliminary clinical study. *Experimental Brain Research*, **202**, 779–85.

Vanneste, S., Plazier, M., van der Loo, E., Ost, J., Van de Heyning, P., and De Ridder, D. (2010b) Burst transcranial magnetic stimulation: which tinnitus characteristics influence the amount of transient tinnitus suppression? *European Journal of Neurology*, **66**, 986–90.

Vanneste, S., Plazier, M., der Loo, E., van de Heyning, P., Congedo, M., and De Ridder, D. (2010c) The neural correlates of tinnitus-related distress. *Neuroimage*, **52**, 470–80.

Vanneste, S., Plazier, M., van der Loo, E., Van de Heyning, P., and De Ridder, D. (2010d) The difference between uni- and bilateral auditory phantom percept. *Clinical Neurophysiology*, **122**, 578–87.

Vanneste, S., Plazier, M., Van de Heyning, P., and De Ridder, D. (2010e) Transcutaneous electrical nerve stimulation (TENS) of upper cervical nerve (C2) for the treatment of somatic tinnitus. *Experimental Brain Research*, **204**, 283–87.

Vanneste, S., Plazier, M., van der Loo, E., Van de Heyning, P., and De Ridder, D. (2010f) The differences in brain activity between narrow band noise and pure tone tinnitus. *PLoS One*, **5**(**10**), e13618.

Varela, F., Lachaux, J.P., Rodriguez, E., and Martinerie, J. (2001) The brainweb: phase synchronization and large-scale integration. *Nature Reviews Neuroscience*, **2**, 229–39.

Vass, Z., Shore, S.E., Nuttall, A.L., Jancsó, G., Brechtelsbauerm P.B., and Miller J.M. (1997) Trigeminal ganglion innervation of the cochlea—a retrograde transport study. *Neuroscience*, **79**, 605–615.

Vass, Z., Shore, S.E., Nuttall, A.L., and Miller, J.M. (1998) Direct evidence of trigeminal innervation of the cochlear blood vessels. *Neuroscience*, **84**, 559–67.

Verhagen, W.I.M., Bom, S.J.H., Huygen, P.L.M., Fransen, E., Van Camp, G., and Cremers, C.W.R.J. (2000) Familial progressive vestibulocochlear dysfunction caused by a COCH mutation (DFNA9) *Archives of Neurology*, **57**, 1045–47.

Vernon, J. and Meikle, M.B. (2003) Tinnitus: clinical measurement. *Otolaryngology Clinics of North America*, **36**, 293–305.

Vernon, J., Schleuning, A., Well, I., and Hughes, F. (1977) A tinnitus clinic. *Ear Nose and Throat Journal*, **56**, 58–71.

Versnel, H., Prijs, V.F., and Schoonhoven, R. (1992) Round-window recorded potential of single-fibre discharge (unit response) in normal and noise-damaged cochleas. *Hearing Research*, **59**, 157–70.

Viani, L.G. (1989) Tinnitus in children with hearing loss. *Journal of Laryngology & Otology*, **3**, 1142–45.

Vicente-Torres, M.A., Dávila, D., Bartolomé, M.V., Carricondo, F., and Gil-Loyzaga, P. (2003) Biochemical evidence for the presence of serotonin transporters in the rat cochlea. *Hearing Research*, **182**, 43–47.

Villa, A.E. and Abeles, M. (1990) Evidence for spatiotemporal firing patterns within the auditory thalamus of the cat. *Brain Research*, **509**, 325–27.

Viskontas, I.V., Knowlton, B.J., Steinmetz, P.N., and Fried, I. (2006) Differences in mnemonic processing by neurons in the human hippocampus and parahippocampal regions. *Journal of Cognitive Neuroscience*, **18**, 1654–62.

Vogler, D.P., Robertson, D., and Mulders, W.H.A.M. (2011) Hyperactivity in the ventral cochlear nucleus after cochlear trauma. *Journal of Neuroscience*, **31**, 6639–45.

Voigt, H.F. and Young, E.D. (1980) Evidence of inhibitory interactions between neurons in dorsal cochlear nucleus. *Journal of Neurophysiology*, **44**, 76–96.

Volcy, M., Sheftell, F.D., Tepper, S.J., Rapoport, A.M., and Bigal, M.E. (2005) Tinnitus in migraine: an allodynic symptom secondary to abnormal cortical functioning? *Headache*, **45**, 1083–87.

Vyleta, N.P. and Smith, S.M. (2011) Spontaneous glutamate release is independent of calcium influx and tonically activated by the calcium-sensing receptor. *Journal of Neuroscience*, **31**, 4593–4606.

Wahlström, B. and Axelsson, A. (1995) The description of tinnitus sounds. In: G.E. Reich, and J.A. Vernon (eds.), *Proceedings of the Fifth International Tinnitus Seminar*, pp. 298–301. Portland, OR: American Tinnitus Association.

Waldvogel, D., Mattle, H.P., Sturzenegger, M., and Schroth, G. (1998) Pulsatile tinnitus—a review of 84 patients. *Journal of Neurology*, **245**, 137–42.

Wallace, M.N., Kitzes, L.M., and Jones, E.G. (1991) Intrinsic inter- and intralaminar connections and their relationship to the tonotopic map in cat primary auditory cortex. *Experimental Brain Research*, **86**, 527–44.

Wallhäusser-Franke, E. (1997) Salicylate evokes c-fos expression in the brain stem: implications for tinnitus. *Neuroreport*, **8**, 725–28.

Wallhäusser-Franke, E., Braun, S., and Langner, G. (1996) Salicylate alters 2-DG uptake in the auditory system: a model for tinnitus? *Neuroreport*, 7, 1585–88.

Wallhausser-Franke, E., Mahlke, C., Oliva, R., Braun, S., Wenz, G., and Langner, G. (2003) Expression of c-fos in auditory and non-auditory brain regions of the gerbil after manipulations that induce tinnitus. *Experimental Brain Research*, **153**, 649–54.

Wang, H., Brozoski, T.J., Ling, L., Hughes, L.F.and Caspary, D.M. (2011) Impact of sound exposure and aging on brain-derived neurotrophic factor and tyrosine kinase B receptors levels in dorsal cochlear nucleus 80 days following sound exposure. *Neuroscience*, **172**, 453–59.

Wang, H.T., Luo, B., Zhou, K.Q., Xu, T.L., and Chen. L. (2006) Sodium salicylate reduces inhibitory postsynaptic currents in neurons of rat auditory cortex. *Hearing Research*, **215**, 77–83.

Wang, H.T., Luo, B., Zhou, K.Q., and Chen. L. (2008) Sodium salicylate suppresses serotonin-induced enhancement of GABAergic spontaneous inhibitory postsynaptic currents in rat inferior colliculus in vitro. *Hearing Research*, **236**, 42–51.

Wang, H., Brozoski, T.J., Turner, J.G., Ling, L., Parrish, J.L., Hughes, L.F., and Caspary, D.M. (2009) Plasticity at glycinergic synapses in dorsal cochlear nucleus of rats with behavioral evidence of tinnitus. *Neuroscience*, **164**, 747–59.

Wang, H.P., Spencer, D., Fellous, J.M., and Sejnowski, T.J. (2010) Synchrony of thalamocortical inputs maximizes cortical reliability. *Science*, **328**, 106–109.

Wang, J., Salvi, R.J., and Powers, N. (1996) Plasticity of response properties of inferior colliculus neurons following acute cochlear damage. *Journal of Neurophysiology*, **75**, 171–83.

Wang, L., Andersson, S., Warner, M., and Gustafsson, J.A. (2001) Morphological abnormalities in the brains of estrogen receptor beta knockout mice. *Proceedings of the National Academy of Science of the United States of America*, **98**, 2792–96.

Wang, J., Ding, D., and Salvi, R.J. (2002) Functional reorganization in chinchilla inferior colliculus associated with chronic and acute cochlear damage. *Hearing Research*, **168**, 238–49.

Wang, X.-J. and Buzsáki, G. (1996) Gamma oscillation by synaptic inhibition in a hippocampal interneuronal network model. *Journal of Neuroscience*, **16**, 6402–6413.

Wangemann, P., Nakaya, K., Wu, T., Maganti, R.J., Itza, E.M., Sanneman, J.D., Harbidge, D.G., Billings, S., and Marcus, D.C. (2007) Loss of cochlear HCO3- secretion causes deafness via endolymphatic acidification and inhibition of Ca^{2+} reabsorption in a Pendred syndrome mouse model. *American Journal of Physiology Renal Physiology*, **292**, F1345–1353.

Ward, L.M. (2011) The thalamic dynamic core theory of conscious experience. *Consciousness and Cognition*, **20**, 464–86.

Ward, L.M. and Baumann, M. (2009) Measuring tinnitus loudness using constrained psychophysical scaling. *American Journal of Audiology*, **18**, 119–28.

Weber, H., Pfadenhauer, K., Stöhr, M., and Rösler, A. (2002) Central hyperacusis with phonophobia in multiple sclerosis. *Multiple Sclerosis*, **8**, 505–509.

Wei, L., Ding, D., and Salvi, R. (2010a) Salicylate-induced degeneration of cochlea spiral ganglion neurons-apoptosis signaling. *Neuroscience*, **168**, 288–99.

Wei, L., Ding, D., Sun, W., Xu-Friedman, M.A., and Salvi, R. (2010b) Effects of sodium salicylate on spontaneous and evoked spike rate in the dorsal cochlear nucleus. *Hearing Research*, **267**, 54–60.

Weinberg, R.J. and Rustioni, A. (1987) A cuneocochlear pathway in the rat. *Neuroscience*, **20**, 209–219.

Weisz, N., Hartmann, T., Dohrmann, K., Schlee, W., and Norena, A. (2006) High-frequency tinnitus without hearing loss does not mean absence of deafferentation. *Hearing Research*, **222**, 108–14.

Weisz, N., Moratti, S., Meinzer, M., Dohrmann, K., and Elbert, T. (2005a) Tinnitus perception and distress is related to abnormal spontaneous brain activity as measured by magnetoencephalography. *PLoS Med*, **2**, e153.

Weisz, N., Müller, S., Schlee, W., Dohrmann, K., Hartmann, T., and Elbert, T. (2007) The neural code of auditory phantom perception. *Journal of Neuroscience*, **27**, 1479–84.

Weisz, N., Voss, S., Berg, P., and Elbert, T. (2004) Abnormal auditory mismatch response in tinnitus sufferers with high-frequency hearing loss is associated with subjective distress level. *BMC Neuroscience*, **5**, 8.

Weisz, N., Wienbruch, C., Dohrmann, K., and Elbert, T. (2005b) Neuromagnetic indicators of auditory cortical reorganization of tinnitus. *Brain*, **128**, 2722–31.

West, A.E., Chen, W.G., Dalva, M.B., Dolmetsch, R.E., Kornhauser, J.M., Shaywitz, A.J., Takasu, M.A., Tao, X., and Greenberg, M.E. (2001) Calcium regulation of neuronal gene expression. *Proceedings of the National Academy of Science of the United States of America*, **98**, 11024–31.

West, R.L., Ward, L.M., and Khosla, R. (2000) Constrained scaling: the effect of learned psychophysical scales on idiosyncratic response bias. *Perception and Psychophysics*, **62**, 137–51.

White, J.A., Burgess, B.J., Hall, R.D., and Nadol, J.B. (2000) Pattern of degeneration of the spiral ganglion cell and its processes in the C57BL/6J mouse. *Hearing Research*, **141**, 12–18.

Whiting, B., Moiseff, A., and Rubio, M.E. (2009) Cochlear nucleus neurons redistribute synaptic AMPA and glycine receptors in response to monaural conductive hearing loss. *Neuroscience*, **163**, 1264–76.

Wienbruch, C., Paul, I., Weisz, N., Elbert, T., and Roberts, L. E. (2006) Frequency organization of the 40-Hz auditory steady-state response in normal hearing and in tinnitus. *Neuroimage*, **33**, 180–94.

Willner, P. (1986) Validation criteria for animal models of human mental disorders: learned helplessness as a paradigm case. *Progress in Neuropsychopharmacology and Biological Psychiatry*, **10**, 677–90.

Willner, P. and Mitchell, P.J. (2002) The validity of animal models of predisposition to depression. *Behavioural Pharmacology*, **13**, 169–88.

Willott, J.F. (1986) Effects of aging, hearing loss, and anatomical location on thresholds of inferior colliculus neurons in C57BL/6 and CBA mice. *Journal of Neurophysiology*, **56**, 391–408.

Willott, J.F. (1991) *Aging and the Auditory System: Anatomy, Physiology and Psychophysics*. San Diego, CA: Singular Publishing Group.

Willott, J.F. (2009) Effects of sex, gonadal hormones, and augmented acoustic environments on sensorineural hearing loss and the central auditory system: insights from research on C57BL/6J mice. *Hearing Research*, **252**, 89–99.

Willott, J.F., Aitkin, L.M., and McFadden, S.L. (1993) Plasticity of auditory cortex associated with sensorineural hearing loss in adult C57BL/6J mice. *Journal of Comparative Neurology*, **329**, 402–11.

Willott, J.F. and Bross, L.S. (1990) Morphology of the octopus cell area of the cochlear nucleus in young and aging C57BL/6J and CBA/J mice. *Journal of Comparative Neurology*, **300**, 61–81.

Willott, J.F. and Bross, L.S. (1996) Morphological changes in the anteroventral cochlear nucleus that accompany sensorineural hearing loss in DBA/2J and C57BL/6J mice. *Brain Research Developmental Brain Research*, **91**, 218–26.

Willott, J.F., Bross, L.S., and McFadden, S.L. (1992) Morphology of the dorsal cochlear nucleus in C57BL/6J and CBA/J mice across the life span. *Journal of Comparative Neurology*, **321**, 666–78.

Willott, J.F., Bross, L.S., and McFadden, S.L. (1994) Morphology of the cochlear nucleus in CBA/J mice with chronic, severe sensorineural cochlear pathology induced during adulthood. *Hearing Research*, **74**, 1–21.

Willott, J.F., Milbrandt, J.C., Bross, L.S., and Caspary, D.M. (1997) Glycine immunoreactivity and receptor binding in the cochlear nucleus of C57BL/6J and CBA/CaJ mice: effects of cochlear impairment and aging. *Journal of Comparative Neurology*, **385**, 405–414.

Willott, J.F., Parham, K., and Hunter, K.P. (1988a) Response properties of inferior colliculus neurons in young and very old CBA/J mice. *Hearing Research*, **37**, 1–14.

Willott, J.F., Parham, K., and Hunter, K.P. (1988b) Response properties of inferior colliculus neurons in middle-aged C57BL/6J mice with presbycusis. *Hearing Research*, **37**, 15–27.

Wilson, P.H. and Henry, J.L. (1998) Tinnitus Cognitions Questionnaire: development and psychometric properties of a measure of dysfunctional cognitions associated with tinnitus. *International Tinnitus Journal*, **4**, 23–30.

Wilson, P.H., Henry, J., Bowen, M., and Haralambous, G. (1991) Tinnitus Reaction Questionnaire: psychometric properties of a measure of distress associated with tinnitus. *Journal of Speech Language and Hearing Research*, **34**, 197–201.

Winer, J.A. (1990) The functional architecture of the medial geniculate body and the primary auditory cortex. In: D.B. Webster, A.N. Popper, R.R. Fay, and R.R. (eds.), *The Mammalian Auditory Pathway: Neuroanatomy*, pp. 222–409. New York: Springer Verlag.

Winer, J.A. (1992) The functional architecture of the medial geniculate body and the primary auditory cortex. In: D.B. Webster, A.N. Popper, and R.R. Fay (eds.), *Springer Handbook of Auditory Research, vol. 1, The Mammalian Auditory Pathway: Neuroanatomy*, pp. 222–409. New York: Springer-Verlag.

Winer, J.A. (2006) Decoding the auditory corticofugal systems. *Hearing Research*, **212**, 1–8.

Winer, J.A. and Lee, C.C. (2007) The distributed auditory cortex. *Hearing Research*, **229**, 3–13.

Winter, I.M., Robertson, D., and Yates, G.K. (1990) Diversity of characteristic frequency rate-intensity functions in guinea pig auditory nerve fibres. *Hearing Research*, **45**, 191–202.

Witsell, D.L., Hannley, M.T., Stinnet, S., and Tucci, D.L. (2007) Treatment of tinnitus with gabapentin: a pilot study. *Otology & Neurotology*, **28**, 11–15.

Wolfart, J., Debay, D., Le Masson, G., Destexhe, A., and Bal, T. (2005) Synaptic background activity controls spike transfer from thalamus to cortex. *Nature Neuroscience*, **8**, 1760–67.

Wright, D.D. and Ryugo, D.K. (1996) Mossy fiber projections from the cuneate nucleus to the cochlear nucleus in the rat. *Journal of Comparative Neurology*, **365**, 159–72.

Wu, G.K., Arbuckle, R., Liu, B.H., Tao, H.W., and Zhang, L.I. (2008) Lateral sharpening of cortical frequency tuning by approximately balanced inhibition. *Neuron*, **58**, 132–43.

Wu, J.L., Chiu, T.W., and Poon, P.W. (2003) Differential changes in Fos-immunoreactivity at the auditory brainstem after chronic injections of salicylate in rats. *Hearing Research*, **176**, 80–93.

Wu, T., Lv, P., Kim, H.J., Yamoah, E.N., and Nuttall, A.L. (2010) Effect of salicylate on KCNQ4 of the guinea pig outer hair cell. *Journal of Neurophysiology*, **103**, 1969–77.

Wutzler, A., Winter, C., Kitzrow, W., Uhl, I., Wolf, R.J., Heinz, A., and Juckel, G. (2008) Loudness dependence of auditory evoked potentials as indicator of central serotonergic neurotransmission: simultaneous electrophysiological recordings and in vivo microdialysis in the rat primary auditory cortex. *Neuropsychopharmacology*, **33**, 3176–81.

Xiao, Z. and Suga, N. (2002) Modulation of cochlear hair cells by the auditory cortex in the mustached bat. *Nature Neuroscience*, **5**, 57–63.

Yan, J. and Ehret, G. (2001) Corticofugal reorganization of the midbrain tonotopic map in mice. *Neuroreport*, **12**, 3313–3316.

Yan, J. and Ehret, G. (2002) Corticofugal modulation of midbrain sound processing in the house mouse. *European Journal of Neuroscience*, **16**, 119–28.

Yan, J., Zhang, Y., and Ehret, G. (2005) Corticofugal shaping of frequency tuning curves in the central nucleus of the inferior colliculus of mice. *Journal of Neurophysiology*, **93**, 71–83.

Yan, W. and Suga, N. (1998) Corticofugal modulation of the midbrain frequency map in the bat auditory system. *Nature Neuroscience*, **1**, 54–58.

Yang, G., Lobarinas, E., Zhang, L., Turner, J., Stolzberg, D., Salvi, R., and Sun, W. (2007) Salicylate induced tinnitus: behavioral measures and neural activity in auditory cortex of awake rats. *Hearing Research*, **226**, 244–53.

Yang, K., Huang, Z.W., Liu, Z.Q., Xiao, B.K., and Peng, J.H. (2009) Long-term administration of salicylate enhances prestin expression in rat cochlea. *International Journal of Audiology*, **48**, 18–23.

Yang, Q., Vernet, M., Orssaud, C., Bonfils, P., Londero, A., and Kapoula, Z. (2010) Central Crosstalk for Somatic Tinnitus: abnormal Vergence Eye Movements. *PLoS One*, **5**, e11845.

Yoo, S.S., Lee, C.U., and Choi, B.G. (2001) Human brain mapping of auditory imagery: event-related functional MRI study. *Neuroreport*, **12**, 3045–49

Yoshimura, H. (2005) The potential of caffeine for functional modification from cortical synapses to neuron networks in the brain. *Current Neuropharmacology*, **3**, 309–36.

Young, C.K. and Eggermont, J.J. (2009) Coupling of mesoscopic brain oscillations: recent advances in analytical and theoretical perspectives. *Progress in Neurobiology*, **89**, 61–78.

Young, E.D., Nelken, I., and Conley, R.A. (1995) Somatosensory effects on neurons in dorsal cochlear nucleus. *Journal of Neurophysiology*, **73**, 743–65.

Young, E.D. and Voigt, H.F. (1982) Response properties of type II and type III units in dorsal cochlear nucleus. *Hearing Research*, **6**, 153–69.

Yu, N., Zhu, M.L., Johnson, B., Liu, Y.P., Jones, R.O., and Zhao, H.B. (2008) Prestin upregulation in chronic salicylate (aspirin) administration: an implication of functional dependence of prestin expression. *Cellular and Molecular Life Sciences*, **65**, 2407–2418.

Yu, X., Wadghiri, Y.Z., Sanes, D.H., and Turnbull, D.H. (2005) In vivo auditory brain mapping in mice with Mn-enhanced MRI. *Nature Neuroscience*, **8**, 961–68.

Yu, X., Zou, J., Babb, J.S., Johnson, G., Sanes, D.H., and Turnbull, D.H. (2008) Statistical mapping of sound-evoked activity in the mouse auditory midbrain using Mn-enhanced MRI. *Neuroimage*, **39**, 223–30.

Zacharek, M.A., Kaltenbach, J.A., Mathog, T.A., and Zhang, J. (2002) Effects of cochlear ablation on noise induced hyperactivity in the hamster dorsal cochlear nucleus: implications for the origin of noise induced tinnitus. *Hearing Research*, **172**, 137–43.

Zeng, C., Nannapaneni, N., Zhou, J., Hughes, L.F., and Shore, S. (2009) Cochlear damage changes the distribution of vesicular glutamate transporters associated with auditory and nonauditory inputs to the cochlear nucleus. *Journal of Neuroscience*, **29**, 4210–4217.

Zeng, F.-G., Tang, Q., Dimitrijevic, A., Starr, A., Larky, J., and Blevins, N.H. (2011) Tinnitus suppression by low-rate electric stimulation and its electrophysiological mechanisms. *Hearing Research*, **277**, 61–66.

Zhang, J.S., Kaltenbach, J.A., Godfrey, D.A., and Wang, J. (2006) Origin of Hyperactivity in the Hamster Dorsal Cochlear Nucleus Following Intense Sound Exposure. *Journal of Neuroscience Research*, **84**, 819–31.

Zhang, J., Zhang, Y., and Zhang, X. (2011) Auditory cortex electrical stimulation suppresses tinnitus in rats. *Journal of the Association for Research in Otolaryngology*, **12**, 185–201.

Zhang, X., Yang, P., Cao, Y., Qin, L. and Sato, Y. (2011) Salicylate induced neural changes in the primary auditory cortex of awake cats. *Neuroscience*, **172**, 232–45.

Zhao, F., Manchaiah, V.K.C., French, D., and Price, S.M. (2010) Music exposure and hearing disorders: an overview. *International Journal of Audiology*, **49**, 54–64

Zheng, J.L. and Gao, W.Q. (1996) Differential damage to auditory neurons and hair cells by ototoxins and neuroprotection by specific neurotrophins in rat cochlear organotypic cultures. *European Journal of Neuroscience*, **8**, 1897–1905.

Zhou, F.M. and Hablitz, J.J. (1999) Activation of serotonin receptors modulates synaptic transmission in rat cerebral cortex. *Journal of Neurophysiology*, **82**, 2989–99.

Zhou, J., Nannapaneni, N., and Shore, S. (2007) Vessicular glutamate transporters 1 and 2 are differentially associated with auditory nerve and spinal trigeminal inputs to the cochlear nucleus. *Journal of Comparative Neurology*, **500**, 777–87.

Zhou, J. and Shore, S. (2004) Projections from the trigeminal nuclear complex to the cochlear nuclei: a retrograde and anterograde tracing study in the guinea pig. *Journal of Neuroscience Research*, **78**, 901–907.

Zhou, J. and Shore, S. (2006) Convergence of spinal trigeminal and cochlear nucleus projections in the inferior colliculus of the guinea pig. *Journal of Comparative Neurology*, **495**, 100–112.

Zwaardemaker, H. (1905) Die physiologisch wahrnehmbaren Energie-wanderungen. *Ergebnisse der Physiologie*, 423–80.

Zwaardemaker, H. (1910) The camera silenta of the physiological laboratory at Utrecht. *Proceedings Royal Netherlands Academy of Arts and Sciences*, **12**, 706–10.

Zöger, S., Svedlund, J., and Holgers, K.M. (2001) Psychiatric disorders in tinnitus patients without severe hearing impairment: 24-month follow-up of patients at an audiological clinic. *Audiology*, **40**, 133–40.

Zöger, S., Svedlund, J., and Holgers, K.M. (2006) The effects of sertraline on severe tinnitus suffering—a randomized, double-blind, placebo-controlled study. *Journal of Clinical Psychopharmacology*, **26**, 32–39.

Author index

Subject index